D1315604

Cracking the
AP®
CHEMISTRY
Exam
2013 Edition

Paul Foglino

PrincetonReview.com

Random House, Inc. New York

The Princeton Review, Inc.
111 Speen Street, Suite 550
Framingham, MA 01701
E-mail: editorialsupport@review.com

ISBN: 978-0-307-94488-7
ISSN: 1092-0102

The Princeton Review is not affiliated with Princeton University.

Editor: Calvin S. Cato
Production Editor: Kathy G. Carter
Production Coordinator: Deborah A. Silvestrini

Printed in the United States of America on partially recycled paper.

10 9 8 7 6 5 4 3 2

2013 Edition

Editorial
Rob Franek, Senior VP, Publisher
Laura Braswell, Senior Editor
Selena Coppock, Senior Editor
Calvin Cato, Editor
Meave Shelton, Editor

Production
Michael Pavese, Publishing Director
Kathy Carter, Project Editor
Michelle Krapf, Editor
Michael Mazzei, Editor
Michael Breslosky, Associate Editor
Stephanie Tantum, Associate Editor
Kristen Harding, Associate Editor
Vince Bonavoglia, Artist
Danielle Joyce, Graphic Designer

Random House Publishing Group
Tom Russell, Publisher
Nicole Benhabib, Publishing Director
Ellen L. Reed, Production Manager
Alison Stoltzfus, Managing Editor

ACKNOWLEDGMENTS

I'd like to thank John Katzman for entrusting me with this project and also my editor, Rebecca Lessem. For her work on past editions, I'd like to thank Rachel Warren, whose guidance and tireless effort helped me to convert chemistry into English. I'd also like to thank Eric Payne, whose rigorous attention to detail helped me to separate chemistry from fantasy. I'd like to thank Tom Meltzer for his advice and Libby O'Connor for her patience.

I'd also like to thank The Princeton Review's editorial and production crew. Thanks also to Robbie Korin, James Karb, and Chris Volpe for their suggestions and corrections.

CONTENTS

Introduction

WHAT IS THE PRINCETON REVIEW?

The Princeton Review is an international test-preparation company with branches in all major U.S. cities and several abroad. In 1981, John Katzman started teaching an SAT prep course in his parents' living room. Within five years, The Princeton Review had become the largest SAT prep program in the country.

Our phenomenal success in improving students' scores on standardized tests is due to a simple, innovative, and radically effective philosophy: Study the test, not just what the test claims to test. This approach has led to the development of techniques for taking standardized tests based on the principles the test writers themselves use to write the tests.

The Princeton Review has found that its methods work not just for cracking the SAT, but for any standardized test. We've already successfully applied our system to the GMAT, LSAT, MCAT, and GRE, to name just a few. Obviously, you need to be well versed in chemistry to do well on the AP Chemistry Exam, but you should remember that any standardized test is partly a measure of your ability to think like the people who write standardized tests. This book will help you brush up on your AP Chemistry and prepare for the exam using our time-tested principle: Crack the system based on how the test is created.

We also offer books and online services that cover an enormous variety of education and career-related topics. If you're interested, check out our website at PrincetonReview.com.

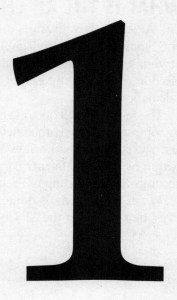

Orientation

WHAT IS THE AP PROGRAM?

The Advanced Placement (AP) Program bridges high schools and colleges by allowing high school students to do college-level work for college credit. The AP Chemistry Exam is one of more than 30 college-level examinations offered every year.

AP courses are offered by more than 10,000 high schools in the United States, Canada, and more than 60 additional countries. More than 3,000 colleges around the world offer college credit to students who perform well on AP tests. The specific score required for credit varies from school to school and from subject to subject.

The AP Program is coordinated by the College Board. The College Board is a national nonprofit organization composed of representatives from various schools and colleges. They see it as their mission to set educational standards.

The College Board appoints a development committee for each of the subjects. The development committee decides what should be covered in an AP course and how it should be covered on the AP test. The AP Chemistry development committee is composed of three high school chemistry teachers, three college professors who teach general chemistry, and an additional college professor who chairs the group. Each member of the development committee serves a three-year term.

The test is administered by the Educational Testing Service (ETS)—the same folks who bring you the SAT. ETS also plays a role in developing the test.

WHAT IS THE AP CHEMISTRY EXAM?

The AP Chemistry Exam is a three-hour-long, two-section test that attempts to cover the material you would learn in a college first-year chemistry course. The first part, which counts for 50 percent of your grade, consists of multiple-choice questions. The second part, which counts for 50 percent of your grade, is composed of free-response questions, such as short essays and problems involving calculations.

The test is offered once every year in May. It's scored in June. The multiple-choice section is scored by computer and the problems and essays are scored by a committee of high school and college teachers. The problems and essays are graded according to a standard set at the beginning of the grading period by the chief faculty consultants. Inevitably, the grading of Section II is never as consistent or accurate as the grading of Section I.

When the grading is done, the results are curved and each student receives a grade based on a five-point scale. For the AP Chemistry Exam, the results break down as follows:

Grade	What It Means	Approximate % of Test Takers Who Get This Score
5	Extremely well qualified	17%
4	Well qualified	18%
3	Qualified	20%
2	Possibly qualified	15%
1	No recommendation	30%

Although standards vary from school to school, it's safe to say that most colleges will give credit for a 5, some will give credit for a 4 or 3, and very few will give credit for a 2.

CRACKING THE MULTIPLE-CHOICE SECTION

THE BASICS

Section I of the test is composed of 75 multiple-choice questions, for which you are allotted 90 minutes. This part is worth 50 percent of your total score.

For this section, you will be given a periodic table of the elements and you may NOT use a calculator. The College Board says that this is because the new scientific calculators not only program and graph but also store information—and they are afraid you'll use this function to cheat!

The first 15 multiple-choice questions, give or take a few, will be formatted with 5 answer choices followed by a series of questions (as shown on the next page).

Questions 1–4

    (A) O_2
    (B) H_2O
    (C) Ni
    (D) Fe
    (E) NaCl

1. This species contains ionic bonds. (E)

2. This species is a gas at standard temperature and pressure. (A)

3. This species is denser as a liquid than as a solid. (B)

4. This species contains a double bond. (A)

These are mostly straightforward, "you know it or you don't" questions. Notice that an answer can be used once, more than once, or not at all.

The rest of the multiple-choice questions are in the standard question-and-answers format shown below.

16. Which of the following species is a gas at standard temperature and pressure?

    (A) O_2
    (B) H_2O
    (C) Ni
    (D) Fe
    (E) NaCl

On the multiple-choice section, you receive 1 point for a correct answer. There is no penalty for leaving a question blank or getting a question wrong.

Your raw score on this section will be just the number of questions you answered correctly. You can make a rough prediction of your overall score from your raw score on the multiple-choice section, assuming that you do about as well on the free-response section.

Roughly speaking,

- if you get a raw score of *at least* **50,** you will probably get a **5**

- if you get a raw score of *at least* **35,** you will probably get at least a **4**

- if you get a raw score of *at least* **25,** you will probably get at least a **3**

PACING

So you can get a 5 with a raw score of 50, a 4 with a raw score of 35, and a 3 with a raw score of 25. That's a pretty generous curve. According to the College Board, the multiple-choice section of the AP Chemistry Exam covers more material than any individual student is expected to know. Nobody is expected to get a perfect or even near-perfect score.

What does that mean to you?

Don't Answer All the Questions!

You can skip every third question and still get a 5. You can skip half the questions and still get a 4. You can skip two out of every three questions and still get a 3. Obviously, you should answer any question that you have a chance of getting right, but you should be aware that the grading curve gives you plenty of slack.

Okay, so you know that you can skip questions. How do you know which questions to skip?

Use the Two-Pass System

Go through the multiple-choice section twice. The first time, do all the questions that you can get answers to immediately. That is, do the questions with little or no math and questions on chemistry topics in which you are well versed. Skip questions on topics that make you uncomfortable. Also, you want to skip the ones that look like number crunchers (even without a calculator, you may still be expected to crunch a few numbers). Circle the questions that you skip in your test booklet so you can find them easily during the second pass. Once you've done all the questions that come easily to you, go back and pick out the tough ones that you have the best shot at.

In general, the questions near the end of the section are tougher than the questions near the beginning. You should keep that in mind, but be aware that each person's experience will be different. If you can do acid-base questions in your sleep, but you'd rather have your teeth drilled than draw a Lewis diagram, you may find questions near the end of the section easier than questions near the beginning.

That's why the Two-Pass System is so handy. By using it, you make sure you get to see all the questions you can get right, instead of running out of time because you got bogged down on questions you couldn't do earlier in the test.

This brings us to another important point.

Don't Turn a Question into a Crusade!

Most people don't run out of time on standardized tests because they work too slowly. Instead, they run out of time because they spend half the test wrestling with two or three particular questions.

You should never spend more than a minute or two on any question. If a question doesn't involve calculation, then either you know the answer, you can make an educated guess, or you don't know the answer. Figure out where you stand on a question, make a decision, and move on.

Any question that requires more than two minutes' worth of calculations probably isn't worth doing. Remember: Skipping a question early in the section is a good thing if it means that you'll have time to get two correct answers later on.

GUESSING

You get one point for every correct answer on the multiple-choice section. Guessing randomly neither helps you nor hurts you. Educated guessing, however, will help you.

Use Process of Elimination (POE) to Find Wrong Answers

There is a fundamental weakness to a multiple-choice test. The test makers must show you the right answer, along with four wrong answers. Sometimes seeing the right answer is all you need. Other times you may not know the right answer, but you may be able to identify one or two of the answers that are clearly wrong. Here is where you should use POE to take an educated guess.

Look at this hypothetical question.

1. Which of the following compounds will produce a
purple solution when added to water?

   (A) Brobogdium rabelide
   (B) Diblythium perjuvenide
   (C) Sodium chloride
   (D) Hynynium gargantuide
   (E) Carbon dioxide

You should have no idea what the correct answer is because three of these compounds are made up, but you do know something about the obviously wrong answers. You know that sodium chloride, choice (C), and carbon dioxide, choice (E), do not turn water purple. So, using POE, you have a one-out-of-three chance at guessing the correct answer. Now the odds are in your favor. Now you should guess.

Guess and Move On

Remember that you're guessing. Pondering the possible differences between brobogdium rabelide and diblythium perjuvenide is a waste of time. Once you've taken POE as far as it will go, pick your favorite letter and move on.

The multiple-choice section is the exact opposite of the free-response section. It's scored by a machine. There's no partial credit. The computer doesn't know, or care if you know, why an answer is correct. All the computer cares about is whether you blackened in the correct oval on your score sheet. You get the same number of points for picking (B) because you know (A) and (E) are wrong and B is a nicer letter than C or D as you would for picking (B) because you fully understood the subtleties of an electrochemical process.

ABOUT CALCULATORS

You will NOT be allowed to use a calculator on this section. That shouldn't worry you. All it means is that there won't be any questions in the section that you'll need a calculator to solve.

Most of the calculation problems will have fairly user-friendly numbers—that is, numbers with only a couple of significant digits, or things like "11.2 liters of gas at STP" or "160 grams of oxygen" or "a temperature increase from 27°C to 127°C." Sometimes these user-friendly numbers will actually point you toward the proper steps to take in your calculations.

Don't be afraid to make rough estimates as you do your calculations. Sometimes knowing that an answer is closer to 50 than to 500 will enable you to pick the correct answer on a multiple-choice test (if the answer choices are far enough apart). Once again, the rule against calculators works in your favor because the College Board will not expect you to do very precise calculations by hand.

There may be a couple of real number-crunching problems on the test. If you can recognize them quickly, these are good ones to skip. There's no point in spending five minutes crunching numbers to get one problem right if that time could be better used in getting three others correct later in the test.

CRACKING THE FREE-RESPONSE SECTION

Section II is composed of a series of six free-response questions, all of which are required. You will be allotted 95 minutes to complete this section, which is worth 50 percent of your total score. You get exactly 55 minutes for Part A, which is composed of three problems requiring calculation, then 40 minutes for Part B, which is composed of two essay questions and one question concerning chemical reactions.

PART A—PROBLEMS

For Part A, you will be given a table of commonly used chemical equations, a table of standard reduction potentials, and a periodic table. You may use a calculator, which you will probably need.

The Equilibrium Problem

This is a multipart question involving calculation to determine some aspect of equilibrium.

This question will be divided into at least four or five parts, with partial credit available for correct answers on each part. The question is worth 20 percent of Section II. The average score for this question is usually less than half credit.

Two Additional Problems

Part A also includes two additional multipart questions that will involve calculation. These questions can come from any topic in the syllabus and one of them may be based in a laboratory setting. The test makers will probably do their best to include concepts from as many different chemistry topics as possible in the different parts of each problem.

These questions, like the equilibrium question, will be divided into at least four or five parts, with partial credit available for correct answers on each part. Each of these questions is worth 20 percent of Section II. The average score for these questions is usually less than half credit.

CRACKING THE PROBLEMS

On Part A, you want to show the graders that you can do chemistry math, so here are some suggestions.

Show every step of your calculations on paper

This section is the opposite of multiple choice. You don't just get full credit for writing the correct answer. You get most of your points on this section for showing the process that got you to the answer. The graders give you partial credit when you show them that you know what you're doing. So even if you can do a calculation in your head, you should set it up and show it on the page.

By showing every step, or explaining what you're doing in words, you insure that you'll get all the partial credit possible, even if you screw up a calculation.

Include units in all your calculations

Scientists like units in calculations. Units make scientists feel secure. You'll get points for including them and you may lose points for leaving them out.

Remember significant figures

You can lose one point per question if your answer is off by more than one significant figure. Without getting too bent out of shape about it, try to remember that a calculation is only as accurate as the least accurate number in it.

The graders will follow your reasoning, even if you've made a mistake

Often, you are asked to use the result of a previous part of a problem in a later part. If you got the wrong answer in part (a) and used it in part (c), you can still get full credit for part (c), as long as your work is correct based on the number that you used. That's important, because it means that botching the first part of a question doesn't necessarily sink the whole question.

Remember the mean!

So let's say that you could complete only parts (a) and (b) on the required equilibrium problem. That's 4 or 5 points out of 10, tops. Are you doomed? Of course not. You're above average. If this test is hard on you, it's probably just as hard on everybody else. Remember: You don't need anywhere near a perfect score to get a 5, and you can leave half the test blank and still get a 4!

PART B—REACTIONS AND ESSAYS

You will not be allowed to use a calculator for Part B, which requires you to write balanced equations for chemical reactions and answer questions about chemical concepts.

Writing Chemical Equations

For this section, you will be given a table of standard reduction potentials and a periodic table and you may not use a calculator.

You will be given 3 sets of chemical reactants and you will be asked to write the appropriate balanced equation for the reaction that will occur for each. Each reaction will be followed by a brief question that will focus on some aspect of the reaction.

Each reaction is worth 5 points; 1 point for reactants, 2 points for products, 1 point for balancing, and 1 point for the question; so you can earn a total of 15 points for this section. This part is worth 10 percent of Section II.

We'll talk about how to approach this section in Chapter 17.

Essays

For the two essay questions, you will be given a table of commonly used chemical equations, a table of standard reduction potentials, and a periodic table. You may not use a calculator for this section, which is fine because there won't be any calculations.

The essays are multipart conceptual questions, with stress on understanding and explaining chemical concepts, rather than on doing calculations. These questions can come from any topic in the syllabus and one of them may be based in a laboratory setting. As in the problems, the test makers will probably try to include questions from as many different chemistry topics as possible. Partial credit will be given for correct answers on each part, and each question is worth 15 percent of Section II.

CRACKING THE ESSAYS

This section is here to test whether you can translate chemistry into English. The term "essay" is a little misleading because all of these questions can be answered in two or three simple sentences, or with a simple diagram or two. Here are some tips for answering the two essay questions on Part II.

Show that you understand the terms used in the question.
If they ask you why sodium and potassium have differing first ionization energies, the first thing you should do is tell them what ionization energy is. That's probably worth the first point of partial credit. Then you should tell them how the differing structures of the atoms make for differing ionization energies. That leads to the next tip.

Take a step-by-step approach.
Grading these tests is hard work. Breaking a question into parts in this way makes it easier on the grader, who must match your response to a set of guidelines he or she has been given that describe how to assign partial and full credit.

Each grader scores each test based on these rough guidelines that are established at the beginning of the grading period. For instance, if a grader has 3 points for the question about ionization energies, the points might be distributed the following way:

> One point for understanding ionization energy.

> One point for explaining the structural difference between sodium and potassium.

> One point for showing how this difference affects the ionization energy.

You can get all three points for this question if the grader thinks that all three concepts are addressed *implicitly* in your answer, but by taking a step by step approach, you improve your chances of *explicitly* addressing the things that a grader has been instructed to look for. Once again, grading these tests is hard work; graders won't know for sure if you understand something unless you tell them.

This leads us to an obvious point.

Write neatly.
Even if writing neatly means working at half-speed. You can't get points for answers if the graders can't understand them. Of course, this applies to the rest of the free-response section as well.

The graders will follow your reasoning, even if you've made a mistake
Just like in the problems section, you might be asked to use the result of a previous part of a problem in a later part. If you decide (incorrectly) that an endothermic reaction in part (a) is exothermic, you can still get full credit in part (c) for your wrong answer about the reaction's spontaneity, as long as your answer in (c) is correct based on an exothermic reaction.

ABOUT THE TOPICS COVERED ON THE TEST
These are the topics covered on the AP Chemistry Exam, as described by the College Board.

  I. **Structure of Matter** (20%)
      A. Atomic theory and atomic structure
          1. Evidence for atomic theory
          2. Atomic masses and how to determine them experimentally
          3. Atomic number and mass number; isotopes
          4. Electron energy levels: atomic spectra, quantum numbers, atomic orbitals
          5. Periodic trends (atomic radii, ionization energies, electron affinities, oxidation states)

B. Bonding
   1. Forces
      a. Types: ionic, covalent, metallic, hydrogen bonding, van der Waals (including London dispersion forces)
      b. Relationships to states, structure, and properties of matter
      c. Polarity, electronegativity
   2. Molecular models
      a. Lewis structures
      b. Valence electrons, hybridization of orbitals, resonance, sigma and pi bonds
      c. VSEPR
   3. Geometry of molecules and ions, structural isomerism of simple organic molecules and coordination complexes, dipole moments, relation of properties to structure
C. Nuclear chemistry: nuclear equations, half-lives, and radioactivity; chemical applications

II. **States of Matter** (20%)
   A. Gases
      1. Ideal gas laws
         a. Equation of state for an ideal gas
         b. Partial pressures, Dalton's law
      2. Kinetic-molecular theory
         a. Interpretation of ideal gas laws on the basis of this theory
         b. Avogadro's hypothesis and the mole concept
         c. Dependence of kinetic energy of molecules on temperature; Graham's law
         d. Deviations from ideal gas laws
   B. Liquids and solids
      1. Liquids and solids and kinetic-molecular theory
      2. Phase diagrams
      3. Changes of state, including critical points and triple points
      4. Structure of solids, lattice energies
   C. Solutions
      1. Types of solutions and factors affecting solubility
      2. Molarity, molality, mole fraction, density
      3. Raoult's law, colligative properties, osmosis
      4. Nonideal behavior

III. **Reactions** (35–40%)
   A. Reaction types
      1. Acid-base reactions; Arrhenius, Brønsted-Lowry, and Lewis theories; coordination complexes; amphoterism
      2. Precipitation reactions
      3. Oxidation-reduction reactions
         a. Oxidation state
         b. The role of the electron in oxidation-reduction
         c. Electrochemistry: electrolytic and galvanic cells; Faraday's laws; standard half-cell potentials; Nernst equation; spontaneity of redox reactions

B. Stoichiometry
   1. Ionic and molecular species present in chemical systems: net ionic equations
   2. Balancing of equations, including redox reactions
   3. Mass and volume relations, using the mole concept in finding empirical formulas and limiting reactants
C. Equilibrium
   1. Dynamic equilibrium, physical and chemical; Le Châtelier's law; equilibrium constants
   2. Quantitative treatment
      a. Equilibrium constants for gaseous reactions: K_p, K_c
      b. Equilibrium constants for reactions in solution
         (1) Constants for acids and bases, pK, pH
         (2) Solubility product constants and their application to precipitation and the dissolution of slightly soluble compounds
         (3) Common ion effect, buffers, hydrolysis
D. Kinetics
   1. Reaction rate
   2. Use of rate laws to determine order of reaction and rate constant from experimental data
   3. Effect of temperature change on rates
   4. Activation energy, catalysts
   5. Reaction mechanisms and rate determining steps
E. Thermodynamics
   1. State functions
   2. First law: enthalpy change; heat of formation; heat of reaction; Hess's law; heats of vaporization and fusion; calorimetry
   3. Second law: entropy; free energy of formation; free energy of reaction; dependence of change in free energy on enthalpy and entropy changes
   4. Relationship of change in free energy to equilibrium constants and electrode potentials

IV. **Descriptive Chemistry** (10–15%)
   A. Chemical reactivity and products of chemical reactions
   B. Relationships in the periodic table: horizontal, vertical, and diagonal with examples from alkali metals, alkaline earth metals, halogens, and the first series of transition elements
   C. Introduction to organic chemistry: hydrocarbons and functional groups (structure, nomenclature, chemical properties)

V. **Laboratory** (5–10%)
   Questions based on experiences and skills students acquire in the laboratory: making observations of chemical reactions and substances, recording data, calculating and interpreting results based on the quantitative data obtained, lab safety, experimental errors

The following list summarizes types of specific chemical calculation problems that may appear on the test:

1. Percentage composition

2. Empirical and molecular formulas from experimental data

3. Molar masses from gas density, freezing-point, and boiling-point measurements

4. Gas laws, including the ideal gas law, Dalton's law, and Graham's law

5. Stoichiometric relations using the concept of the mole; titration calculations

6. Mole fractions; molar and molal solutions

7. Faraday's law of electrolysis

8. Equilibrium constants and their applications, including their use for simultaneous equilibria

9. Standard electrode potentials and their use; Nernst equation

10. Thermodynamic and thermochemical calculations

11. Kinetics calculations

Preparing for the AP Chemistry Exam

THE MONTHS BEFORE THE TEST

START YOUR REVIEW EARLY

Try to spend from a half hour to an hour reviewing chemistry three or four times a week. Leave your chemistry books around so you can leaf through them when there's a really bad sitcom on TV between two shows that you like. If you study consistently, even for very short sessions over a period of a few months, you'll find that by the time test week arrives, you'll know the material and you won't have to study for six hours a day for the last six days—that doesn't work very well anyway.

READ THIS BOOK AND DO ALL OF THE QUESTIONS IN IT

This book covers all of the important information required for the AP Chemistry Exam, and the questions in the book test AP Chemistry material in AP Chemistry style. If you do well on the questions at the end of each chapter and on the practice tests at the end of the book, you will do well on the AP Chemistry Exam.

GET SOME REAL AP CHEMISTRY EXAMS

The College Board releases some test material after the test has been given. It releases a multiple-choice section every five years or so and the free-response section every year, along with the guidelines that were used to grade it. This is the most valuable study resource you have. Remember: You're not just studying chemistry; you're studying the AP Chemistry Exam.

Ask your teacher or guidance counselor if he or she has copies of released exams. Many teachers keep them on file and use them every year for in-class tests and practice material. The free-response sections with grading guidelines are especially useful; if you know what the graders are looking for, it's much easier to give it to them.

If real AP material is not available at your school, you can buy it directly from the College Board and ETS. Check out their website at www.collegeboard.com, or call them at **609-771-7300** and ask them to send you a catalog of what's most currently available.

USE THE INTERNET

If you have a computer or you know someone who does, get on the Internet and try typing "Advanced Placement Chemistry" (or some variation) into a couple of different search engines. Here's what you should find:

- **The College Board home page:** This site gives general information about the test and some test-taking tips. This is one place to go to find out if there have been any last-minute changes to the test.

- **AP students and teachers:** Some AP Chemistry classes have home pages. It might be worth a look to see what other AP classes around the country are up to.

GET A SECOND TEXTBOOK

Don't get a new one. New science textbooks are way too expensive, and you've probably had to buy one already, not to mention what you shelled out for this book.

Go to one of those slightly stale-smelling used book stores that has books piled from floor to ceiling. Somewhere in there, they should have a first-year chemistry textbook that was printed within the last 20 years. It will probably cost less than ten bucks. You should buy it. Don't worry if it's old; first-year chemistry hasn't changed all that much since you were born.

The reason that it's good to have a second book is that all textbook writers have strengths and weaknesses. An author who can make you understand thermodynamics may leave you totally confused when it comes to kinetics. You don't have to read everything in both books; just use the index to see what the second author has to say about any topic that the first author can't help you understand. Sometimes a second point of view is all you need.

TEACH SOMEBODY ELSE

The best way to really learn something is to explain it to someone else. Work with the other people in your class whenever you can. When someone else explains something to you, you're learning. When you explain something to someone else, you're learning even more.

THE WEEK BEFORE THE TEST

MAINTAIN YOUR USUAL ROUTINE

Go to sleep at your usual time. Don't start a strange new diet. You can step up your studying a bit, but if you've been studying with any consistency in the last few months, you probably won't have to. Don't try to cram the night before the test; it's a waste of time and effort.

REVIEW THIS BOOK

In assembling the information presented in this book, everything was pared away that was not absolutely necessary to cover for you to do well on the AP Chemistry Exam. So if it's in this book, then you need to know it. If it's not in this book, you can get a 5 without it.

REVIEW OLD AP CHEMISTRY EXAMS

By now you should know the science pretty well. Practice the test. Read the directions on the test carefully so you will already know them on the day of the test. You should know exactly what you're supposed to do on each section long before you sit for the test. Look for the themes and topics that come up on every test; the more familiar the AP test seems to you when you take it, the easier it will be.

THE DAY OF THE TEST

EAT BREAKFAST

Food gives you energy, and you'll need it for the test.

TAKE EVERYTHING YOU NEED

You need number-two pencils for the multiple-choice section because that's what the grading machines read. You can use either pen or pencil for the free-response section. You need an eraser.

YOU WILL NEED A WATCH

Without getting obsessive, you should keep track of the time as you do both sections. Never trust a proctor to do this for you.

WEAR COMFORTABLE CLOTHING

You don't want anything to distract you from the business at hand.

TAKE A SNACK

A piece of fruit or an energy bar during the break provides a handy energy boost.

RELAX

If you're well prepared, the test is simply an opportunity to show it.

3

Atomic Structure and the Periodic Table

HOW OFTEN DOES ATOMIC STRUCTURE APPEAR ON THE EXAM?

In the multiple-choice section, this topic appears in about 7 out of 75 questions. In the free-response section, this topic appears almost every year.

THE PERIODIC TABLE

The most important tool you will use on this test is the Periodic Table of the Elements.

Periodic Table of the Elements

1 H 1.0																		2 He 4.0
3 Li 6.9	4 Be 9.0											5 B 10.8	6 C 12.0	7 N 14.0	8 O 16.0	9 F 19.0	10 Ne 20.2
11 Na 23.0	12 Mg 24.3											13 Al 27.0	14 Si 28.1	15 P 31.0	16 S 32.1	17 Cl 35.5	18 Ar 39.9
19 K 39.1	20 Ca 40.1	21 Sc 45.0	22 Ti 47.9	23 V 50.9	24 Cr 52.0	25 Mn 54.9	26 Fe 55.8	27 Co 58.9	28 Ni 58.7	29 Cu 63.5	30 Zn 65.4	31 Ga 69.7	32 Ge 72.6	33 As 74.9	34 Se 79.0	35 Br 79.9	36 Kr 83.8
37 Rb 85.5	38 Sr 87.6	39 Y 88.9	40 Zr 91.2	41 Nb 92.9	42 Mo 95.9	43 Tc (98)	44 Ru 101.1	45 Rh 102.9	46 Pd 106.4	47 Ag 107.9	48 Cd 112.4	49 In 114.8	50 Sn 118.7	51 Sb 121.8	52 Te 127.6	53 I 126.9	54 Xe 131.3
55 Cs 132.9	56 Ba 137.3	57 *La 138.9	72 Hf 178.5	73 Ta 180.9	74 W 183.9	75 Re 186.2	76 Os 190.2	77 Ir 192.2	78 Pt 195.1	79 Au 197.0	80 Hg 200.6	81 Tl 204.4	82 Pb 207.2	83 Bi 209.0	84 Po (209)	85 At (210)	86 Rn (222)
87 Fr (223)	88 Ra 226.0	89 †Ac 227.0	104 Unq (261)	105 Unp (262)	106 Unh (263)	107 Uns (262)	108 Uno (265)	109 Une (267)									

*Lanthanide Series:	58 Ce 140.1	59 Pr 140.9	60 Nd 144.2	61 Pm (145)	62 Sm 150.4	63 Eu 152.0	64 Gd 157.3	65 Tb 158.9	66 Dy 162.5	67 Ho 164.9	68 Er 167.3	69 Tm 168.9	70 Yb 173.0	71 Lu 175.0
†Actinide Series:	90 Th 232.0	91 Pa (231)	92 U 238.0	93 Np (237)	94 Pu (244)	95 Am (243)	96 Cm (247)	97 Bk (247)	98 Cf (251)	99 Es (252)	100 Fm (257)	101 Md (258)	102 No (259)	103 Lr (260)

The periodic table gives you very basic but very important information about each element.

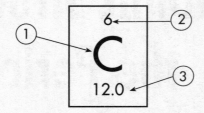

1. This is the **symbol** for the element; carbon, in this case. On the test, the symbol for an element is used interchangeably with the name of the element.

2. This is the **atomic number** of the element. The atomic number is the same as the number of protons in the nucleus of an element; it is also the same as the number of electrons surrounding the nucleus of an element in its neutral state.

3. This is the **molar mass** of the element. It's also called the **atomic weight.**

The identity of an atom is determined by the number of protons contained in its nucleus. The nucleus of an atom also contains neutrons. The mass number of an atom is the sum of its neutrons and protons. Atoms of an element with different numbers of neutrons are called isotopes; for instance, carbon-12, which contains 6 protons and 6 neutrons, and carbon-14, which contains 6 protons and 8 neutrons, are isotopes of carbon. The molar mass given on the periodic table is the average of the mass numbers of all known isotopes weighted by their percent abundance.

The molar mass of an element will give you a pretty good idea of the most common isotope of that element. For instance, the molar mass of carbon is 12.0 and about 99 percent of the carbon in existence is carbon-12.

The horizontal rows of the periodic table are called **periods.**

The vertical columns of the periodic table are called **groups.**

ELECTRONS

QUANTUM NUMBERS

The positions of the electrons in relation to the nucleus are described by their **quantum numbers**. Each electron has four quantum numbers that apply to its **shell, subshell, orbital,** and **spin.**

Shells: $n = 1, 2, 3...$

In a hydrogen atom, the principal quantum number, or shell, of an electron determines its average distance from the nucleus as well as its energy. So electrons in shells with higher values are farther away on average from the nucleus and will have more energy and less stability than electrons in shells with lower values.

Subshells: $l = 0, 1, 2...$

The angular momentum quantum number, or subshell, describes the shape of an electron's orbital.

- The first shell ($n = 1$) has one subshell: s, or $l = 0$.

- The second shell ($n = 2$) has two subshells: s ($l = 0$), and p ($l = 1$).

- The third shell ($n = 3$) has three subshells: s ($l = 0$), p ($l = 1$), and d ($l = 2$).

The orbitals of s subshells are spherical, while the orbitals of p subshells are dumbbell shaped.

Orbitals: $m_l = ...-1, 0, +1...$

The magnetic quantum number, or orbital, describes the orientation of the orbital in space. Roughly, that means it describes whether the path of the electron lies mostly on the x, y, or z axis of a three-dimensional grid.

- The s subshell ($l = 0$) has one orbital: $m_l = 0$.

- The p subshell ($l = 1$) has three orbitals: $m_l = -1$, $m_l = 0$, and $m_l = +1$.

- The d subshell ($l = 2$) has five orbitals: $m_l = -2, -1, 0, 1$, and 2.

Spin: $m_s = +\dfrac{1}{2}, -\dfrac{1}{2}$

Each orbital can contain two electrons: one with a positive spin and one with a negative spin.

Here are a couple of graphical ways of looking at quantum numbers.

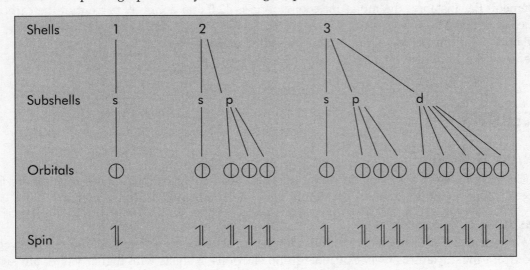

	The Quantum Numbers for levels 1–4		The number of electrons in the states	
n	l	m	In the subshell	In the level
1	0(s)	0	2	2
2	0 (s)	0	2	8
	1 (p)	−1, 0, +1	6	
3	0 (s)	0	2	18
	1 (p)	−1, 0, +1	6	
	2 (d)	−2, −1, 0, +1, +2	10	
4	0 (s)	0	2	32
	1 (p)	−1, 0, +1	6	
	2 (d)	−2, −1, 0, +1, +2	10	
	3 (f)	−3, −2, −1, 0, +1, +2, +3	14	

THE AUFBAU PRINCIPLE

The **Aufbau principle** states that when building up the electron configuration of an atom, electrons are placed in orbitals, subshells, and shells in order of increasing energy.

THE PAULI EXCLUSION PRINCIPLE

The **Pauli exclusion principle** states that within an atom, no two electrons can have the same set of quantum numbers. So, each electron in any atom has its own distinct set of four quantum numbers.

QUANTUM NUMBERS AND THE PERIODIC TABLE

You can use the periodic table to tell the first two quantum numbers of the valence electrons of any element. Note that n = principal quantum number.

You can also tell the order in which shells and subshells are filled by following the table from left to right across each period.

You should note that after the third period, the filling of subshells becomes more complicated. Notice, for instance, that the $4s$ subshell fills before the $3d$ subshell.

HUND'S RULE

Hund's rule says that when an electron is added to a subshell, it will always occupy an empty orbital if one is available. Electrons always occupy orbitals singly if possible and pair up only if no empty orbitals are available.

Watch how the 2p subshell fills as we go from boron to neon.

	1s	2s	2p
Boron	⇅	⇅	↑
Carbon	⇅	⇅	↑ ↑
Nitrogen	⇅	⇅	↑ ↑ ↑
Oxygen	⇅	⇅	⇅ ↑ ↑
Fluorine	⇅	⇅	⇅ ⇅ ↑
Neon	⇅	⇅	⇅ ⇅ ⇅

DIAMAGNETISM AND PARAMAGNETISM

Diamagnetism and **paramagnetism** are types of magnetism. **Diamagnetic** elements have all of their electrons spin paired. So, diamagnetic elements are elements with all of their subshells completed.

Some diamagnetic elements are

Helium	$1s^2$
Beryllium	$1s^2 2s^2$
Neon	$1s^2 2s^2 2p^6$

Most of the elements do not have all of their electrons spin paired and are called **paramagnetic** elements.

Paramagnetic elements are strongly affected by magnetic fields, whereas diamagnetic elements are not very strongly affected.

Molecules can also be diamagnetic or paramagnetic, depending on the pairing of electrons in their molecular orbitals, but the same basic rule holds: Paramagnetic molecules are affected by magnetic fields, and diamagnetic molecules are not.

ELECTRONS AND ENERGY

The positively charged nucleus is always pulling at the negatively charged electrons around it, and the electrons have potential energy that increases with their distance from the nucleus. It works the same way that the gravitational potential energy of a brick on the third floor of a building is greater than the gravitational potential energy of a brick nearer to ground level.

The energy of electrons, however, is **quantized.** That's important. It means that electrons can exist only at specific energy levels, separated by specific intervals. It's kind of like if the brick in the building could be placed only on the first, second, or third floor of the building, but not in-between.

The quantized energy of an electron in a hydrogen atom can be found if you know its principal quantum number or shell.

Energy of an Electron

$$E_n = \frac{-2.178 \times 10^{-18}}{n^2} \text{ joules}$$

E_n = the energy of the electron

n = the principal quantum number of the electron

When atoms absorb energy in the form of **electromagnetic radiation,** electrons jump to higher energy levels. When electrons drop from higher to lower energy levels, atoms give off energy in the form of electromagnetic radiation.

The relationship between the change in energy level of an electron and the electromagnetic radiation absorbed or emitted is given below.

Energy and Electromagnetic Radiation

$$\Delta E = hv = \frac{hc}{\lambda}$$

ΔE = energy change

h = Planck's constant, 6.63×10^{-34} joule-sec

v = frequency of the radiation

λ = wavelength of the radiation

c = the speed of light, 3.00×10^8 m/sec $(c = \lambda f)$

The energy level changes for the electrons of a particular atom are always the same, so atoms can be identified by their emission and absorption spectra.

NAMES AND THEORIES

Dalton's Elements

In the early 1800s, John Dalton presented some basic ideas about atoms that we still use today. He was the first to say that there are many different kinds of atoms, which he called elements. He said that these elements combine to form compounds and that these compounds always contain the same ratios of elements. Water (H_2O), for instance, always has two hydrogen atoms for every oxygen atom. He also said that atoms are never created or destroyed in chemical reactions.

Development of the Periodic Table

In 1869, Dmitri Mendeleev and Lothar Meyer independently proposed arranging the elements into early versions of the periodic table, based on the trends of the known elements.

Thomson's Experiment

In the late 1800s, J. J. Thomson watched the deflection of charges in a cathode ray tube and put forth the idea that atoms are composed of positive and negative charges. The negative charges were called electrons, and Thomson guessed that they were sprinkled throughout the positively charged atom like chocolate chips sprinkled throughout a blob of cookie dough.

Millikan's Experiment

Robert Millikan was able to calculate the charge on an electron by examining the behavior of charged oil drops in an electric field.

The Plum Pudding Model of an Atom

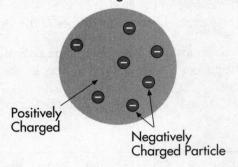

Positively Charged

Negatively Charged Particle

Rutherford's Experiment

In the early 1900s, Ernest Rutherford fired alpha particles at gold foil and observed how they were scattered. This experiment led him to conclude that all of the positive charge in an atom was concentrated in the center and that an atom is mostly empty space. This led to the idea that an atom has a positively charged nucleus, which contains most of the atom's mass, and that the tiny, negatively charged electrons travel around this nucleus.

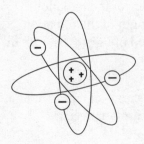

Quantum Theory

Max Planck figured out that electromagnetic energy is quantized. That is, for a given frequency of radiation (or light), all possible energies are multiples of a certain unit of energy, called a quantum (mathematically, that's $E = hv$). So, energy changes do not occur smoothly but rather in small but specific steps.

The Bohr Model

Neils Bohr took the quantum theory and used it to predict that electrons orbit the nucleus at specific, fixed radii, like planets orbiting the Sun. The Bohr model worked for atoms and ions with one electron but not for more complex atoms.

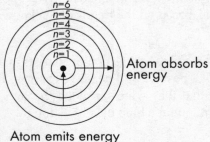

Atom absorbs energy

Atom emits energy

The Heisenberg Uncertainty Principle

Werner Heisenberg said that it is impossible to know both the position and momentum of an electron at a particular instant. In terms of atomic structure, this means that electron orbitals do not represent specific orbits like those of planets. Instead, an electron orbital is a probability function describing the possibility that an electron will be found in a region of space.

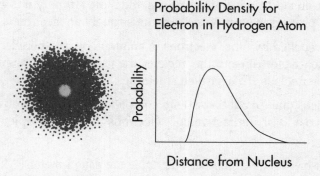

Probability Density for
Electron in Hydrogen Atom

Probability

Distance from Nucleus

The de Broglie Hypothesis

Louis de Broglie said that all matter has wave characteristics. This is important because sometimes the behavior of electrons is better described in terms of waves than particles.

There is a simple relationship between an electron's wave and particle characteristics.

The de Broglie Equation
$$\lambda = \frac{h}{mv}$$
λ = wavelength associated with a particle m = mass of the particle v = speed of the particle $mv = p$ = momentum of the particle h = Planck's constant, 6.63×10^{-34} joule-sec

De Broglie's hypothesis is useful for very small particles, such as electrons. For larger particles, the wavelength becomes too small to be of interest.

PERIODIC TRENDS

You can make predictions about certain behavior patterns of an atom and its electrons based on the position of the atom in the periodic table. All the periodic trends can be understood in terms of three basic rules.

1. Electrons are attracted to the protons in the nucleus of an atom.
   a. The closer an electron is to the nucleus, the more strongly it is attracted.
   b. The more protons in a nucleus, the more strongly an electron is attracted.

2. Electrons are repelled by other electrons in an atom. So, if other electrons are between a valence electron and the nucleus, the valence electron will be less attracted to the nucleus. That's called shielding.

3. Completed shells (and to a lesser extent, completed subshells) are very stable. Atoms prefer to add or subtract valence electrons to create complete shells if possible.

The atoms in the left-hand side of the periodic table are called metals. Metals give up electrons when forming bonds. Most of the elements in the table are metals. The elements in the upper right-hand portion of the table are called nonmetals. Nonmetals generally gain electrons when forming bonds. The metallic character of the elements decreases as you move from left to right across the periodic table. The elements in the borderline between metal and nonmetal, such as silicon and arsenic, are called metalloids.

ATOMIC RADIUS

The atomic radius is the approximate distance from the nucleus of an atom to its valence electrons.

Moving from Left to Right Across a Period (Li to Ne, for Instance), Atomic Radius Decreases

Moving from left to right across a period, protons are added to the nucleus, so the valence electrons are more strongly attracted to the nucleus; this decreases the atomic radius. Electrons are also being added, but they are all in the same shell at about the same distance from the nucleus, so there is not much of a shielding effect.

Moving Down a Group (Li to Cs, for Instance), Atomic Radius Increases

Moving down a group, shells of electrons are added to the nucleus. Each shell shields the more distant shells from the nucleus and the valence electrons get farther away from the nucleus. Protons are also being added, but the shielding effect of the negatively charged electron shells cancels out the added positive charge.

Cations (Positively Charged Ions) Are Smaller than Atoms

Generally, when electrons are removed from an atom to form a cation, the outer shell is lost, making the cation smaller than the atom. Also, when electrons are removed, electron-electron repulsions are reduced, allowing all of the remaining valence electrons to move closer to the nucleus.

Anions (Negatively Charged Ions) Are Larger than Atoms

When an electron is added to an atom, forming an anion, electron-electron repulsions increase, causing the valence electrons to move farther apart, which increases the radius.

IONIZATION ENERGY

Electrons are attracted to the nucleus of an atom, so it takes energy to remove an electron. The energy required to remove an electron from an atom is called the first ionization energy. Once an electron has been removed, the atom becomes a positively charged ion. The energy required to remove the next electron from the ion is called the second ionization energy, and so on.

Moving from Left to Right Across a Period, Ionization Energy Increases

Moving from left to right across a period, protons are added to the nucleus, which increases its positive charge. For this reason, the negatively charged valence electrons are more strongly attracted to the nucleus, which increases the energy required to remove them. Electrons are also being added, and the shielding effect provided by the filling of the s subshell causes a slight deviation in the trend in moving from Group 2A to Group 3A. There is also a slight deviation when the electrons in the p subshell start to pair up, so oxygen has a slightly lower first ionization energy than nitrogen does.

Moving Down a Group, Ionization Energy Decreases

Moving down a group, shells of electrons are added to the nucleus. Each inner shell shields the more distant shells from the nucleus, reducing the pull of the nucleus on the valence electrons and making them easier to remove. Protons are also being added, but the shielding effect of the negatively charged electron shells cancels out the added positive charge.

The Second Ionization Energy Is Greater than the First Ionization Energy

When an electron has been removed from an atom, electron-electron repulsion decreases and the remaining valence electrons move closer to the nucleus. This increases the attractive force between the electrons and the nucleus, increasing the ionization energy.

As Electrons Are Removed, Ionization Energy Increases Gradually Until a Shell Is Empty, Then Makes a Big Jump

- For each element, when the valence shell is empty, the next electron must come from a shell that is much closer to the nucleus, making the ionization energy for that electron much larger than for the previous ones.

- For Na, the second ionization energy is much larger than the first.

- For Mg, the first and second ionization energies are comparable, but the third is much larger than the second.

- For Al, the first three ionization energies are comparable, but the fourth is much larger than the third.

ELECTRON AFFINITY

Electron affinity is a measure of the change in energy of an atom when an electron is added to it. When the addition of an electron makes the atom more stable, energy is given off. This is true for most of the elements. When the addition of an electron makes the atom less stable, energy must be put in; that's because the added electron must be placed in a higher energy level, making the element less stable. This is the case for elements with full subshells, like the alkaline earths and the noble gases.

When moving from left to right across a period, the energy given off when an electron is added increases. Electron affinities don't change very much moving down a group.

ELECTRONEGATIVITY

Electronegativity refers to how strongly the nucleus of an atom attracts the electrons of other atoms in a bond. Electronegativities of elements are estimated based on ionization energies and electron affinities, and they basically follow the same trends.

- Moving from left to right across a period, electronegativity increases.
- Moving down a group, electronegativity decreases.

The various periodic trends, which don't include the noble gases, are summarized in the diagram below.

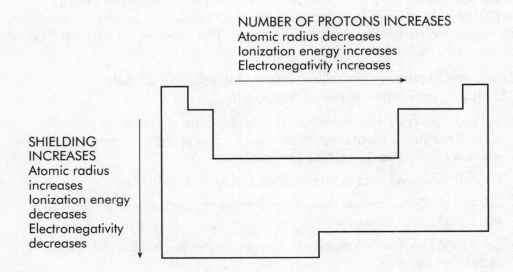

NUMBER OF PROTONS INCREASES
Atomic radius decreases
Ionization energy increases
Electronegativity increases

SHIELDING INCREASES
Atomic radius increases
Ionization energy decreases
Electronegativity decreases

CHAPTER 3 QUESTIONS

MULTIPLE-CHOICE QUESTIONS

Questions 1–4

    (A) C
    (B) N
    (C) O
    (D) F
    (E) Ne

1. This is the most electronegative element.

2. The nuclear decay of an isotope of this element is used to measure the age of archaeological artifacts.

3. All of the electrons in this element are spin-paired.

4. This element, present as a diatomic gas, makes up most of the earth's atmosphere.

Questions 5–7

    (A) Hg
    (B) Si
    (C) Cu
    (D) Zn
    (E) Ag

5. This element is commonly used in the manufacture of semiconductors.

6. This element is a liquid at room temperature.

7. After oxygen, this is by far the most common element in the earth's crust.

8. What is the most likely electron configuration for a sodium ion in its ground state?

    (A) $1s^2\,2s^22p^5$
    (B) $1s^2\,2s^22p^6$
    (C) $1s^2\,2s^22p^6\,3s^1$
    (D) $1s^2\,2s^22p^5\,3s^2$
    (E) $1s^2\,2s^22p^6\,3s^2$

9. Which of the following statements is true regarding sodium and chlorine?

    (A) Sodium has greater electronegativity and a larger first ionization energy.
    (B) Sodium has a larger first ionization energy and a larger atomic radius.
    (C) Chlorine has a larger atomic radius and a greater electronegativity.
    (D) Chlorine has greater electronegativity and a larger first ionization energy.
    (E) Chlorine has a larger atomic radius and a larger first ionization energy.

10. Which of the following could be the quantum numbers (n, l, m_l, m_s) for the valence electron in a potassium atom in its ground state?

    (A) $3, 0, 0, \dfrac{1}{2}$

    (B) $3, 1, 1, \dfrac{1}{2}$

    (C) $4, 0, 0, \dfrac{1}{2}$

    (D) $4, 1, 1, \dfrac{1}{2}$

    (E) $4, 2, 1, \dfrac{1}{2}$

11. Which of the following elements is diamagnetic?

   (A) H
   (B) Li
   (C) Be
   (D) B
   (E) C

12. Which of the following rules states that no two electrons in an atom can have the same set of quantum numbers?

   (A) Hund's rule
   (B) The Heisenberg uncertainty principle
   (C) The Pauli exclusion principle
   (D) The de Broglie hypothesis
   (E) The Bohr model

13. Which of the following is true of the alkali metal elements?

   (A) They usually take the +2 oxidation state.
   (B) They have oxides that act as acid anhydrides.
   (C) They form covalent bonds with oxygen.
   (D) They are generally found in nature in compounds.
   (E) They have relatively large first ionization energies.

14. Which of the following nuclei has 3 more neutrons than protons? (Remember: The number before the symbol indicates atomic mass.)

   (A) ^{11}B
   (B) ^{37}Cl
   (C) ^{24}Mg
   (D) ^{70}Ga
   (E) ^{19}F

15. Which of the following ions has the smallest ionic radius?

   (A) O^{2-}
   (B) F^-
   (C) Na^+
   (D) Mg^{2+}
   (E) Al^{3+}

16. Which of the following is an impossible set of quantum numbers (n, l, m_l, m_s)?

   (A) $4, 0, 0, \dfrac{1}{2}$

   (B) $4, 0, 1, \dfrac{1}{2}$

   (C) $4, 1, 0, \dfrac{1}{2}$

   (D) $4, 1, 1, \dfrac{1}{2}$

   (E) $4, 2, 1, \dfrac{1}{2}$

17. Which of the following represents the energy of the single electron in a hydrogen atom when it is in the $n = 4$ state?

   (A) $\left(\dfrac{-2.178 \times 10^{-18}}{2} \right)$ joules

   (B) $\left(\dfrac{-2.178 \times 10^{-18}}{4} \right)$ joules

   (C) $\left(\dfrac{-2.178 \times 10^{-18}}{8} \right)$ joules

   (D) $\left(\dfrac{-2.178 \times 10^{-18}}{16} \right)$ joules

   (E) $\left(\dfrac{-2.178 \times 10^{-18}}{64} \right)$ joules

18. When an electron in a hydrogen atom makes the transition from the $n = 4$ state to the $n = 2$ state, blue light with a wavelength of 434 nm is emitted. Which of the following expressions gives the energy released by the transition?

(A) $\dfrac{\left(6.63 \times 10^{-34}\right)\left(3.00 \times 10^{8}\right)}{\left(4.34 \times 10^{-7}\right)}$ joules

(B) $\dfrac{\left(6.63 \times 10^{-34}\right)\left(4.34 \times 10^{-7}\right)}{\left(3.00 \times 10^{8}\right)}$ joules

(C) $\dfrac{\left(6.63 \times 10^{-34}\right)}{\left(4.34 \times 10^{-7}\right)\left(3.00 \times 10^{8}\right)}$ joules

(D) $\dfrac{\left(4.34 \times 10^{-7}\right)}{\left(6.63 \times 10^{-34}\right)\left(3.00 \times 10^{8}\right)}$ joules

(E) $\left(6.63 \times 10^{-34}\right)\left(4.34 \times 10^{-7}\right)$ joules

19. The ionization energies for an element are listed in the table below.

First	Second	Third	Fourth	Fifth
8 eV	15 eV	80 eV	109 eV	141 eV

Based on the ionization energy table, the element is most likely to be

(A) sodium.
(B) magnesium.
(C) aluminum.
(D) silicon.
(E) phosphorous.

20. A researcher listed the first five ionization energies for a silicon atom in order from first to fifth. Which of the following lists corresponds to the ionization energies for silicon?

(A) 780 kJ, 13,675 kJ, 14,110 kJ, 15,650 kJ, 16,100 kJ
(B) 780 kJ, 1,575 kJ, 14,110 kJ, 15,650 kJ, 16,100 kJ
(C) 780 kJ, 1,575 kJ, 3,220 kJ, 15,650 kJ, 16,100 kJ
(D) 780 kJ, 1,575 kJ, 3,220 kJ, 4,350 kJ, 16,100 kJ
(E) 780 kJ, 1,575 kJ, 3,220 kJ, 4,350 kJ, 5,340 kJ

ESSAYS

1. Explain each of the following in terms of atomic and molecular structures and/or forces.

    (a)   The first ionization energy for magnesium is greater than the first ionization energy for calcium.

    (b)   The first and second ionization energies for calcium are comparable, but the third ionization energy is much greater.

    (c)   Solid sodium conducts electricity, but solid sodium chloride does not.

    (d)   The first ionization energy for aluminum is lower than the first ionization energy for magnesium.

2. Silicon is a nonmetal with four valence electrons.

    (a)   Write the ground state electron configuration for silicon.

    (b)   Which fundamental atomic theory is violated by the following list of quantum numbers representing silicon's valence electrons?

n	l	m_l	m_s
3	0	0	$-\dfrac{1}{2}$
3	0	0	$-\dfrac{1}{2}$
3	1	1	$-\dfrac{1}{2}$
3	1	1	$-\dfrac{1}{2}$

    (c)   Which fundamental atomic theory is violated by the following list of quantum numbers representing silicon's valence electrons?

n	l	m_l	m_s
3	0	0	$-\dfrac{1}{2}$
3	0	0	$+\dfrac{1}{2}$
3	1	-1	$-\dfrac{1}{2}$
3	1	-1	$+\dfrac{1}{2}$

    (d)   Will a lone silicon atom be diamagnetic or paramagnetic? Justify your answer.

3. Use your knowledge of the periodic table of the elements to answer the following questions.

   (a) Explain the trend in electronegativity from P to S to Cl.

   (b) Explain the trend in electronegativity from Cl to Br to I.

   (c) Explain the trend in atomic radius from Li to Na to K.

   (d) Explain the trend in atomic radius from Al to Mg to Na.

4. Use your knowledge of atomic theory to answer the following questions.

   (a) State the Heisenberg uncertainty principle.

   (b) The absorption spectrum of a hydrogen atom contains dark bands at specific wavelengths. The emission spectrum of a hydrogen atom contains bright bands at the same wavelengths. Explain what causes these bright and dark bands at specific wavelengths.

   (c) Explain why the addition of an electron to a chlorine atom is an exothermic process and the addition of an electron to a magnesium atom is an endothermic process.

   (d) Explain how the valence electron configuration of sulfur is consistent with the existence of Na_2S and SF_6.

CHAPTER 3 ANSWERS AND EXPLANATIONS

MULTIPLE-CHOICE QUESTIONS

1. **D** Fluorine, which needs one electron to complete its outer shell, is the most electronegative element, with an electronegativity of 4.0.

2. **A** In carbon dating, the ratio of carbon-14 to carbon-12 in an organic artifact is used to determine the age of the artifact.

3. **E** Neon's second shell is complete, so all of its electrons are spin-paired.

4. **B** Nitrogen gas (N_2) makes up 78 percent of the earth's atmosphere. Oxygen is next at 21 percent.

5. **B** Silicon (Si) is used in semiconductor technology because it has properties that lie in-between metals and nonmetals.

6. **A** Mercury (Hg) is unusual among metals in that it is a liquid at room temperature. Its melting point is –39 degrees Celsius.

7. **B** Silicon makes up 26 percent of the earth's crust by weight (oxygen makes up 50 percent). In fact, compounds including silicon and oxygen make up nearly all rocks and soils.

8. **B** Neutral sodium in its ground state has the electron configuration shown in choice (C). Sodium forms a bond by giving up its one valence electron and becoming a positively charged ion with the same electron configuration as neon.

9. **D** As we move from left to right across the periodic table within a single period (from sodium to chlorine), we add protons to the nuclei, which progressively increases the pull of each nucleus on its electrons. So chlorine will have a higher first ionization energy, greater electronegativity, and a smaller atomic radius.

10. **C** Potassium's valence electron is in the $4s$ subshell. That means that $n = 4$, $l = 0$, $m_l = 0$, and $m_s = \frac{1}{2}$ or $-\frac{1}{2}$.

11. **C** Beryllium's electrons are paired up in the completed orbitals of the $1s$ and $2s$ subshells. Choice (E) is wrong because, according to Hund's rule, carbon's two $2p$ electrons must be placed in different orbitals.

12. **C** The Pauli exclusion principle states that no two electrons in an atom can have the same set of quantum numbers.

    About the other answers:

    (A) Hund's rule states that within a subshell, electrons will be placed in empty orbitals while they are available and will start to pair up in orbitals only when no more empty orbitals are available.

    (B) The Heisenberg uncertainty principle states that it is impossible to know with certainty both the position and momentum of a particle at the same moment.

    (D) The de Broglie hypothesis relates the wave and particle properties of matter, using the following equation:

    $$\text{Wavelength} = \frac{h}{\text{momentum}}$$

    h is Planck's constant, 6.63×10^{-34} joule-sec

    (E) The Bohr model of the hydrogen atom (which was disproved by the Heisenberg uncertainty principle) states that electrons orbit the nucleus in fixed, quantized circular orbits.

13. **D**   The alkali metals (Li, Na, K...) are extremely reactive and are found in nature almost exclusively in compounds.

As for the other answers:

Choice (A) is wrong because the alkali metals take the +1 oxidation state. Choice (B) is wrong because alkali metal oxides are basic anhydrides; that is, they form basic solutions in water. Choice (C) is wrong because they form ionic bonds with oxygen. Choice (E) is wrong because they have only one valence electron, so they have relatively small first ionization energies.

14. **B**   The atomic mass is the sum of the neutrons and protons in any atom's nucleus. Since atomic number, which indicates the number of protons, is unique to each element we can subtract this from the weight to find the number of neutrons. The atomic number of B is 5; hence, the ^{11}B has 6 neutrons, only 1 in excess. The atomic number of Cl is 17, so ^{37}Cl has 20 neutrons, 3 in excess. The atomic number of Mg is 12, so ^{24}Mg has the same number of neutrons as protons. The element Ga has an atomic number of 31, meaning there are 39 neutrons in the given nucleus and the atomic number for F is 9, meaning 10 neutrons in ^{19}F.

15. **E**   All of the ions listed have the same electron configuration as neutral neon. Al^{3+} has the most protons, so its electrons will experience greater attractive force from the nucleus, resulting in the smallest ions.

16. **B**   The m_l quantum number (orbital) can't be larger than the l quantum number (subshell), so $4, 0, 1, \frac{1}{2}$ is impossible because 1 is greater than 0.

17. **D**   When you know the principal quantum number, you must use the formula for the energy of an electron, which is shown below.

$$E_n = \left(\frac{-2.178 \times 10^{-18}}{n^2} \right)$$

The variable n stands for the principal quantum number; in this case, $n = 4$.

18. **A**   When you know the wavelength emitted, you must use the formula for the energy of an electron transition, which is shown below. Remember: A nanometer is equal to 10^{-9} meters.

$$E_n = \frac{hc}{\lambda} = \frac{\left(6.63 \times 10^{-34} \right)\left(3.00 \times 10^8 \right)}{\left(4.34 \times 10^{-7} \right)}$$

19. **B**   The ionization energy will show a large jump when enough electrons have been removed to leave a stable shell. In this case, the jump occurs between the second and third electrons removed, so the element is stable after two electrons are removed. Magnesium (Mg) is the only element on the list with exactly two valence electrons.

20. **D**   The ionization energy will show a large jump when enough electrons have been removed to leave a stable shell. Silicon has four valence electrons, so we would expect to see a large jump in ionization energy after the fourth electron has been removed. That's choice (D).

ESSAYS

1.   (a) Ionization energy is the energy required to remove an electron from an atom. The outermost electron in Ca is at the $4s$ energy level. The outermost electron in Mg is at the $3s$ level. The outermost electron in Ca is at a higher energy level and is more shielded from the nucleus, making it easier to remove.

(b) Calcium has two electrons in its outer shell. The second ionization energy will be larger than the first but still comparable because both electrons are being removed from the same energy level. The third electron is much more difficult to remove because it is being removed from a lower energy level, so it will have a much higher ionization energy than the other two.

(c) Solid sodium exhibits metallic bonding, in which the positively charged sodium ions are held together by a sea of mobile, delocalized electrons. These electrons move freely from nucleus to nucleus, making solid sodium a good conductor.

Sodium chloride exhibits ionic bonding, in which positively charged sodium ions and negatively charged chlorine ions hold fixed places in a crystal lattice. The electrons are localized around particular nuclei and are not free to move about the lattice. This makes solid sodium chloride a bad conductor of electricity.

(d) The valence electron to be removed from magnesium is located in the completed $3s$ subshell, while the electron to be removed from aluminum is the lone electron in the $3p$ subshell. It is easier to remove the electron from the higher-energy $3p$ subshell than from the lower energy (completed) $3s$ subshell, so the first ionization energy is lower for aluminum.

2. (a) $1s^22s^22p^63s^23p^2$

    (b) The Pauli exclusion principle is violated. The Pauli exclusion principle states that no two electrons can have the same set of quantum numbers.

    (c) Hund's rule is violated. Hund's rule states that within an energy level, electrons will be placed in empty orbitals while they are available and will start to pair up in orbitals only when no more empty orbitals are available.

    (d) A lone silicon atom will be paramagnetic and will be affected by a magnetic field. If all the electrons in an atom are spin-paired, the atom is diamagnetic. If any of the electrons are not spin-paired, the atom is paramagnetic. Silicon has two electrons in the $3p$ subshell. They will be placed in different orbitals, so they will not be spin-paired.

3. (a) Electronegativity is the pull of the nucleus of one atom on the electrons of other atoms; it increases from P to S to Cl because nuclear charge increases. This is because as you from move left to right across the periodic table, atomic radii decrease in size. Increasing nuclear charge means that Cl has the most positively charged nucleus of the three and will exert the greatest pull on the electrons of other atoms.

    (b) Electronegativity is the pull of the nucleus of one atom on the electrons of other atoms; it decreases from Cl to Br to I because electron shells are added. The added electron shells shield the nucleus, causing it to have less of an effect on the electrons of other atoms. Therefore, iodine will exert the least pull on the electrons of other atoms.

    (c) Atomic radius increases from Li to Na to K because electrons are being added in higher energy levels, which are farther away from the nucleus; therefore, the K atom is the largest of the three.

    (d) Atomic radius increases from Al to Mg to Na because protons are being removed from the nucleus while the energy levels of the valence electrons remains unchanged. If there are fewer positive charges in the nucleus, the electrons of Na will be less attracted to the nucleus and will remain farther away.

4. (a) A straightforward statement of the Heisenberg uncertainty principle is as follows: It is impossible to know with certainty both the position and momentum of a particle at one moment.

    (b) When a hydrogen atom absorbs energy, its electrons jump to higher energy levels. The absorbed energy shows up as a dark area on the absorption spectrum.

    A hydrogen atom gives off energy when its electrons jump back down to lower energy levels. Electromagnetic waves, which are emitted at these jumps, show up as bright areas on the emission spectrum.

In an atom, energy is quantized, which means that electrons can exist only at specific energy levels. When an electron jumps from one energy level to another, it will always emit or absorb exactly the same amount of energy, and because $\Delta E = h\nu$, for a particular energy change, radiation of the same frequency will always be emitted or absorbed.

(c) When an electron is added to chlorine, the chlorine ion created has a complete valence shell, which is an extremely stable, low energy configuration. When something becomes more stable, energy is given off, making the process exothermic.

When an electron is added to a magnesium atom, it must be placed by itself in the $3p$ energy level. Adding an electron in a higher energy level makes the magnesium ion created more energetic and less stable, which means that the process is endothermic.

(d) Sulfur has six valence electrons in its outer shell. In Na_2S, sulfur gains two electrons to give its outer shell a complete octet. In SF_6, sulfur uses sp^3d^2 hybridization to share all six of its valence electrons with fluorine atoms.

Bonding

HOW OFTEN DOES BONDING APPEAR ON THE EXAM?

In the multiple-choice section, this topic appears in about 9 out of 75 questions. In the free-response section, this topic appears every year.

COULOMB'S LAW

All bonds occur because of electrostatic attractions. Atoms stick together to form molecules, and atoms and molecules stick together to form liquids or solids because the negatively charged electrons of one atom are attracted to the positively charged nucleus of another atom.

Electrostatic forces are governed by Coulomb's law, and the entire study of bonding comes down to understanding how Coulomb's law applies to different chemical situations.

Let's take a look at Coulomb's law.

Coulomb's Law

Attractive force is proportional to $\dfrac{(+q)(-q)}{r^2}$

$+q$ = magnitude of the positive charge
$-q$ = magnitude of the negative charge
$\ r$ = distance between the charges

You should be able to extrapolate two things from the formula above.

- Bigger charges mean stronger bonds; smaller charges mean weaker bonds.

- Charges close together mean stronger bonds; charges far apart mean weaker bonds.

BONDS

Atoms join to form molecules because atoms like to have a full outer shell of electrons. This usually means having eight electrons in the outer shell. So atoms with too many or too few electrons in their valence shells will find one another and pass the electrons around until all the atoms in the molecule have stable outer shells. Sometimes an atom will give up electrons completely to another atom, forming an ionic bond. Sometimes atoms share electrons, forming covalent bonds.

IONIC BONDS

An ionic solid is held together by the electrostatic attractions between ions that are next to one another in a lattice structure. They often occur between metals and nonmetals. In an ionic bond, electrons are not shared. Instead, one atom gives up electrons and becomes a positively charged ion while the other atom accepts electrons and becomes a negatively charged ion.

The two ions in an ionic bond are held together by electrostatic forces. In the diagram below, a sodium atom has given up its single valence electron to a chlorine atom, which has seven valence electrons and uses the electron to complete its outer shell (with eight). The two atoms are then held together by the positive and negative charges on the ions.

$$\left[\text{Na}\right]^+ \left[:\ddot{\underset{..}{\text{Cl}}}:\right]^-$$

The same electrostatic attractions that hold together the ions in a molecule of NaCl hold together a crystal of NaCl, so there is no real distinction between the molecules and the solid. Ionic bonds are strong, and substances held together by ionic bonds have high melting and boiling points.

Coulomb's law states that more highly charged ions will form stronger bonds than less highly charged ions and smaller ions will form stronger bonds than larger ions. So an ionic bond composed of ions with +2 and –2 charges will be stronger than a bond composed of ions with +1 and –1 charges. Also from Coulomb's law, we know that the smaller the ions in an ionic bond, the stronger the bond. This is because a small ionic radius allows the charges to get closer together and increases the force between them.

In an ionic solid, each electron is localized around a particular atom, so electrons do not move around the lattice; this makes ionic solids poor conductors of electricity. Ionic liquids, however, do conduct electricity because the ions themselves are free to move about in the liquid phase, although the electrons are still localized around particular atoms. Salts are held together by ionic bonds.

COVALENT BONDS

In a covalent bond, two atoms share electrons. Each atom counts the shared electrons as part of its valence shell. In this way, both atoms achieve complete outer shells.

In the diagram below, two fluorine atoms, each of which has seven valence electrons and needs one electron to complete its valence shell, form a covalent bond. Each atom donates an electron to the bond, which is considered to be part of the valence shell of both atoms.

$$:\ddot{F}\cdot \ + \ \cdot\ddot{F}: \ \Rightarrow \ :\ddot{F}\!:\!\ddot{F}:$$

The number of covalent bonds an atom can form is the same as the number of electrons in its valence shell.

The first covalent bond formed between two atoms is called a sigma (σ) bond. All single bonds are sigma bonds. If additional bonds between the two atoms are formed, they are called pi (π) bonds. The second bond in a double bond is a pi bond and the second and third bonds in a triple bond are also pi bonds. Double and triple bonds are stronger and shorter than single bonds, but they are not twice or triple the strength.

Summary of Multiple Bonds			
Bond type:	Single	Double	Triple
Bond designation:	One sigma (σ)	One sigma (σ) and one pi (π)	One sigma (σ) and two pi (π)
Bond order:	One	Two	Three
Bond length:	Longest	Intermediate	Shortest
Bond energy:	Least	Intermediate	Greatest

Single bonds have one sigma (σ) bond and a bond order of one.
The single bond has the longest bond length and the least bond energy.

Polarity

In the F_2 molecule shown above, the two fluorine atoms share the electrons equally, but that's not usually the case in molecules. Usually, one of the atoms (the more electronegative one) will exert a stronger pull on the electrons in the bond—not enough to make the bond ionic, but enough to keep the electrons on one side of the molecule more than on the other side. This gives the molecule a dipole. That is, the side of the molecule where the electrons spend more time will be negative and the side of the molecule where the electrons spend less time will be positive.

Dipole Moment

The polarity of a molecule is measured by the dipole moment. The larger the dipole moment, the more polar the molecule. The greater the charge at the ends of the dipole and the greater the distance between the charges, the greater the value of the dipole moment.

LEWIS DOT STRUCTURES

Drawing Lewis Dot Structures

At some point on the test, you'll be asked to draw the Lewis structure for a molecule or polyatomic ion. Here's how to do it.

1. Count the valence electrons in the molecule or polyatomic ion; refer to page 18 for a periodic table.

2. If a polyatomic ion has a negative charge, add electrons equal to the charge of the total in (1). If a polyatomic ion has a positive charge, subtract electrons equal to the charge of the electrons from the total in (1).

3. Draw the skeletal structure of the molecule and place two electrons (or a single bond) between each pair of bonded atoms. If the molecule contains three or more atoms, the least electronegative atom will usually occupy the central position.

4. Add electrons to the surrounding atoms until each has a complete outer shell.

5. Add the remaining electrons to the central atom.

6. Look at the central atom.

    (a) If the central atom has fewer than eight electrons, remove an electron pair from an outer atom and add another bond between that outer atom and the central atom. Do this until the central atom has a complete octet.

    (b) If the central atom has a complete octet, you are finished.

    (c) If the central atom has more than eight electrons, that's okay too.

Let's find the Lewis structure for the CO_3^{2-} ion.

1. Carbon has 4 valence electrons; oxygen has 6.
   $4 + 6 + 6 + 6 = 22$

2. The ion has a charge of –2, so add 2 electrons.
   $22 + 2 = 24$

3. Carbon is the central atom.

4. Add electrons to the oxygen atoms.

5. We've added all 24 electrons, so there's nothing left to put on the carbon atom.

6. (a) We need to give carbon a complete octet, so we take an electron pair away from one of the oxygens and make a double bond instead. Place a bracket around the model and add a charge of negative two.

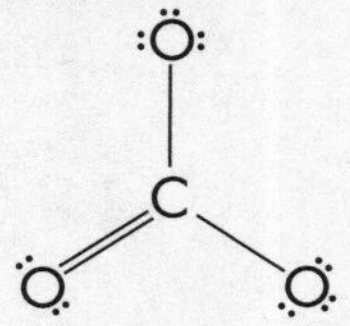

Resonance Forms

When we put a double bond into the CO_3^{2-} ion, we place it on any one of the oxygen atoms, as shown below.

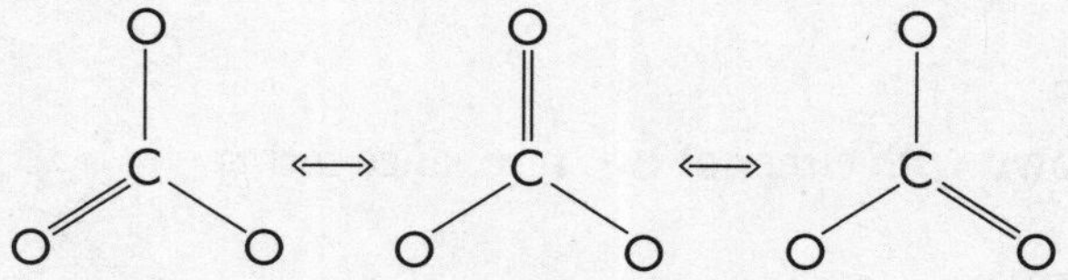

All three resonance forms are considered to exist simultaneously, and the strength and lengths of all three bonds are the same: somewhere between the strength and length of a single bond and a double bond.

Incomplete Octets

Some atoms can have a complete outer shell with less than eight electrons; for example, hydrogen can have a maximum of two electrons, and beryllium can be stable with only four valence electrons, as in BeH_2.

$$H - Be - H$$

Boron can be stable with only six valence electrons, as in BF_3.

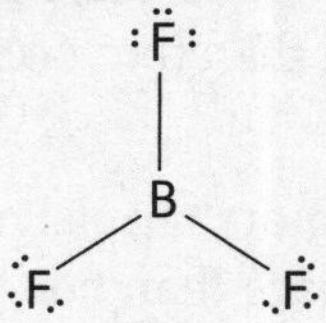

Expanded Octets

In molecules that have d subshells available, the central atom can have more than eight valence electrons.

Here are some examples.

PCl_5

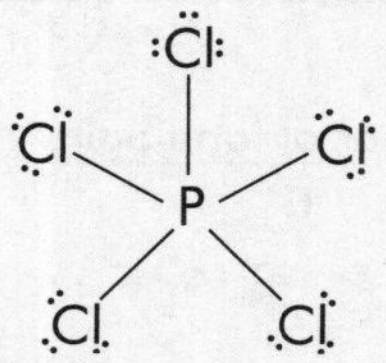

SF_4

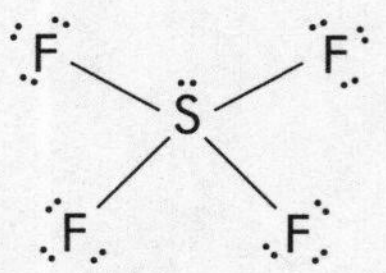

SF_6

Odd Numbers of Electrons

Molecules almost always have an even number of electrons, allowing electrons to be paired, but there are exceptions, usually involving nitrogen.

NO

NO_2

Note that NO_2 can be shown with either of two resonance forms.

MOLECULAR GEOMETRY

Electrons repel one another, so when atoms come together to form a molecule, the molecule will assume the shape that keeps its different electron pairs as far apart as possible. When we predict the geometries of molecules using this idea, we are using the **valence shell electron-pair repulsion (VSEPR) model.**

In a molecule with more than two atoms, the shape of the molecule is determined by the number of electron pairs on the central atom. The central atom forms **hybrid orbitals,** each of which has a standard shape. Variations on the standard shape occur depending on the number of bonding pairs and lone pairs of electrons on the central atom.

Here are some things you should remember when dealing with the VSEPR model.

- Double and triple bonds are treated in the same way as single bonds in terms of predicting overall geometry for a molecule; however, multiple bonds have slightly more repulsive strength and will therefore occupy a little more space than single bonds.

- Lone electron pairs have a little more repulsive strength than bonding pairs, so lone pairs will occupy a little more space than bonding pairs.

The tables on the following pages show the different hybridizations and geometries that you might see on the test.

If the central atom has **2** electron pairs, then it has *sp* hybridization and its basic shape is **linear.**

Number of lone pairs	Geometry	Examples
0	B—A—B	$BeCl_2$
	linear	CO_2

If the central atom has **3** electron pairs, then it has sp^2 hybridization and its basic shape is **trigonal planar**; its bond angles are about 120°.

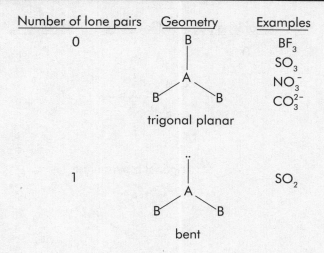

Number of lone pairs	Geometry	Examples
0	trigonal planar	BF_3 SO_3 NO_3^- CO_3^{2-}
1	bent	SO_2

If the central atom has **4** electron pairs, then it has sp^3 hybridization and its basic shape is **tetrahedral**; its bond angles are about 109.5°.

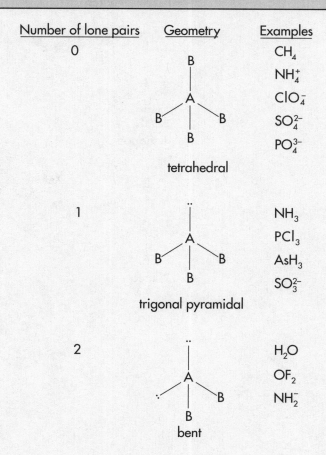

Number of lone pairs	Geometry	Examples
0	tetrahedral	CH_4 NH_4^+ ClO_4^- SO_4^{2-} PO_4^{3-}
1	trigonal pyramidal	NH_3 PCl_3 AsH_3 SO_3^{2-}
2	bent	H_2O OF_2 NH_2^-

If the central atom has **5** electron pairs, then it has dsp^3 hybridization and its basic shape is **trigonal bipyramidal.**

Number of lone pairs	Geometry	Examples
0		PCl_5
		PF_5

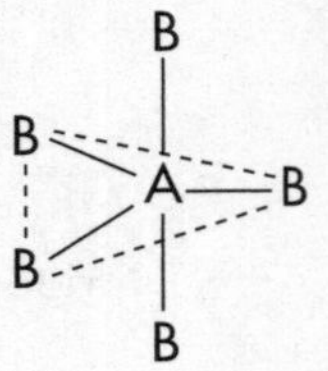

trigonal bipyramidal

| 1 | | SF_4 |
| | | IF_4^+ |

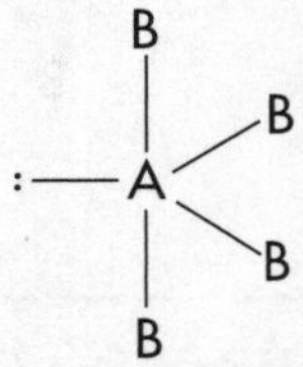

folded square, seesaw,
distorted tetrahedron

| 2 | | ClF_3 |
| | | ICl_3 |

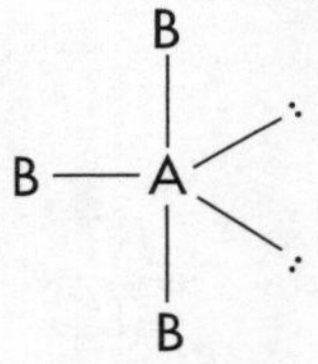

T-shaped

| 3 | | XeF_2 |
| | | I_3^- |

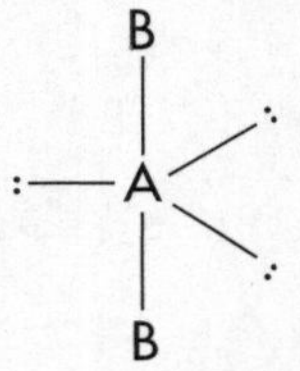

linear

> If the central atom has **6** electron pairs, then it has d^2sp^3 hybridization and its basic shape is **octahedral.**

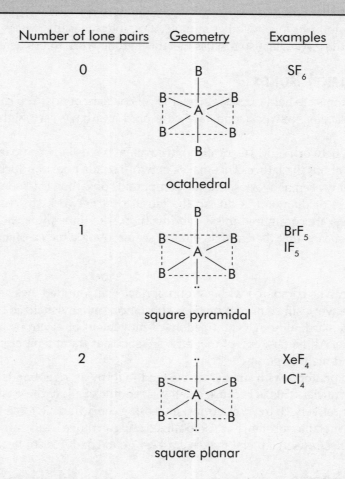

Number of lone pairs	Geometry	Examples
0	octahedral	SF_6
1	square pyramidal	BrF_5 IF_5
2	square planar	XeF_4 ICl_4^-

In trigonal bipyrimidal shapes, place the lone pairs in axial position first. In octahedral shapes, place lone pairs in equatorial position first.

BONDS BETWEEN MOLECULES—ATTRACTIVE FORCES IN SOLIDS AND LIQUIDS

Sometimes the bonds that hold the atoms or ions in liquids and solids together are the same strong bonds that hold the atoms or ions together in molecules. Intermolecular forces exist only with covalently bonded molecules. We will also discuss metallic and network forces here.

NETWORK (COVALENT) BONDS

In a network solid, atoms are held together in a lattice of covalent bonds. You can visualize a network solid as one big molecule. Network solids are very hard and have very high melting and boiling points.

The electrons in a network solid are localized in covalent bonds between particular atoms, so they are not free to move about the lattice. This makes network solids poor conductors of electricity.

The most commonly seen network solids are compounds of carbon (diamond) and silicon (SiO_2— quartz). The hardness of diamond is due to the tetrahedral network structure formed by carbon atoms whose electrons are configured in sp^3 hybridization. The three-dimensional complexity of the tetrahedral network means that there are no natural seams along which a diamond can be broken.

METALLIC BONDS

Metallic substances can be compared with a group of nuclei surrounded by a sea of mobile electrons. As with ionic and network substances, a metallic substance can be visualized as one large molecule. Most metals are very hard, although the freedom of movement of electrons in metals makes them malleable and ductile. All metals, except mercury, are solids at room temperature, and most metals have high boiling and melting points.

Metals composed of atoms with smaller nuclei tend to form stronger bonds than metals made up of atoms with larger nuclei. This is because smaller sized nuclei allow the positively charged nuclei to be closer to the negatively charged electrons, increasing the attractive force from Coulomb's law.

The electrons in a metallic substance are delocalized and can move freely throughout the substance. The freedom of the electrons in a metal makes it a very good conductor of heat and electricity.

VAN DER WAALS FORCES

Dipole–Dipole Forces

Dipole–dipole forces occur between neutral, polar molecules: The positive end of one polar molecule is attracted to the negative end of another polar molecule.

Molecules with greater polarity will have greater dipole–dipole attraction, so molecules with larger dipole moments tend to have higher melting and boiling points. Dipole–dipole attractions are relatively weak, however, and these substances melt and boil at very low temperatures. Most substances held together by dipole–dipole attraction are gases or liquids at room temperature.

London Dispersion Forces

London dispersion forces occur between neutral, nonpolar molecules. These very weak attractions occur because of the random motions of electrons on atoms within molecules. At a given moment, a nonpolar molecule might have more electrons on one side than on the other, giving it an instantaneous polarity. For that fleeting instant, the molecule will act as a very weak dipole.

Since London dispersion forces depend on the random motions of electrons, molecules with more electrons will experience greater London dispersion forces. So among substances that experience only London dispersion forces, the one with more electrons will generally have higher melting and boiling points. London dispersion forces are even weaker than dipole–dipole forces, so substances that experience only London dispersion forces melt and boil at extremely low temperatures and tend to be gases at room temperature.

HYDROGEN BONDS

Hydrogen bonds are similar to dipole–dipole attractions. In a hydrogen bond, the positively charged hydrogen end of a molecule is attracted to the negatively charged end of another molecule containing an extremely electronegative element (fluorine, oxygen, or nitrogen—F, O, N).

Hydrogen bonds are much stronger than dipole–dipole forces because when a hydrogen atom gives up its lone electron to a bond, its positively charged nucleus is left virtually unshielded. Substances that have hydrogen bonds, like water and ammonia, have higher melting and boiling points than substances that are held together by dipole–dipole forces.

Water is less dense as a solid than as a liquid because its hydrogen bonds force the molecules in ice to form a crystal structure, which keeps them farther apart than they are in the liquid form.

CHAPTER 4 QUESTIONS

MULTIPLE-CHOICE QUESTIONS

Questions 1–4

(A) Metallic bonding
(B) Network covalent bonding
(C) Hydrogen bonding
(D) Ionic bonding
(E) London dispersion forces

1. Solids exhibiting this kind of bonding are excellent conductors of heat.

2. This kind of bonding is the reason that water is more dense than ice.

3. This kind of bonding exists between atoms with very different electronegativities.

4. The stability exhibited by diamonds is due to this kind of bonding.

Questions 5–7

(A) CH_4
(B) NH_3
(C) $NaCl$
(D) N_2
(E) H_2

5. This substance undergoes ionic bonding.

6. This molecule contains two pi (π) bonds.

7. This substance undergoes hydrogen bonding.

Questions 8–10

(A) BF_3
(B) CO_2
(C) H_2O
(D) CF_4
(E) PH_3

8. The central atom in this molecule forms sp^2 hybrid orbitals.

9. This molecule has a tetrahedral structure.

10. This molecule has a linear structure.

11. A liquid whose molecules are held together by which of the following forces would be expected to have the lowest boiling point?

   (A) Ionic bonds
   (B) London dispersion forces
   (C) Hydrogen bonds
   (D) Metallic bonds
   (E) Network bonds

12. Hydrogen bonding would be seen in a sample of which of the following substances?

   (A) CH_4
   (B) H_2
   (C) H_2O
   (D) HI
   (E) All of the above

13. Which of the following species does NOT have a tetrahedral structure?

   (A) CH_4
   (B) NH_4^+
   (C) SF_4
   (D) $AlCl_4^-$
   (E) CBr_4

14. Which form of orbital hybridization can form molecules with shapes that are either trigonal pyramidal or tetrahedral?

   (A) sp
   (B) sp^2
   (C) sp^3
   (D) d^2sp
   (E) dsp^3

15. The six carbon atoms in a benzene molecule are shown in different resonance forms as three single bonds and three double bonds. If the length of a single carbon–carbon bond is 154 pm and the length of a double carbon–carbon bond is 133 pm, what length would be expected for the carbon–carbon bonds in benzene?

   (A) 126 pm
   (B) 133 pm
   (C) 140 pm
   (D) 154 pm
   (E) 169 pm

16. Which of the following could be the Lewis structure for sulfur trioxide?

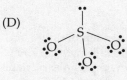

17. In which of the following species does the central atom NOT form sp^2 hybrid orbitals?

   (A) SO_2
   (B) BF_3
   (C) NO_3^-
   (D) SO_3
   (E) PCl_3

18. A molecule whose central atom has d^2sp^3 hybridization can have which of the following shapes?

   I. Tetrahedral
   II. Square pyramidal
   III. Square planar

   (A) I only
   (B) III only
   (C) I and II only
   (D) II and III only
   (E) I, II, and III

19. Which of the following molecules will have a Lewis dot structure with exactly one unshared electron pair on the central atom?

    (A) H_2O
    (B) PH_3
    (C) PCl_5
    (D) CH_2Cl_2
    (E) $BeCl_2$

20. Which of the following lists of species is in order of increasing boiling points?

    (A) H_2, N_2, NH_3
    (B) N_2, NH_3, H_2
    (C) NH_3, H_2, N_2
    (D) NH_3, N_2, H_2
    (E) H_2, NH_3, N_2

21. Solid NaCl melts at a temperature of 800°C, while solid NaBr melts at 750°C. Which of the following is an explanation for the higher melting point of NaCl?

    (A) A chlorine ion has less mass than a bromine ion.
    (B) A chlorine ion has a greater negative charge than a bromine ion.
    (C) A chlorine ion has a lesser negative charge than a bromine ion.
    (D) A chlorine ion is smaller than a bromine ion.
    (E) A chlorine ion is larger than a bromine ion.

22. Which of the compounds listed below would require the greatest energy to separate it into ions in the gaseous state?

    (A) NaCl
    (B) NaI
    (C) MgO
    (D) Na_2O
    (E) $MgCl_2$

23. Which sample of the following compounds contains both ionic and covalent bonds?

    (A) H_2O_2
    (B) CH_3Cl
    (C) C_2H_5OH
    (D) $NaNO_3$
    (E) NH_2OH

24. Which of the molecules listed below has the largest dipole moment?

    (A) Cl_2
    (B) HCl
    (C) SO_3
    (D) NO
    (E) N_2

25. Which of the following statements about boiling points is (are) correct?

    I. H_2O boils at a higher temperature than CO_2.
    II. Ar boils at a higher temperature than He.
    III. Rb boils at a higher temperature than Na.

    (A) I only
    (B) I and II only
    (C) I and III only
    (D) II and III only
    (E) I, II, and III

ESSAYS

1. Use the principles of bonding and molecular structure to explain the following statements.

   (a) The boiling point of argon is –186°C, whereas the boiling point of neon is –246°C.

   (b) Solid sodium melts at 98°C, but solid potassium melts at 64°C.

   (c) More energy is required to break up a CaO(s) crystal into ions than to break up a KF(s) crystal into ions.

   (d) Molten KF conducts electricity, but solid KF does not.

2. The carbonate ion CO_3^{2-} is formed when carbon dioxide, CO_2, reacts with slightly basic cold water.

   (a) (i) Draw the Lewis electron dot structure for the carbonate ion. Include resonance forms if they apply.

   (ii) Draw the Lewis electron dot structure for carbon dioxide.

   (b) Describe the hybridization of carbon in the carbonate ion.

   (c) (i) Describe the relative lengths of the three C–O bonds in the carbonate ion.

   (ii) Compare the average length of the C–O bonds in the carbonate ion to the average length of the C–O bonds in carbon dioxide.

3.

Substance	Boiling Point (°C)	Bond Length (Å)	Bond Strength (kcal/mol)
H_2	–253°	0.75	104.2
N_2	–196°	1.10	226.8
O_2	–182°	1.21	118.9
Cl_2	–34°	1.99	58.0

   (a) Explain the differences in the properties given in the table above for each of the following pairs.

   (i)   The bond strengths of N_2 and O_2

   (ii) The bond lengths of H_2 and Cl_2

   (iii) The boiling points of O_2 and Cl_2

   (b) Use the principles of molecular bonding to explain why H_2 and O_2 are gases at room temperature, while H_2O is a liquid at room temperature.

4. H_2S, SO_4^{2-}, XeF_2, ICl_4^-

   (a) Draw a Lewis electron dot diagram for each of the molecules listed above.

   (b) Use the valence shell electron-pair repulsion (VSEPR) model to predict the geometry of each of the molecules.

5. Use the principles of bonding and molecular structure to explain the following statements.

    (a)   The angle between the N–F bonds in NF_3 is smaller than the angle between the B–F bonds in BF_3.

    (b) $I_2(s)$ is insoluble in water, but it is soluble in carbon tetrachloride.

    (c)   Diamond is one of the hardest substances on Earth.

    (d)   HCl has a lower boiling point than either HF or HBr.

CHAPTER 4 ANSWERS AND EXPLANATIONS

MULTIPLE-CHOICE QUESTIONS

1. **A** In metallic bonding, nuclei are surrounded by a sea of mobile electrons. The electrons' freedom to move allows them to conduct heat and electricity.

2. **C** When ice forms, the hydrogen bonds join the molecules in a lattice structure, which forces them to remain farther apart than they were in the liquid form. Because the molecules are farther apart in the solid than in the liquid, the solid (ice) is less dense than the liquid.

3. **D** Electronegativity is a measure of how much pull an atom exerts on another atom's electrons. If the difference in electronegativities is large enough (greater than 1.7) then the more electronegative atom will simply take an electron away from the other atom. The two atoms will then be held together by electrostatic attraction (the atom that has gained an electron becomes negative and the atom that has lost an electron becomes positive). That's an ionic bond.

4. **B** The carbon atoms in diamond are held together by a network of covalent bonds. The carbon atoms form sp^3 hybrid orbitals, resulting in a tetrahedral structure, which is very stable and has no simple breaking points.

5. **C** NaCl is composed of two elements that have very different electronegativities than each other, so Na gives up an electron to Cl and the two are held together by electrostatic attraction in an ionic bond.

6. **D** N_2 contains a triple bond, so it has one sigma (σ) bond and two pi (π) bonds.

7. **B** Hydrogen bonding occurs between hydrogen atoms of one molecule and electronegative elements (F, O, or N) of another molecule. So in ammonia, hydrogens from one ammonia molecule will form bonds with nitrogens from another ammonia molecule.

8. **A** In BF_3, boron forms three bonds with fluorine atoms and has no unbonded valence electrons, so it must form sp^2 hybrid orbitals.

9. **D** CF_4 forms a tetrahedral structure as shown in the diagram below. The central carbon atom is hybridized sp^3.

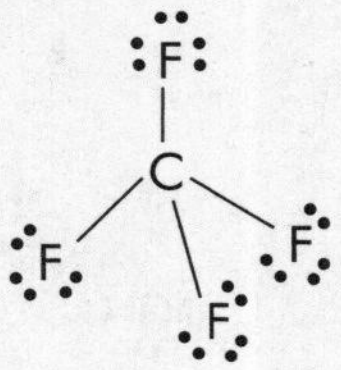

10. **B** CO_2 forms a linear structure as shown in the diagram below. The central carbon atom is sp hybridized.

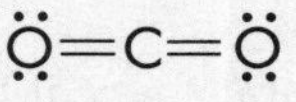

11. **B** A liquid with a low boiling point must be held together by weak bonds. London dispersion forces are the weakest kind of intermolecular forces.

12. **C** Hydrogen bonding specifically describes the attraction experienced by a hydrogen atom in one molecule to an extremely electronegative element (F, O, or N) in another molecule. So in water, a hydrogen atom in one water molecule will be attracted to an oxygen atom in another water molecule.

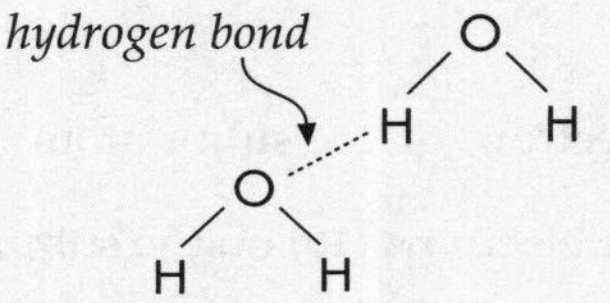

13. **C** SF_4 has 34 valence electrons distributed in the Lewis dot structure and shape shown below.

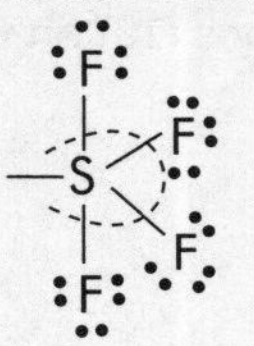

In this molecule, sulfur forms dsp^3 hybrid orbitals, which have a trigonal bipyramid structure. Because SF_4 has one unshared electron pair, the molecule takes the "seesaw" or "folded square" shape.

Choices (A) and (B), CH_4 and NH_4^+, each have 8 valence electrons distributed in the same Lewis dot structure and shape, shown below for NH_4^+.

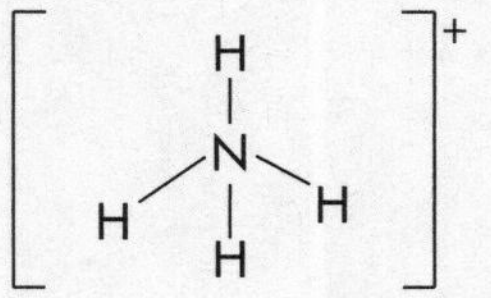

In these molecules, the central atom forms sp^3 hybrid orbitals, which have a tetrahedral structure. There are no unshared electron pairs on the central atom, so the molecules are tetrahedral.

Choices (D) and (E), $AlCl_4^-$ and CBr_4, each have 32 valence electrons distributed in the same Lewis dot structure and shape, shown below for $AlCl_4^-$.

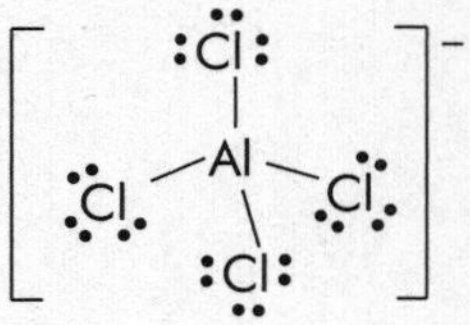

In these molecules, the central atom forms sp^3 hybrid orbitals, which have a tetrahedral structure. There are no unshared electron pairs on the central atom, so the molecules are tetrahedral.

14. **C** The sp^3 hybrid orbitals take a tetrahedral shape if the central atom has no unshared electron pairs (CH_4, for instance). If the central atom has one unshared electron pair, the molecule takes the trigonal pyramid shape (NH_3, for instance).

15. **C** Resonance is used to describe a situation that lies between single and double bonds, so the bond length would also be expected to be in between that of single and double bonds.

16. **A** Choice (A) has the correct number of valence electrons (6 + 6 + 6 + 4 + 2 = 24), and 8 valence electrons on each atom.

    About the other choices:

    (B) There are only 6 valence electrons on the sulfur atom.

    (C) There are too many valence electrons (10) on the sulfur atom.

    (D) There are too many valence electrons (26).

    (E) There are too many valence electrons (10) on the sulfur atom and not enough (6) on one of the oxygen atoms.

17. **E** PCl_3 is the only one that doesn't form sp^2 hybrid orbitals, forming sp^3 orbitals instead. The Lewis dot structures for all of the choices are shown below (note that boron does not need an octet).

(A)

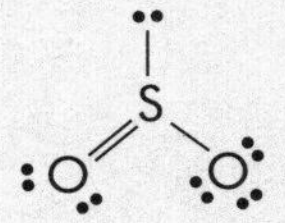

(B)

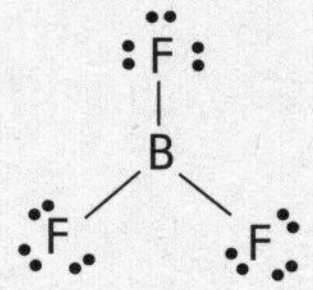

(C)

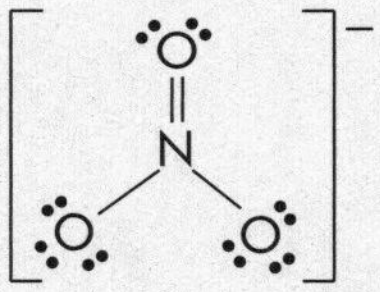

(D)

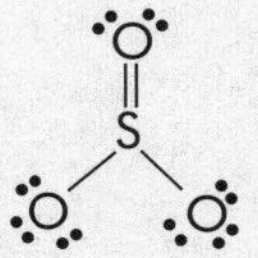

(E)

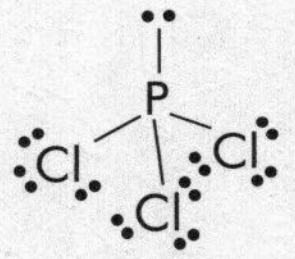

18. **D** A molecule with d^2sp^3 hybridization has octahedral structure if the central atom has no unbonded electrons (SF_6, for instance).

If the central atom has one unbonded electron pair, the molecule is square pyramidal (IF_5, for instance).

If the central atom has two unbonded electron pairs, the molecule is square planar (XeF_4, for instance).

A molecule with d^2sp^3 hybridization can never be tetrahedral.

19. **B** The Lewis dot structures for the answer choices are shown below. Only PH₃ has a single unshared electron pair on its central atom.

(A)

(B)

(C)

(D)

(E)

20. **A** H₂ experiences only van der Waals forces and has the lowest boiling point.

N₂ also experiences only van der Waals forces, but it is larger than H₂ and has more electrons, so it has stronger van der Waals interactions with other molecules.

NH₃ is polar and undergoes hydrogen bonding, so it has the strongest intermolecular interactions and the highest boiling point.

21. **D** The greater the bond strength, the higher the melting point. Both NaCl and NaBr are held together by ionic bonds, and the strength of ionic bonds depends on the strength of the charges and the sizes of the ions. The strength of the charges doesn't matter in this case because both chlorine and bromine ions have charges of –1. The reason for NaCl's higher melting point is that chlorine ions are smaller than bromine ions, so the ions can get closer together.

22. **C** The energy required to separate an ionic compound into gaseous ions is called the lattice energy. The most important factor in determining the lattice energy is the strength of the charges on the two ions. In MgO, the Mg ion is +2 and the O ion is –2. All of the other choices listed contain +1 or –1 ions, so their lattice energies will not be as large.

23. **D** A sample of NaNO₃ contains an Na⁺ ion and an NO₃⁻ polyatomic ion, which are held together with ionic bonds. The atoms in the NO₃⁻ ion are held together by covalent bonds. All of the bonds in the other choices are covalent bonds.

24. **B** The bond that holds HCl together is a covalent bond with a large polarity. The bond that holds NO together is also polar covalent, but its polarity is very small because N and O are so close together on the periodic table. Cl_2 and N_2 are nonpolar because they share electrons equally, and SO_3 is nonpolar because it is symmetrical (trigonal planar). Nonpolar molecules have dipole moments of zero.

25. **B** Groups I and II are listed in order of increasing melting points. Remember: Stronger bonds or intermolecular forces mean higher melting points.

   I. H_2O has hydrogen bonding, which makes its intermolecular forces much stronger than those of CO_2, which is nonpolar and exhibits only London dispersion forces. So water has a higher boiling point than CO_2.

   II. These are nonpolar atoms, so they are held together by weak London dispersion forces. Because London dispersion forces depend on the random movement of electrons, the more electrons in a compound the stronger the London dispersion forces. Ar has more electrons than He, so it has stronger London dispersion forces and a higher boiling point.

   III. These are held together by metallic bonding. In general, the smaller the nucleus of a metal, the stronger the metallic bonds. Rb has a larger nucleus than Na, so it has weaker metallic bonds and a lower boiling point, so this statement is not correct.

ESSAYS

1. (a) Molecules of noble gases in the liquid phase are held together by London dispersion forces, which are weak interactions brought about by instantaneous polarities in nonpolar atoms and molecules.

   Atoms with more electrons are more easily polarized and experience stronger London dispersion forces. Argon has more electrons than neon, so it experiences stronger London dispersion forces and boils at a higher temperature.

   (b) Sodium and potassium are held together by metallic bonds and positively charged ions in a delocalized sea of electrons.

   Potassium is larger than sodium, so the electrostatic attractions that hold the atoms together act at a greater distance, reducing the attractive force and resulting in its lower melting point.

   (c) Both CaO(s) and KF(s) are held together by ionic bonds in crystal lattices.

   Ionic bonds are held together by an electrostatic force, which can be determined by using Coulomb's law.

   $$F = k\frac{Q_1 Q_2}{r^2}$$

   CaO is more highly charged, with Ca^{2+} bonded to O^{2-}. So for CaO, Q_1 and Q_2 are +2 and –2.

   KF is not as highly charged, with K^+ bonded to F^-. So for KF, Q_1 and Q_2 are +1 and –1.

   CaO is held together by stronger forces and is more difficult to break apart.

   (d) KF is composed of K^+ and F^- ions. In the liquid (molten) state, these ions are free to move and can thus conduct electricity.

   In the solid state, the K^+ and F^- ions are fixed in a crystal lattice and their electrons are localized around them, so there is no charge that is free to move and thus no conduction of electricity.

2. (a) (i)   (ii)

(b) The central carbon atom forms three sigma bonds with oxygen atoms and has no free electron pairs, so its hybridization must be sp^2.

(c) (i) All three bonds will be the same length because no particular resonance form is preferred over the others. The actual structure is an average of the resonance structures.

(ii) The C–O bonds in the carbonate ion have resonance forms between single and double bonds, while the C–O bonds in carbon dioxide are both double bonds.

The bonds in the carbonate ion will be shorter than single bonds and longer than double bonds, so the carbonate bonds will be longer than the carbon dioxide bonds.

3. (a) (i) The bond strength of N_2 is larger than the bond strength of O_2 because N_2 molecules have triple bonds and O_2 molecules have double bonds. Triple bonds are stronger and shorter than double bonds.

(ii) The bond length of H_2 is smaller than the bond length of Cl_2 because hydrogen is a smaller atom than chlorine, allowing the hydrogen nuclei to be closer together.

(iii) Liquid oxygen and liquid chlorine are both nonpolar substances that experience only London dispersion forces of attraction. These forces are greater for Cl_2 because it has more electrons (which makes it more polarizable), so Cl_2 has a higher boiling point than O_2.

(b) H_2 and O_2 are both nonpolar molecules that experience only London dispersion forces, which are too weak to form the bonds required for a substance to be liquid at room temperature.

H_2O is a polar substance whose molecules form hydrogen bonds with each other. Hydrogen bonds are strong enough to form the bonds required in a liquid at room temperature.

4. (a)

(b) H_2S has two bonds and two free electron pairs on the central S atom. The greatest distance between the electron pairs is achieved by tetrahedral arrangement. The electron pairs at two of the four corners will cause the molecule to have a bent shape, like water.

SO_4^{2-} has four bonds around the central S atom and no free electron pairs. The four bonded pairs will be farthest apart when they are arranged in a tetrahedral shape, so the molecule is tetrahedral.

XeF_2 has two bonds and three free electron pairs on the central Xe atom. The greatest distance between the electron pairs can be achieved by a trigonal bipyramidal arrangement. The three free electron pairs will occupy the equatorial positions, which are 120 degrees apart, to minimize repulsion. The two F atoms are at the poles, so the molecule is linear.

ICl_4^- has four bonds and two free electron pairs on the central I atom. The greatest distance between the electron pairs can be achieved by an octahedral arrangement. The two free electron pairs will be opposite each other to minimize repulsion. The four Cl atoms are in the equatorial positions, so the molecule is square planar.

5. (a) BF_3 has three bonds on the central B atom and no free electron pairs, so the structure of BF_3 is trigonal planar, with each of the bonds 120 degrees apart.

NF_3 has three bonds and one free electron pair on the central N atom. The four electron pairs are pointed toward the corners of a tetrahedron, 109.5 degrees apart. The added repulsion from the free electron pair causes the N–F bonds to be even closer together, and the angle between them is more like 107 .

(b) Polar solvents are best at dissolving polar solutes. Nonpolar solvents are best at dissolving nonpolar solutes.

$I_2(s)$ is nonpolar, so it dissolves better in carbon tetrachloride, CCl_4, which is nonpolar, than in water, H_2O, which is polar.

(c) The carbon atoms in diamond are bonded together in a tetrahedral network, with each carbon atom bonded to three other carbon atoms. The tetrahedral structure of the network bonds does not leave any seams along which the diamond can be broken, so a diamond behaves as one big molecule with no weaknesses.

(d) HBr and HCl are polar molecules. In liquid form, both substances are held together by dipole–dipole interactions. These interactions are stronger for molecules with more electrons, so HBr has stronger intermolecular bonds and a higher boiling point.

HF has a higher boiling point than HCl because HF undergoes hydrogen bonding, while HCl does not; this causes HF to remain a liquid at higher temperatures than HCl, although HF is a polar molecule with fewer electrons than HCl.

5

Stoichiometry and Chemical Equations

HOW OFTEN DO STOICHIOMETRY AND CHEMICAL EQUATIONS APPEAR ON THE EXAM?

In the multiple-choice section, this topic appears in about 7 out of 75 questions.
In the free-response section, this topic appears almost every year.

SOME MATH

SIGNIFICANT FIGURES

When you do calculations on the AP Chemistry Exam, you'll be expected to present your answers with the proper number of significant figures, so let's review the rules.

- Nonzero digits and zeros between nonzero digits *are* significant.

245	3 significant figures
7.907	4 significant figures
907.08	5 significant figures

- Zeros to the *left* of the first nonzero digit in a number are *not* significant.

0.005	1 significant figure
0.0709	3 significant figures

- Zeros at the *end of a number* to the *right* of the decimal point *are* significant.

12.000	5 significant figures
0.080	2 significant figures
1.0	2 significant figures

- Zeros at the *end of a number greater than 1* are *not* significant, unless their significance is indicated by the presence of a decimal point.

1,200	2 significant figures
1,200.	4 significant figures
10	1 significant figure
10.	2 significant figures

- The coefficients of a balanced equation and numbers obtained by counting objects are infinitely significant. So if a balanced equation calls for 3 moles of carbon, we can think of it as $3.\overline{00}$ moles of carbon.

- When multiplying and dividing, the result should have the same number of significant figures as the number in the calculation with the smallest number of significant figures.

  $$0.352 \times 0.90876 = 0.320$$
  $$864 \times 12 = 1.0 \times 10^4$$
  $$7 \div 0.567 = 10$$

- When adding and subtracting, the result should have the same number of decimal places as the number in the calculation with the smallest number of decimal places.

  $$26 + 45.88 + 0.09534 = 72$$

The whole point is that the result of a calculation cannot be more accurate than the least accurate number in the calculation.

LOGARITHMS

Let's review some basic facts about logarithms.

If $10^x = y$, then $\log y = x$

If $e^x = y$, then $\ln y = x$

$e = 2.7183$

$\ln y = 2.303 \log y$

$\log(ab) = \log a + \log b$

$\log\left(\dfrac{a}{b}\right) = \log a - \log b$

MOLES

The mole (**Avogadro's number**) is the most important number in chemistry, serving as a bridge that connects all the different quantities that you'll come across in chemical calculations. The coefficients in chemical reactions tell you about the reactants and products in terms of moles, so most of the stoichiometry questions you'll see on the test will be exercises in converting between moles and grams, liters, molarities, and other units.

MOLES AND MOLECULES

The definition of Avogadro's number gives you the information you need to convert between moles and individual molecules and atoms.

$$1 \text{ mole} = 6.022 \times 10^{23} \text{ molecules}$$

$$\text{Moles} = \frac{\text{molecules}}{\left(6.022 \times 10^{23}\right)}$$

MOLES AND GRAMS

Moles and grams can be related by using the atomic weights given in the periodic table. Atomic weights on the periodic table are given in terms of atomic mass units (amu), but an amu is the same as a gram per mole, so if 1 carbon atom weighs 12 amu, then 1 mole of carbon atoms weighs 12 grams.

You can use the relationship between amu and g/mol to convert between grams and moles by using the following equation:

$$\text{Moles} = \frac{\text{grams}}{\text{molar mass}}$$

MOLES AND GASES

We'll talk more about the ideal gas equation in Chapter 6, but for now, you should know that you can use it to calculate the number of moles of a gas if you know some of the gas's physical properties. All you need to remember at this point is that in the equation $PV = nRT$, n stands for moles of gas.

$$\text{Moles} = \frac{PV}{RT}$$

P = pressure (atm)
V = volume (L)
T = temperature (K)
R = the gas constant, 0.0821 L-atm/mol-K

The equation above gives the general rule for finding the number of moles of a gas. Many gas problems will take place at STP, or standard temperature and pressure, where $P = 1$ atmosphere and $T = 273$ K. At STP, the situation is much simpler and you can convert directly between the volume of a gas and the number of moles. That's because at STP, one mole of gas always occupies 22.4 liters.

$$\text{Moles} = \frac{\text{liters}}{(22.4 \, \text{L} / \text{mol})}$$

MOLES AND SOLUTIONS

We'll talk more about molarity and molality in Chapter 9, but for now you should realize that you can use the equations that define these common measures of concentration to find the number of moles of solute in a solution. Just rearrange the equations to isolate moles of solute.

Moles = (molarity)(liters of solution)
Moles = (molality)(kilograms of solvent)

PERCENT COMPOSITION

To solve many problems on the exam, you will need to use percent composition, or mass percents. Percent composition is the percent by mass of each element that makes up a compound. It is calculated by dividing the mass of each element or component in a compound by the total molar mass for the substance.

EMPIRICAL AND MOLECULAR FORMULAS

You will also need to know how to determine the empirical and molecular formula of a compound given masses or mass percents of the components of that compound. Remember that the empirical formula represents the simplest ratio of one element to another in a compound (e.g., CH_2O), while the molecular formula represents the actual formula for the substance (e.g., $C_6H_{12}O_6$).

CHEMICAL EQUATIONS

BALANCING CHEMICAL EQUATIONS

Normally, balancing a chemical equation is a trial-and-error process. You start with the most complicated-looking compound in the equation and work from there. There is, however, an old Princeton Review SAT trick that you may want to try if you see a balancing equation question on the multiple-choice section. The trick is called **backsolving**.

It works like this: To make a balancing equation question work in a multiple-choice format, one of the answer choices is the correct coefficient for one of the species in the reaction. So instead of starting blind in the trial-and-error process, you can insert the answer choices one by one to see which one works. You probably won't have to try all five, and if you start in the middle and the number doesn't work, it might be obviously too small or large, eliminating other choices before you have to try them. Let's try it.

$$...NH_3 + ...O_2 \rightarrow ...N_2 + ...H_2O$$

1. If the equation above were balanced with lowest whole number coefficients, the coefficient for NH_3 would be

   (A) 1
   (B) 2
   (C) 3
   (D) 4
   (E) 5

Start at (C) because it's the middle number. If there are 3 NH_3's, then there can't be a whole number coefficient for N_2, so (C) is wrong, and so are the other odd number answers, (A) and (E).

Try (D).

If there are 4 NH_3's, then there must be 2 N_2's and 6 H_2O's.

If there are 6 H_2O's, then there must be 3 O_2's, and the equation is balanced with lowest whole number coefficients.

Backsolving is more efficient than the methods that you're accustomed to. If you use the answer choices that you're given, you streamline the trial-and-error process and allow yourself to use POE as you work on the problem.

CHEMICAL EQUATIONS AND CALCULATIONS

Many of the stoichiometry problems on the test will be formatted in the following way: You will be given a balanced chemical equation and told that you have some number of grams (or liters of gas, or molar concentration, and so on) of reactant. Then you will be asked what number of grams (or liters of gas, or molar concentration, and so on) of products are generated.

In these cases, follow this simple series of steps.

1. Convert whatever quantity you are given into moles.

2. If you are given information about two reactants, you may have to use the equation coefficients to determine which one is the limiting reagent. Remember: The limiting reagent is not necessarily the reactant that you have the least of; it is the reactant that runs out first.

3. Use the balanced equation to determine how many moles of the desired product are generated.

4. Convert moles of product to the desired unit.

Let's try one.

$$2 \, HBr(aq) + Zn(s) \rightarrow ZnBr_2(aq) + H_2(g)$$

2. A piece of solid zinc weighing 98 grams was added to a solution containing 324 grams of HBr. What is the volume of H_2 produced at standard temperature and pressure if the reaction above runs to completion?

    (A) 11 liters
    (B) 22 liters
    (C) 34 liters
    (D) 45 liters
    (E) 67 liters

1. Convert whatever quantity you are given into moles.

$$\text{Moles of Zn} = \frac{\text{grams}}{\text{molar mass}} = \frac{(98 \, g)}{(65.4 \, g/mol)} = 1.5 \, mol$$

$$\text{Moles of HBr} = \frac{\text{grams}}{\text{molar mass}} = \frac{(324 \, g)}{(80.9 \, g/mol)} = 4.0 \, mol$$

2. Use the balanced equation to find the limiting reagent.
From the balanced equation, 2 moles of HBr are used for every mole of Zn that reacts, so when 1.5 moles of Zn react, 3 moles of HBr are consumed, and there will be HBr left over when all of the Zn is gone. That makes Zn the limiting reagent.

3. Use the balanced equation to determine how many moles of the desired product are generated.

1 mole of H_2 is produced for every mole of Zn consumed, so if 1.5 moles of Zn are consumed, then 1.5 moles of H_2 are produced.

4. Convert moles of product to the desired unit.

The H_2 gas is at STP, so we can convert directly from moles to volume.
Volume of H_2 = (moles)(22.4 L/mol) = (1.5 mol)(22.4 L/mol) = 33.6 L $\cong$ 34 L

So (C) is correct.

Let's try another one using the same reaction.

$$2 \, HBr(aq) + Zn(s) \rightarrow ZnBr_2(aq) + H_2(g)$$

3. A piece of solid zinc weighing 13.1 grams was placed in a container. A 0.10-molar solution of HBr was slowly added to the container until the zinc was completely dissolved. What was the volume of HBr solution required to completely dissolve the solid zinc?

(A) 1.0 L
(B) 2.0 L
(C) 3.0 L
(D) 4.0 L
(E) 5.0 L

1. Convert whatever quantity you are given into moles.

$$\text{Moles of Zn} = \frac{\text{grams}}{\text{MW}} = \frac{(13.1 \text{ g})}{(65.4 \text{ g / mol})} = 0.200 \text{ mol}$$

2. Use the balanced equation to find the limiting reagent.

3. Use the balanced equation to determine how many moles of the desired product are generated.

In this case, we're using the balanced reaction to find out how much of one reactant is required to consume the other reactant. It's a slight variation on the process described in (2) and (3).

We can see from the balanced equation that it takes 2 moles of HBr to react completely with 1 mole of Zn, so it will take 0.400 moles of HBr to react completely with 0.200 moles of Zn.

4. Convert moles of product to the desired unit.
Moles of HBr = (molarity)(volume)

$$\text{Volume of HBr} = \frac{\text{moles}}{\text{molarity}} = \frac{(0.400 \text{ mol})}{(0.10 \text{ mol/L})} = 4.0 \text{ L}$$

So (D) is correct.

When you perform calculations, always include units. Including units in your calculations will help you (and the person scoring your test) keep track of what you are doing. Including units will also get you partial credit points on the free-response section.

CHAPTER 5 QUESTIONS

MULTIPLE-CHOICE QUESTIONS

Questions 1–3

    (A) Moles
    (B) Liters
    (C) Grams
    (D) Atmospheres
    (E) Volts

1. One mole of solid zinc has a mass of 65.39 of these.

2. These units can be calculated by dividing a quantity by 6.02×10^{23}.

3. Four grams of helium gas occupy 22.4 of these at standard temperature and pressure.

4. What is the mass ratio of fluorine to boron in a boron trifluoride molecule?

    (A) 1.8 to 1
    (B) 3.0 to 1
    (C) 3.5 to 1
    (D) 5.3 to 1
    (E) 6.0 to 1

5. A hydrocarbon sample with a mass of 6 grams underwent combustion, producing 11 grams of carbon dioxide. If all of the carbon initially present in the compound was converted to carbon dioxide, what was the percent of carbon, by mass, in the hydrocarbon sample?

    (A) 25%
    (B) 33%
    (C) 50%
    (D) 66%
    (E) 75%

6. What is the mass of oxygen in 148 grams of calcium hydroxide ($Ca(OH)_2$)?

    (A) 16 grams
    (B) 24 grams
    (C) 32 grams
    (D) 48 grams
    (E) 64 grams

7. An ion containing only oxygen and chlorine is 31% oxygen by mass. What is its empirical formula?

    (A) ClO^-
    (B) ClO_2^-
    (C) ClO_3^-
    (D) ClO_4^-
    (E) Cl_2O^-

8. A sample of propane, C_3H_8, was completely burned in air at STP. The reaction occurred as shown below.

$$C_3H_8 + O_2 \rightarrow 3\ CO_2 + 4\ H_2O$$

If 67.2 liters of CO_2 were produced and all of the carbon in the CO_2 came from the propane, what was the mass of the propane sample?

    (A) 11 grams
    (B) 22 grams
    (C) 33 grams
    (D) 44 grams
    (E) 55 grams

9. What is the percent composition by mass of the elements in the compound $NaNO_3$?

    (A) Na 20%, N 20%, O 60%
    (B) Na 23%, N 14%, O 48%
    (C) Na 23%, N 14%, O 63%
    (D) Na 27%, N 16%, O 57%
    (E) Na 36%, N 28%, O 36%

10.
$$CaCO_3(s) \rightarrow CaO(s) + CO_2(g)$$

A sample of pure $CaCO_3$ was heated and decomposed according to the reaction given above. If 28 grams of CaO were produced by the reaction, what was the initial mass of $CaCO_3$?

    (A) 14 grams
    (B) 25 grams
    (C) 42 grams
    (D) 50 grams
    (E) 84 grams

11. The composition of a typical glass used in bottles is 12.0% Na_2O, 12.0% CaO, and 76.0% SiO_2. Which of the following lists the three compounds in order of greatest to least number of moles present in a typical sample of bottle glass?

    (A) SiO_2, CaO, Na_2O
    (B) SiO_2, Na_2O, CaO
    (C) Na_2O, SiO_2, CaO
    (D) Na_2O, CaO, SiO_2
    (E) CaO, Na_2O, SiO_2

12. The concentration of sodium chloride in sea water is about 0.5 molar. How many grams of $NaCl$ are present in 1 kg of sea water?

    (A) 30 grams
    (B) 60 grams
    (C) 100 grams
    (D) 300 grams
    (E) 600 grams

13. A sample of a hydrate of $CuSO_4$ with a mass of 250 grams was heated until all the water was removed. The sample was then weighed and found to have a mass of 160 grams. What is the formula for the hydrate?

    (A) $CuSO_4 \cdot 10\,H_2O$
    (B) $CuSO_4 \cdot 7\,H_2O$
    (C) $CuSO_4 \cdot 5\,H_2O$
    (D) $CuSO_4 \cdot 2\,H_2O$
    (E) $CuSO_4 \cdot H_2O$

14. A compound containing only sulfur and oxygen is 50% sulfur by weight. What is the empirical formula for the compound?

    (A) SO
    (B) SO_2
    (C) SO_3
    (D) S_2O
    (E) S_3O

15. $2\,Na(s) + 2\,H_2O(l) \rightarrow 2\,NaOH(aq) + H_2(g)$

    Elemental sodium reacts with water to form hydrogen gas as shown above. If a sample of sodium reacts completely to form 20 liters of hydrogen gas, measured at standard temperature and pressure, what was the mass of the sodium?

    (A) 5 grams
    (B) 10 grams
    (C) 20 grams
    (D) 30 grams
    (E) 40 grams

16. $ZnSO_3(s) \rightarrow ZnO(s) + SO_2(g)$

    What is the STP volume of SO_2 gas produced by the above reaction when 145 grams of $ZnSO_3$ are consumed?

    (A) 23 liters
    (B) 36 liters
    (C) 45 liters
    (D) 56 liters
    (E) 90 liters

17. $...CN^- + ...OH^- \rightarrow ...CNO^- + ...H_2O + ...e^-$

    When the half reaction above is balanced, what is the coefficient for OH^- if all the coefficients are reduced to the lowest whole number?

    (A) 1
    (B) 2
    (C) 3
    (D) 4
    (E) 5

18. $...MnO_4^- + ...I^- + ...H_2O \rightarrow ...MnO_2 + ...IO_3^- + ...OH^-$

    The oxidation-reduction reaction above is to be balanced with lowest whole number coefficients. What is the coefficient for OH^-?

    (A) 1
    (B) 2
    (C) 3
    (D) 4
    (E) 5

19. $CaCO_3(s) + 2 H^+(aq) \rightarrow Ca^{2+}(aq) + H_2O(l) + CO_2(g)$

If the reaction above took place at standard temperature and pressure and 150 grams of $CaCO_3(s)$ were consumed, what was the volume of $CO_2(g)$ produced at STP?

(A) 11 L
(B) 22 L
(C) 34 L
(D) 45 L
(E) 56 L

20. A gaseous mixture at 25 C contained 1 mole of CH_4 and 2 moles of O_2 and the pressure was measured at 2 atm. The gases then underwent the reaction shown below.

$$CH_4(g) + 2 O_2(g) \rightarrow CO_2(g) + 2 H_2O(g)$$

What was the pressure in the container after the reaction had gone to completion and the temperature was allowed to return to 25°C?

(A) 1 atm
(B) 2 atm
(C) 3 atm
(D) 4 atm
(E) 5 atm

21. A sample of a hydrate of $BaCl_2$ with a mass of 61 grams was heated until all the water was removed. The sample was then weighed and found to have a mass of 52 grams. What is the formula for the hydrate?

(A) $BaCl_2 \bullet 5 H_2O$
(B) $BaCl_2 \bullet 4 H_2O$
(C) $BaCl_2 \bullet 3 H_2O$
(D) $BaCl_2 \bullet 2 H_2O$
(E) $BaCl_2 \bullet H_2O$

22. A hydrocarbon was found to be 20% hydrogen by weight. If 1 mole of the hydrocarbon has a mass of 30 grams, what is its molecular formula?

(A) CH
(B) CH_2
(C) CH_3
(D) C_2H_4
(E) C_2H_6

23. $...CuFeS_2 + ...O_2 \rightarrow Cu_2S + ...FeO + ...SO_2$

When the half-reaction above is balanced, what is the coefficient for O_2 if all the coefficients are reduced to the lowest whole number?

(A) 2
(B) 3
(C) 4
(D) 6
(E) 8

24. A hydrocarbon contains 75% carbon by mass. What is the empirical formula for the compound?

(A) CH_2
(B) CH_3
(C) CH_4
(D) C_2H_5
(E) C_3H_8

25. When chlorine gas is combined with fluorine gas, a compound is formed that is 38% chlorine and 62% fluorine. What is the empirical formula of the compound?

(A) ClF
(B) ClF_2
(C) ClF_3
(D) ClF_5
(E) ClF_7

PROBLEMS

1. A 10.0 gram sample containing calcium carbonate and an inert material was placed in excess hydrochloric acid. A reaction occurred producing calcium chloride, water, and carbon dioxide.

   (a) Write the balanced equation for the reaction.

   (b) When the reaction was complete, 900 milliliters of carbon dioxide gas were collected at 740 mmHg and 30°C. How many moles of calcium carbonate were consumed in the reaction?

   (c) If all of the calcium carbonate initially present in the sample was consumed in the reaction, what percent by mass of the sample was due to calcium carbonate?

   (d) If the inert material was silicon dioxide, what was the molar ratio of calcium carbonate to silicon dioxide in the original sample?

2. A gaseous hydrocarbon sample is completely burned in air, producing 1.80 liters of carbon dioxide at standard temperature and pressure and 2.16 grams of water.

   (a) What is the empirical formula for the hydrocarbon?

   (b) What was the mass of the hydrocarbon consumed?

   (c) The hydrocarbon was initially contained in a closed 1.00 liter vessel at a temperature of 32°C and a pressure of 760 millimeters of mercury. What is the molecular formula of the hydrocarbon?

   (d) Write the balanced equation for the combustion of the hydrocarbon.

3. The table below shows three common forms of copper ore.

Ore #	Empirical Formula	Percent by Weight		
		Copper	Sulfur	Iron
1	Cu_2S	?	?	0
2	?	34.6	34.9	30.5
3	?	55.6	28.1	16.3

   (a) What is the percent by weight of copper in Cu_2S?

   (b) What is the empirical formula of ore #2?

   (c) If a sample of ore #3 contains 11.0 grams of iron, how many grams of sulfur does it contain?

   (d) Cu can be extracted from Cu_2S by the following process:

   $$3\,Cu_2S + 3\,O_2 \rightarrow 3\,SO_2 + 6\,Cu$$

   If 3.84 grams of O_2 are consumed in the process, how many grams of Cu are produced?

4. $2\,Mg(s) + 2\,CuSO_4(aq) + H_2O(l) \rightarrow 2\,MgSO_4(aq) + Cu_2O(s) + H_2(g)$

(a)   If 1.46 grams of $Mg(s)$ are added to 500 milliliters of a 0.200-molar solution of $CuSO_4$, what is the maximum molar yield of $H_2(g)$?

(b)   When all of the limiting reagent has been consumed in (a), how many moles of the other reactant (not water) remain?

(c)   What is the mass of the Cu_2O produced in (a)?

(d)   What is the value of $[Mg^{2+}]$ in the solution at the end of the experiment? (Assume that the volume of the solution remains unchanged.)

CHAPTER 5 ANSWERS AND EXPLANATIONS

MULTIPLE-CHOICE QUESTIONS

1. **C** The units for atomic weight are grams/mole.

2. **A** A mole is equal to 6.02×10^{23}.

3. **B** Four grams of helium is a mole. A mole of gas occupies 22.4 liters at STP.

4. **D** The empirical formula of boron trifluoride is BF_3.

   Grams = (moles)(MW)

   Grams of boron = (1 mol)(10.8 g/mol) = 10.8 g

   Grams of fluorine = (3 mol)(19.0 g/mol) = 57.0 g

   So the mass ratio is about 57 to 11, which is about 5.3 to 1.

5. **C** Moles = $\dfrac{\text{grams}}{\text{MW}}$

   Moles of $CO_2 = \dfrac{(11g)}{(44g/mol)} = \dfrac{1}{4}$ mol

   If $\dfrac{1}{4}$ mole of CO_2 was produced, then $\dfrac{1}{4}$ mole of C was consumed.

   Grams = (moles)(MW)

   Grams of carbon = ($\dfrac{1}{4}$ mol)(12 g/mol) = 3 g

   So the percent by mass of carbon was $\dfrac{3}{6} = \dfrac{1}{2} = 50\%$

6. **E** $Moles = \dfrac{grams}{MW}$

Moles of calcium hydroxide = $\dfrac{(148\ g)}{(74\ g/mol)} = 2\ moles$

Every mole of $Ca(OH)_2$ contains 2 moles of oxygen.

So there are (2)(2) = 4 moles of oxygen
Grams = (moles)(MW)
So grams of oxygen = (4 mol)(16 g/mol) = 64 grams

7. **A** Assume that we have 100 grams of the compound. That means that we have 31 grams of oxygen and 69 grams of chlorine.

$Moles = \dfrac{grams}{MW}$

Moles of oxygen = $\dfrac{(31g)}{(16\ g/mol)}$ = slightly less than 2 mol

Moles of chlorine = $\dfrac{(69\ g)}{(35.5\ g/mol)}$ = slightly less than 2 mol

So the ratio of chlorine to oxygen is 1 to 1, and the empirical formula is ClO^-.

8. **D** $Moles = \dfrac{liters}{(22.4\ L/mol)}$

Moles of CO_2 = $\dfrac{67.2\ L}{(22.4\ L/mol)}$ = 3 mol

According to the balanced equation, if 3 moles of CO_2 were produced, 1 mole of C_3H_8 was consumed.
Grams = (moles)(MW)
So grams of C_3H_8 = (1 mol)(44 g/mol) = 44 grams

9. **D** The molecular weight of $NaNO_3$ is (23) + (14) + (3)(16) = 85 g/mol.

We can get the answer using pretty rough estimates.

The percent by mass of Na = $\dfrac{(23)}{(85)}$ = between 25% ($\dfrac{1}{4}$) and 33% ($\dfrac{1}{3}$)

The percent by mass of N = $\dfrac{(14)}{(85)}$ = between 10% ($\dfrac{1}{10}$) and 20% ($\dfrac{1}{5}$)

The percent by mass of O = $\dfrac{(48)}{(85)}$ = between 50% ($\dfrac{1}{2}$) and 60% ($\dfrac{3}{5}$)

You can use POE to get choice (D).

10. **D** Moles = $\dfrac{grams}{MW}$

Moles of CaO = $\dfrac{(28g)}{(56 g/mol)}$ = 0.50 mol

From the balanced equation, if 0.50 mol of CaO was produced, then 0.50 mol of $CaCO_3$ was consumed.

Grams = (moles)(MW)

Grams of $CaCO_3$ = (0.50 mol)(100 g/mol) = 50 g

11. **A** The molecular weights of the three compounds are as follows:

Na_2O – 62 g/mol

CaO – 56 g/mol

SiO_2 – 60 g/mol

Because the molecular weights are close together, we can safely say that SiO_2, which makes up a much greater percentage by mass of bottle glass than the other two, will have far and away the most moles in a sample. So the answer must be (A) or (B).

Remember that (grams) = (moles)(MW). Because a sample of bottle glass will have the same number of grams of Na_2O and CaO, the one with the smaller molecular weight must have the greater number of moles. So there must be more moles of CaO than Na_2O.

12. **A** First, you have to remember that 1 liter of water has a mass of 1 kg.

Moles = (molarity)(liters)

Moles of NaCl = (0.5 M)(1 L) = 0.5 moles

Grams = (moles)(MW)

Grams of NaCl = (0.5 mol)(59 g/mol) = 30 g

13. **C** The molecular weight of $CuSO_4$ is 160 g/mol, so we have only 1 mole of the hydrate. The lost mass was due to water, so 1 mole of the hydrate must have contained 90 grams of H_2O.

Moles = $\dfrac{grams}{MW}$

Moles of water = $\dfrac{(90\ g)}{(18\ g/mol)}$ = 5 moles

So if 1 mole of hydrate contains 5 moles of H_2O, then the formula for the hydrate must be $CuSO_4 \bullet 5\ H_2O$.

14. **B** You might be able to do this one in your head just from knowing that sulfur's molecular weight is twice as large as oxygen's. If not, let's say you have 100 grams of the compound. So you have 50 grams of sulfur and 50 grams of oxygen.

Moles = $\dfrac{grams}{MW}$

Moles of sulfur = $\dfrac{(50\ g)}{(32\ g/mol)}$ = a little more than 1.5

Moles of oxygen = $\dfrac{(50\ g)}{(16\ g/mol)}$ = a little more than 3

The molar ratio of O to S is 2 to 1, so the empirical formula must be SO_2.

15. **E** Moles $= \dfrac{\text{liters}}{(22.4\,\text{L/mol})}$

Moles of $H_2 = \dfrac{(20\,\text{L})}{(22.4\,\text{L/mol})} =$ about 0.9 moles

From the balanced equation, for every mole of H_2 produced, 2 moles of Na are consumed, so 1.8 moles of Na are consumed.

Grams $=$ (moles)(MW)
Grams of Na $=$ (1.8 mol)(23 g/mol) $=$ about 40 grams
You don't really have to do the math, because you can get the answer by using rough estimates.

16. **A** Moles $= \dfrac{\text{grams}}{\text{MW}}$

Moles of $ZnSO_3 = \dfrac{(145\,\text{g})}{(145\,\text{g/mol})} = 1$ mole

From the balanced equation, when 1 mole of $ZnSO_3$ is consumed, 1 mole of SO_2 will be produced. So about 1 mole of SO_2 is produced.

Liters $=$ (moles)(22.4 L/mol)
Liters of $SO_2 =$ (about 1 mol)(22.4 L) $= 23$ liters

17. **B** Use trial and error or backsolve.

Start with (C). If there are 3 OH^-, there can't be a whole number coefficient for H_2O, so (C) is wrong. You should also be able to see that the answer can't be an odd number, so (A) and (E) are also wrong.
Try (D). If there are 4 OH^-, then there are 2 H_2O.
That leaves 2 more Os on the product side, so there must be 2 CNO^-.
If there are 2 CNO^- then there are 2 CN^-.
These are all whole numbers, but they are not the lowest whole numbers, so (D) is wrong.
If we divide all the coefficients by 2, we get the lowest whole number coefficients. That leaves us with 2 OH^-, which is choice (B).
By the way, N^{5-} (in CN^-) is oxidized to N^{3-} (in CNO^-), so there are 2 e^-.

18. **B** Use trial and error or backsolve.

Start at (C).

If there are 3 OH^-, there can't be a whole number coefficient for H_2O, so (C) is wrong. Also, the answer can't be an odd number, so (A) and (E) are wrong.

Notice that you don't have to test both of the remaining answers. If the one you pick works, you're done. If the one you pick doesn't work, then the one that's left must be correct. With a choice of only two answers, pick the one that looks easier to work with.

Try (B) because it's smaller. If there are 2 OH^-, then there is 1 H_2O. If you put in 1 for I^- and IO_3^- and 2 for MnO_4^- and MnO_2, the equation is balanced. So (B) is correct.

19. **C** $\text{Moles} = \dfrac{\text{grams}}{\text{MW}}$

$\text{Moles of } CaCO_3 = \dfrac{(150\text{ g})}{(100\text{ g/mol})} = 1.5\text{ moles}$

From the balanced equation, for every mole of $CaCO_3$ consumed, one mole of CO_2 is produced. So 1.5 moles of CO_2 are produced.
At STP, volume of gas = (moles)(22.4 L)
So volume of CO_2 = (1.5)(22.4) = 34 L

20. **B** All of the reactants are consumed in the reaction and the temperature doesn't change, so the pressure will change only if the number of moles of gas changes over the course of the reaction. The number of moles of gas (3 moles) doesn't change in the balanced equation, so the pressure will remain the same (2 atm) at the end of the reaction as at the beginning.

21. **D** The molecular weight of $BaCl_2$ is 208 g/mol, so we can figure out how many moles of the hydrate we have.

$\text{Moles} = \dfrac{\text{grams}}{\text{MW}}$

$\text{Moles of hydrate} = \dfrac{(52\text{ g})}{(208\text{ g/mol})} = \dfrac{1}{4}\text{ mole}$

The lost mass was due to water, so 1 mole of the hydrate must have contained 9 grams of H_2O.

$\text{Moles} = \dfrac{\text{grams}}{\text{MW}}$

$\text{Moles of water} = \dfrac{(9\text{ g})}{(18\text{ g/mol})} = \dfrac{1}{2}\text{ moles}$

So if $\dfrac{1}{4}$ mole of hydrate contains $\dfrac{1}{2}$ mole of H_2O, there must be 2 moles of H_2O for every mole of hydrate, and the formula for the hydrate must be $BaCl_2 \bullet 2\ H_2O$.

22. **E** Let's say we have 100 grams of the compound.

$\text{Moles} = \dfrac{\text{grams}}{\text{MW}}$

So moles of carbon = $\dfrac{(80\text{ g})}{(12\text{ g/mol})} = 6.7\text{ moles}$

and moles of hydrogen = $\dfrac{(20\text{ g})}{(1\text{ g/mol})} = 20\text{ moles}$

According to our rough calculation, there are about three times as many moles of hydrogen in the compound as there are moles of carbon, so the empirical formula is CH_3.

The molar mass for the empirical formula is 15 g/mol, so we need to double the moles of each element to get a compound with a molar mass of 30 g/mol. That makes the molecular formula of the compound C_2H_6.

23. **C** Backsolving doesn't work so well in this case because there are two different compounds that contain oxygen on the right side of the equation, which makes the process kind of confusing. Instead, let's just try plugging in values for the most complicated compound in the equation, $CuFeS_2$.

What if there's 1 $CuFeS_2$? That's impossible because there are 2 Cu's on the right.

What if there are 2 $CuFeS_2$'s? Then the right side has 1 Cu_2S to balance the Cu and 2 FeO to balance the Fe. The right side must also have 3 SO_2 to balance the S.

Now there are 8 O's on the right, so there must be 4 O_2's on the left, and the equation is balanced.

$$2\ CuFeS_2 + 4\ O_2 \rightarrow 1\ Cu_2S + 2\ FeO + 3\ SO_2$$

24. **C** Let's say we have 100 grams of the compound.

$$Moles = \frac{grams}{MW}$$

So moles of carbon = $\frac{(75\ g)}{(12\ g/mol)}$ = 6 moles

and moles of hydrogen = $\frac{(25\ g)}{(1\ g/mol)}$ = 25 moles

According to our rough calculation, there are about four times as many moles of hydrogen in the compound as there are moles of carbon, so the empirical formula is CH_4.

25. **C** Let's say we have 100 grams of the compound.

$$Moles = \frac{grams}{MW}$$

Moles of chlorine = $\frac{(38\ g)}{(35.5\ g/mol)}$ = about 1 mole

Moles of fluorine = $\frac{(62\ g)}{(19\ g/mol)}$ = about 3 moles

According to our rough calculation, there are about three times as many moles of fluorine in the compound as there are moles of chlorine, so the empirical formula is ClF_3.

PROBLEMS

1. (a) $CaCO_3 + 2\,HCl \rightarrow CaCl_2 + H_2O + CO_2$

   (b) Use the ideal gas equation to find the number of moles of CO_2 produced. Remember to convert to the proper units.

   $$n = \frac{PV}{RT} = \frac{\left(\frac{740}{760}\,atm\right)(0.900\,L)}{(0.0821\,)(303\,K)} = 0.035\,moles$$

   From the balanced equation, for every mole of CO_2 produced, 1 mole of $CaCO_3$ was consumed.

   So 0.035 moles of $CaCO_3$ were consumed.

   (c) We know the number of moles of $CaCO_3$, so we can find the mass.
   Grams = (moles)(MW)
   Grams of $CaCO_3$ = (0.035 mol)(100 g/mol) = 3.50 grams

   $$Percent\ by\ mass = \frac{mass\ of\ CaCO_3}{mass\ of\ sample} \times 100 = \frac{(3.50\,g)}{(10.0\,g)} \times 100 = 35\%$$

   (d) Mass of SiO_2 = 10.0 g – 3.5 g = 6.5 g

   $$Moles = \frac{grams}{MW}$$

   $$Moles\ of\ SiO_2 = \frac{(6.5\,g)}{(60\,g/mol)} = 0.11\,mol$$

   $$Molar\ ratio = \frac{moles\ of\ CaCO_3}{moles\ of\ SiO_2} = \frac{(0.035)}{(0.11)} = 0.32$$

2. (a) All of the hydrogen in the water and all of the carbon in the carbon dioxide must have come from the hydrocarbon.

   $$Moles\ of\ H_2O = \frac{(2.16\,g)}{(18.0\,g/mol)} = 0.120\,moles$$

   Every mole of water contains 2 moles of hydrogen, so there are 0.240 moles of hydrogen.

   $$Moles\ of\ CO_2 = \frac{(1.80\,L)}{(22.4\,L/mol)} = 0.080\,moles$$

   Every mole of CO_2 contains 1 mole of carbon, so there are 0.080 moles of carbon.

   There are three times as many moles of hydrogen as there are moles of carbon, so the empirical formula of the hydrocarbon is CH_3.

   (b) In (a), we found the number of moles of hydrogen and carbon consumed, so we can find the mass of the hydrocarbon.

   Grams = (moles)(MW)
   Grams of H = (0.240 mol)(1.01 g/mol) = 0.242 g
   Grams of C = (0.080 mol)(12.01 g/mol) = 0.961 g
   Grams of hydrocarbon = (0.242) + (0.961) = 1.203 g

(c) First let's find the number of moles of hydrocarbon from the ideal gas law. Don't forget to convert to the appropriate units (760 mmHg = 1 atm, 32°C = 305 K).

$$n = \frac{PV}{RT} = \frac{(1.00 \text{ atm})(1.00 \text{ L})}{(0.0821\ )(305 \text{ K})} = 0.040 \text{ moles}$$

Now we can use the mass we found in (b) to find the molecular weight of the hydrocarbon.

$$MW = \frac{\text{grams}}{\text{moles}} = \frac{(1.203 \text{ g})}{(0.040 \text{ mol})} = 30.1 \text{ g/mole}$$

CH_3 would have a molecular weight of 15, so we can just double the empirical formula to get the molecular formula, which is C_2H_6.

(d) $2\ C_2H_6 + 7\ O_2 \rightarrow 4\ CO_2 + 6\ H_2O$

3. (a) First find the molecular weight of Cu_2S.

MW of Cu_2S = (2)(63.6) + (1)(32.1) = 159.3% by mass of Cu = $\dfrac{\text{mass of Cu}}{\text{mass of } Cu_2S} \times 100$

$= \dfrac{(2)(63.6)}{(159.3)} \times 100 = 79.8\%$

(b) Assume that we have 100 grams of ore #2. So we have 34.6 g of Cu, 30.5 g of Fe, and 34.9 g of S. To get the empirical formula, we need to find the number of moles of each element.

$$\text{Moles} = \frac{\text{grams}}{MW}$$

$$\text{Moles of Cu} = \frac{(34.6 \text{ g})}{(63.6 \text{ g/mol})} = 0.544 \text{ moles of Cu}$$

$$\text{Moles of Fe} = \frac{(30.5 \text{ g})}{(55.9 \text{ g/mol})} = 0.546 \text{ moles of Fe}$$

$$\text{Moles of S} = \frac{(34.9 \text{ g})}{(32.1 \text{ g/mol})} = 1.09 \text{ moles of S}$$

So the molar ratio of Cu:Fe:S is 1:1:2, and the empirical formula for ore #2 is $CuFeS_2$.

(c) You can use the ratio of the percents by weight.

$$\text{Mass of S} = \frac{\% \text{ by mass of S}}{\% \text{ by mass of Fe}} \times (\text{mass of Fe}) = \frac{(28.1\,\%)}{(16.3\,\%)} (11.0 \text{ g}) = 19.0 \text{ g}$$

(d) First find the moles of O_2 consumed.

$$\text{Moles} = \frac{\text{grams}}{MW}$$

$$\text{Moles of } O_2 = \frac{(3.84 \text{ g})}{(32.0 \text{ g/mol})} = 0.120 \text{ moles}$$

From the balanced equation, for every 3 moles of O_2 consumed, 6 moles of Cu are produced, so the number of moles of Cu produced will be twice the number of moles of O_2 consumed. So 0.240 moles of Cu are produced.

Grams = (moles)(MW)

Grams of Cu = (0.240 mol)(63.6 g/mol) = 15.3 grams

4. (a) We need to find the limiting reagent. There's plenty of water, so it must be one of the other two reactants.

$$\text{Moles} = \frac{\text{grams}}{\text{MW}}$$

$$\text{Moles of Mg} = \frac{(1.46 \text{ g})}{(24.3 \text{ g/mol})} = 0.060 \text{ moles}$$

Moles = (molarity)(volume)
Moles of $CuSO_4$ = (0.200 M)(0.500 L) = 0.100 moles

From the balanced equation, Mg and $CuSO_4$ are consumed in a 1:1 ratio, so we'll run out of Mg first. Mg is the limiting reagent, and we'll use it to find the yield of H_2.

From the balanced equation, 1 mole of H_2 is produced for every 2 moles of Mg consumed, so the number of moles of H_2 produced will be half the number of moles of Mg consumed.

$$\text{Moles of } H_2 = \frac{1}{2}(0.060 \text{ mol}) = 0.030 \text{ moles}$$

(b) Mg is the limiting reagent, so some $CuSO_4$ will remain. From the balanced equation, Mg and $CuSO_4$ are consumed in a 1:1 ratio, so when 0.060 moles of Mg are consumed, 0.060 moles of $CuSO_4$ are also consumed.

Moles of $CuSO_4$ remaining = (0.100 mol) − (0.060 mol) = 0.040 moles

(c) From the balanced equation, 1 mole of Cu_2O is produced for every 2 moles of Mg consumed, so the number of moles of Cu_2O produced will be half the number of moles of Mg consumed.

$$\text{Moles of } Cu_2O = \frac{1}{2}(0.060 \text{ mol}) = 0.030 \text{ moles}$$

Grams = (moles)(MW)
Grams of Cu_2O = (0.030 mol)(143 g/mol) = 4.29 grams

(d) All of the Mg consumed ends up as Mg^{2+} ions in the solution.

$$\text{Molarity} = \frac{\text{moles}}{\text{liters}}$$

$$[Mg^{2+}] = \frac{(0.060 \text{ mol})}{(0.500 \text{ L})} = 0.120 \text{ } M$$

6

Gases

HOW OFTEN DO GASES APPEAR ON THE EXAM?

In the multiple-choice section, this topic appears in about 5 out of 75 questions. In the free-response section, this topic appears almost every year.

STANDARD TEMPERATURE AND PRESSURE (STP)

You should be familiar with standard temperature and pressure (STP), which comes up fairly often in problems involving gases.

At STP: Pressure = 1 atmosphere = 760 millimeters of mercury (mmHg)

Temperature = 0°C = 273 K

At STP, 1 mole of gas occupies 22.4 liters

KINETIC MOLECULAR THEORY

Most of the gas problems you will see on the test will assume that gases behave in what is called an ideal manner. For ideal gases, the following assumptions can be made:

- The kinetic energy of an ideal gas is directly proportional to its absolute temperature: The greater the temperature, the greater the average kinetic energy of the gas molecules.

The Total Kinetic Energy of a Gas Sample
$$KE = \frac{3}{2}nRT$$ R = the gas constant; 8.31 joules/mol-K T = absolute temperature (K) n = number of moles (mol)

The Average Kinetic Energy of a Single Gas Molecule
$$KE = \frac{1}{2}mv^2$$ m = mass of the molecule (kg) v = speed of the molecule (meters/sec) KE is measured in joules

- If several different gases are present in a sample at a given temperature, all the gases will have the same average kinetic energy. That is, the average kinetic energy of a gas depends only on the absolute temperature, not on the identity of the gas.

- The volume of an ideal gas particle is insignificant when compared with the volume in which the gas is contained.

- There are no forces of attraction between the gas molecules in an ideal gas.

- Gas molecules are in constant motion, colliding with one another and with the walls of their container.

THE IDEAL GAS EQUATION

You can use the ideal gas equation to calculate any of the four variables relating to the gas, provided that you already know the other three.

The Ideal Gas Equation
$$PV = nRT$$ P = the pressure of the gas (atm) V = the volume of the gas (L) n = the number of moles of gas T = the absolute temperature of the gas (K) R = the gas constant, 0.0821 L-atm/mol-K

You can also manipulate the ideal gas equation to figure out how changes in each of its variables affect the other variables.

$$\frac{P_1 V_1}{T_1} = \frac{P_2 V_2}{T_2}$$

P = the pressure of the gas (atm)
V = the volume of the gas (L)
T = the absolute temperature of the gas (K)

You should be comfortable with the following simple relationships:

- If the volume is constant: As pressure increases, temperature increases; as temperature increases, pressure increases.

- If the temperature is constant: As pressure increases, volume decreases; as volume increases, pressure decreases. That's Boyle's law.

- If the pressure is constant: As temperature increases, volume increases; as volume increases, temperature increases. That's Charles's law.

DALTON'S LAW

Dalton's law states that the total pressure of a mixture of gases is just the sum of all the partial pressures of the individual gases in the mixture.

Dalton's Law
$P_{total} = P_a + P_b + P_c + ...$

You should also note that the partial pressure of a gas is directly proportional to the number of moles of that gas present in the mixture. So if 25 percent of the gas in a mixture is helium, then the partial pressure due to helium will be 25 percent of the total pressure.

Partial Pressure
$P_a = (P_{total})(X_a)$
$X_a = \dfrac{\text{moles of gas A}}{\text{total moles of gas}}$

GRAHAM'S LAW

Part of the first assumption of kinetic molecular theory was that all gases at the same temperature have the same average kinetic energy. Knowing this, we can find the average speed of a gas molecule at a given temperature.

$$u_{rms} = \sqrt{\frac{3kT}{m}} = \sqrt{\frac{3RT}{M}}$$

u_{rms} = average speed of a gas molecule (meters/sec)
T = absolute temperature (K)
m = mass of the gas molecule (kg)
M = molecular weight of the gas (kg/mol)
k = Boltzmann's constant, 1.38×10^{-23} joule/K
R = the gas constant, 8.31 joules/mol-K

By the way, you may have noticed that Boltzmann's constant, k, and the gas constant, R, differ only by a factor of Avogadro's number, N_A, the number of molecules in a mole. That is, $R = kN_A$.

Knowing that the average kinetic energy of a gas molecule is dependent only on the temperature, we can compare the average speeds (and the rates of effusion) of two different gases in a sample. The equation used to do this is called Graham's law.

Graham's Law

$$\frac{r_1}{r_2} = \sqrt{\frac{M_2}{M_1}}$$

r = rate of effusion of a gas or average speed of the molecules of a gas
M = molecular weight

You should note that Graham's law states that at a given temperature, lighter molecules move faster than heavier molecules.

VAN DER WAALS EQUATION

At low temperature and/or high pressure, gases behave in a less-than-ideal manner. That's because the assumptions made in kinetic molecular theory become invalid under conditions where gas molecules are packed too tightly together.

Two things happen when gas molecules are packed too tightly.

- *The volume of the gas molecules becomes significant.*
  The ideal gas equation does not take the volume of gas molecules into account, so the actual volume of a gas under nonideal conditions will be larger than the volume predicted by the ideal gas equation.

- *Gas molecules attract one another and stick together.*
  The ideal gas equation assumes that gas molecules never stick together. When a gas is packed tightly together, van der Waals forces (dipole–dipole attractions and London dispersion forces) become significant, causing some gas molecules to stick together. When gas molecules stick together, there are fewer particles bouncing around and creating pressure, so the real pressure in a nonideal situation will be smaller than the pressure predicted by the ideal gas equation.

The van der Waals equation adjusts the ideal gas equation to take nonideal conditions into account.

Van der Waals Equation
$$(P + \frac{n^2 a}{V^2})(V - nb) = nRT$$ P = the pressure of the gas (atm) V = the volume of the gas (L) n = the number of moles of gas (mol) T = the absolute temperature of the gas (K) R = the gas constant, 0.0821 L-atm/mol-K a = a constant, different for each gas, that takes into account the attractive forces between molecules b = a constant, different for each gas, that takes into account the volume of each molecule

DENSITY

You may be asked about the density of a gas. The density of a gas is measured in the same way as the density of a liquid or solid: in mass per unit volume.

Density of a Gas
$$D = \frac{m}{V}$$ D = density m = mass of gas, usually in grams V = volume occupied by a gas, usually in liters

CHAPTER 6 QUESTIONS

MULTIPLE-CHOICE QUESTIONS

Questions 1–5

    (A) H_2
    (B) He
    (C) O_2
    (D) N_2
    (E) CO_2

1. This is the most plentiful gas in the earth's atmosphere.

2. A 1 mole sample of this gas occupying 1 liter will have the greatest density.

3. At a given temperature, this gas will have the greatest rate of effusion.

4. The molecules of this nonpolar gas contain polar bonds.

5. The molecules of this gas contain triple bonds.

6. The temperature of a sample of an ideal gas confined in a 2.0 L container was raised from 27°C to 77°C. If the initial pressure of the gas was 1,200 mmHg, what was the final pressure of the gas?

    (A) 300 mmHg
    (B) 600 mmHg
    (C) 1,400 mmHg
    (D) 2,400 mmHg
    (E) 3,600 mmHg

7. A sealed container containing 8.0 grams of oxygen gas and 7.0 of nitrogen gas is kept at a constant temperature and pressure. Which of the following is true?

    (A) The volume occupied by oxygen is greater than the volume occupied by nitrogen.
    (B) The volume occupied by oxygen is equal to the volume occupied by nitrogen.
    (C) The volume occupied by nitrogen is greater than the volume occupied by oxygen.
    (D) The density of nitrogen is greater than the density of oxygen.
    (E) The average molecular speeds of the two gases are the same.

8. A gas sample contains 0.1 mole of oxygen and 0.4 moles of nitrogen. If the sample is at standard temperature and pressure, what is the partial pressure due to nitrogen?

    (A) 0.1 atm
    (B) 0.2 atm
    (C) 0.5 atm
    (D) 0.8 atm
    (E) 1.0 atm

9. A mixture of gases contains 1.5 moles of oxygen, 3.0 moles of nitrogen, and 0.5 moles of water vapor. If the total pressure is 700 mmHg, what is the partial pressure of the nitrogen gas?

(A) 70 mmHg
(B) 210 mmHg
(C) 280 mmHg
(D) 350 mmHg
(E) 420 mmHg

10. A mixture of helium and neon gases has a total pressure of 1.2 atm. If the mixture contains twice as many moles of helium as neon, what is the partial pressure due to neon?

(A) 0.2 atm
(B) 0.3 atm
(C) 0.4 atm
(D) 0.8 atm
(E) 0.9 atm

11. Nitrogen gas was collected over water at 25°C. If the vapor pressure of water at 25°C is 23 mmHg, and the total pressure in the container is measured at 781 mmHg, what is the partial pressure of the nitrogen gas?

(A) 23 mmHg
(B) 46 mmHg
(C) 551 mmHg
(D) 735 mmHg
(E) 758 mmHg

12. When 4.0 moles of oxygen are confined in a 24-liter vessel at 176°C, the pressure is 6.0 atm. If the oxygen is allowed to expand isothermally until it occupies 36 liters, what will be the new pressure?

(A) 2 atm
(B) 3 atm
(C) 4 atm
(D) 8 atm
(E) 9 atm

13. A gas sample is confined in a 5-liter container. Which of the following will occur if the temperature of the container is increased?

   I. The kinetic energy of the gas will increase.
  II. The pressure of the gas will increase.
 III. The density of the gas will increase.

(A) I only
(B) II only
(C) I and II only
(D) I and III only
(E) I, II, and III

14. A 22.0 gram sample of an unknown gas occupies 11.2 liters at standard temperature and pressure. Which of the following could be the identity of the gas?

(A) CO_2
(B) SO_3
(C) O_2
(D) N_2
(E) He

15. A gaseous mixture at a constant temperature contains O_2, CO_2, and He. Which of the following lists the three gases in order of increasing average molecular speeds?

(A) O_2, CO_2, He
(B) O_2, He, CO_2
(C) He, CO_2, O_2
(D) He, O_2, CO_2
(E) CO_2, O_2, He

16. Which of the following conditions would be most likely to cause the ideal gas laws to fail?

   I. High pressure
  II. High temperature
 III. Large volume

(A) I only
(B) II only
(C) I and II only
(D) I and III only
(E) II and III only

17. Which of the following expressions is equal to the density of helium gas at standard temperature and pressure?

(A) $\dfrac{1}{22.4}$ g/L

(B) $\dfrac{2}{22.4}$ g/L

(C) $\dfrac{1}{4}$ g/L

(D) $\dfrac{4}{22.4}$ g/L

(E) $\dfrac{4}{4}$ g/L

18. An ideal gas is contained in a 5.0 liter chamber at a temperature of 37°C. If the gas exerts a pressure of 2.0 atm on the walls of the chamber, which of the following expressions is equal to the number of moles of the gas? The gas constant, R, is 0.08 (L-atm)/(mol-K).

(A) $\dfrac{(2.0)(5.0)}{(0.08)(37)}$ moles

(B) $\dfrac{(2.0)(0.08)}{(5.0)(37)}$ moles

(C) $\dfrac{(2.0)(0.08)}{(5.0)(310)}$ moles

(D) $\dfrac{(2.0)(310)}{(0.08)(5.0)}$ moles

(E) $\dfrac{(2.0)(5.0)}{(0.08)(310)}$ moles

19. Which of the following gases would be expected to have a rate of effusion that is one-third as large as that of H_2?

(A) O_2
(B) N_2
(C) He
(D) H_2O
(E) CO_2

20. In an experiment $H_2(g)$ and $O_2(g)$ were completely reacted, above the boiling point of water, according to the following equation in a sealed container of constant volume and temperature:

$$2H_2(g) + O_2(g) \rightarrow 2H_2O(g)$$

If the initial pressure in the container before the reaction is denoted as P_i, which of the following expressions gives the final pressure, assuming ideal gas behavior?

(A) P_i
(B) $2\,P_i$
(C) $(3/2)P_i$
(D) $(2/3)P_i$
(E) $(1/2)P_i$

21. Nitrogen gas was collected over water at a temperature of 40°C, and the pressure of the sample was measured at 796 mmHg. If the vapor pressure of water at 40°C is 55 mmHg, what is the partial pressure of the nitrogen gas?

(A) 55 mmHg
(B) 741 mmHg
(C) 756 mmHg
(D) 796 mmHg
(E) 851 mmHg

22. A balloon occupies a volume of 1.0 liter when it contains 0.16 grams of helium at 37°C and 1 atm pressure. If helium is added to the balloon until it contains 0.80 grams while pressure and temperature are kept constant, what will be the new volume of the balloon?

(A) 0.50 liters
(B) 1.0 liters
(C) 2.0 liters
(D) 4.0 liters
(E) 5.0 liters

23. An ideal gas fills a balloon at a temperature of 27°C and 1 atm pressure. By what factor will the volume of the balloon change if the gas in the balloon is heated to 127 C at constant pressure?

(A) $\dfrac{27}{127}$

(B) $\dfrac{3}{4}$

(C) $\dfrac{4}{3}$

(D) $\dfrac{2}{1}$

(E) $\dfrac{127}{27}$

24. A gas sample with a mass of 10 grams occupies 6.0 liters and exerts a pressure of 2.0 atm at a temperature of 26°C. Which of the following expressions is equal to the molecular mass of the gas? The gas constant, R, is 0.08 (L-atm)/(mol-K).

(A) $\dfrac{(10)(0.08)(299)}{(2.0)(6.0)}$ g/mol

(B) $\dfrac{(299)(0.08)}{(10)(2.0)(6.0)}$ g/mol

(C) $\dfrac{(2.0)(6.0)(299)}{(10)(0.08)}$ g/mol

(D) $\dfrac{(10)(2.0)(6.0)}{(299)(0.08)}$ g/mol

(E) $\dfrac{(2.0)(6.0)}{(10)(299)(0.08)}$ g/mol

25. Which of the following assumptions is (are) valid based on kinetic molecular theory?

   I. Gas molecules have negligible volume.
   II. Gas molecules exert no attractive forces on one another.
   III. The temperature of a gas is directly proportional to its kinetic energy.

(A) I only
(B) III only
(C) I and III only
(D) II and III only
(E) I, II, and III

PROBLEMS

1.

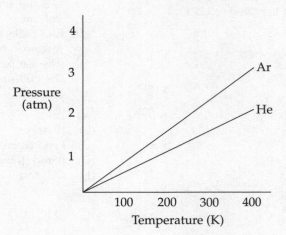

The graph above shows the changes in pressure with changing temperature of gas samples of helium and argon confined in a closed 2-liter vessel.

(a) What is the total pressure of the two gases in the container at a temperature of 200 K?

(b) How many moles of helium are contained in the vessel?

(c) How many molecules of helium are contained in the vessel?

(d) What is the ratio of the average speeds of the helium atoms to the average speeds of the argon atoms?

(e) If the volume of the container were reduced to 1 liter at a constant temperature of 300 K, what would be the new pressure of the helium gas?

2. $$2 \, KClO_3(s) \rightarrow 2 \, KCl(s) + 3 \, O_2(g)$$

The reaction above took place, and 1.45 liters of oxygen gas were collected over water at a temperature of 29°C and a pressure of 755 millimeters of mercury. The vapor pressure of water at 29°C is 30.0 millimeters of mercury.

(a) What is the partial pressure of the oxygen gas collected?

(b) How many moles of oxygen gas were collected?

(c) What would be the dry volume of the oxygen gas at a pressure of 760 millimeters of mercury and a temperature of 273 K?

(d) What was the mass of the $KClO_3$ consumed in the reaction?

ESSAYS

3. Equal molar quantities of two gases, O_2 and H_2O, are confined in a closed vessel at constant temperature.

    (a)   Which gas, if any, has the greater partial pressure?

    (b)   Which gas, if any, has the greater density?

    (c)   Which gas, if any, has the greater concentration?

    (d)   Which gas, if any, has the greater average kinetic energy?

    (e)   Which gas, if any, will show the greater deviation from ideal behavior?

    (f)   Which gas, if any, has the greater average molecular speed?

4. SF_6, H_2O, and CO_2 are all greenhouse gases because they absorb infrared radiation in the atmosphere. As such they are of interest to many researchers.

    (a)   Of the three gases, which deviates the most from ideal gas behavior?

    (b)   If all three ideal gases are held at constant temperature, which gas would have the highest average molecular speed?

    (c)   A container was with equal amounts H_2O and CO_2 and the pressure was measured. Assuming ideal behavior, if SF_6 was then added such that the final pressure was four times that before its addition, the final number of moles of SF_6 is how many times that of H_2O?

    (d)   If a container held equal molar amounts of all three greenhouse gases, but was then compressed decreasing the volume within, which of the gases has the highest, final partial pressure assuming ideal behavior?

    (e)   Gas molecules exert pressure through collisions with the walls of the containers they are in. If a container is filled with equal amounts of each of the greenhouse gases, which gas undergoes the fewest collisions with the wall of the container per unit time?

CHAPTER 6 ANSWERS AND EXPLANATIONS

MULTIPLE-CHOICE QUESTIONS

1. **D** Nitrogen gas makes up about 78 percent of the gas in the earth's atmosphere.

2. **E** Density is a measure of grams per liter. CO_2 has the greatest molecular mass (44), so 1 mole of CO_2 will have the most mass in 1 liter and, therefore, the greatest density.

3. **A** According to Graham's law, the lighter the gas, the greater the rate of effusion. H_2 is the lightest gas (MW = 2), so it will have the greatest rate of effusion.

4. **E** CO_2 is the only gas listed that has bonds between atoms of differing electronegativity. The carbon–oxygen bonds in CO_2 are polar, although the linear geometry of the molecule makes the molecule nonpolar overall.

5. **D** N_2 is the only gas listed whose atoms are held together in a triple bond.

6. **C** Remember to convert Celsius to Kelvin (°C + 273 = K).

   27°C = 300 K and 77°C = 350 K.

   From the relationship $\dfrac{P_1}{T_1} = \dfrac{P_2}{T_2}$ we get

   $$\frac{(1,200\ \text{mm Hg})}{(300\ \text{K})} = \frac{P_2}{(350\ \text{K})}$$

   So P_2 = 1,400 mmHg

7. **B** 8.0 grams of oxygen and 7.0 grams of nitrogen both equal 0.25 moles. Avogadro's law states equal volumes contain equal moles. This makes (A) and (C) wrong.

   (D) This is reversed: The density of oxygen gas would be greater than the density of nitrogen.

   (E) From Graham's law, if the kinetic energies of the two gases are the same, then the molecules of the less massive gas must be moving faster, on average.

8. **D** If the gases are at STP, then the total pressure must be 1.0 atmosphere.

   If $\dfrac{4}{5}$ of the gas in the sample is nitrogen, then from Dalton's law, $\dfrac{4}{5}$ of the pressure must be due to the nitrogen.

   So the partial pressure due to nitrogen is $\dfrac{4}{5}$ (1.0 atm) = 0.8 atm.

9. **E** From Dalton's law, the partial pressure of a gas depends on the number of moles of the gas that are present.

   The total number of moles of gas present is

   1.5 + 3.0 + 0.5 = 5.0 total moles

   If there are 3 moles of nitrogen, then $\dfrac{3}{5}$ of the pressure must be due to nitrogen.

   $(\dfrac{3}{5})$(700 mmHg) = 420 mmHg

10. **C** From Dalton's law, the partial pressure of a gas depends on the number of moles of the gas that are present. If the mixture has twice as many moles of helium as neon, then the mixture must be $\frac{1}{3}$ neon. So $\frac{1}{3}$ of the pressure must be due to neon.

$(\frac{1}{3})(1.2 \text{ atm}) = 0.4 \text{ atm}$.

11. **E** From Dalton's law, the partial pressures of nitrogen and water vapor must add up to the total pressure in the container. The partial pressure of water vapor in a closed container will be equal to the vapor pressure of water, so the partial pressure of nitrogen is

781 mmHg – 23 mmHg = 758 mmHg

12. **C** From the gas laws, we know that with constant temperature

$P_1V_1 = P_2V_2$

Solving for P_2, we get

$P_2 = \dfrac{P_1V_1}{V_2} = \dfrac{(6.0 \text{ atm})(24 \text{ L})}{(36 \text{ L})} = 4.0 \text{ atm}$

13. **C** From kinetic molecular theory, we know that kinetic energy is directly proportional to temperature (think of the expression $KE = \frac{3}{2}kT$), so (I) is true.

From the gas laws, we know that at constant volume, an increase in temperature will bring about an increase in pressure (think of $\dfrac{P_1}{T_1} = \dfrac{P_2}{T_2}$), so (II) is true.

The density of a gas is equal to mass per unit volume, which is not changed by changing temperature, so (III) is not true.

14. **A** Use the relationship

$\text{Moles} = \dfrac{\text{liters}}{22.4 \text{ L/mol}}$

$\text{Moles of unknown gas} = \dfrac{11.2 \text{ L}}{22.4 \text{ L/mol}} = 0.500 \text{ moles}$

$\text{MW} = \dfrac{\text{grams}}{\text{mole}}$

$\text{MW of unknown gas} = \dfrac{22.0 \text{ g}}{0.500 \text{ mole}} = 44.0 \text{ grams/mole}$

That's the molecular weight of CO_2.

15. **E** According to Graham's law, at a given temperature, heavier molecules will have lower average speeds and lighter molecules will have higher speeds.

Helium is the lightest of the three molecules (MW = 4), oxygen is next (MW = 32), and carbon dioxide is the heaviest (MW = 44).

16. **A** The ideal gas laws fail under conditions where gas molecules are packed too tightly together. This can happen as a result of high pressure (I), low temperature (not listed as a choice), or small volume (not listed as a choice).

17. **D** Density is measured in grams per liter. One mole of helium gas has a mass of 4 grams and occupies a volume of 22.4 liters at STP, so the density of helium gas at STP is

$\dfrac{4}{22.4}$ g/L.

18. **E** From the ideal gas equation, we know that $PV = nRT$.

Remember to convert 37°C to 310 K.

Then we just rearrange the equation to solve for n.

$n = \dfrac{PV}{RT} = \dfrac{(2.0)(5.0)}{(0.08)(310)}$ moles

19. **D** Remember Graham's law, which relates the average speeds of different gases (and thus their rates of effusion) to their molecular weights.

$\dfrac{v_1}{v_2} = \sqrt{\dfrac{MW_2}{MW_1}}$

To get a velocity and rate of effusion that is one-third as large as H_2's, we need a molecule with a molecular weight that is nine times as large. The molecular weight of H_2 is 2, so we need a molecule with a molecular weight of 18. That's H_2O.

20. **D** As long as the gases act ideally, the total pressure in the container will be a function only of the number of gas moles present so long as volume and temperature are held constant. The relationship between moles and pressure before and after the reaction, from the ideal gas equation ($PV = nRT$), will be $P_i/n_i = P_f/n_f$ where n_i, P_f, n_f are the initial moles, the final pressure, and the final moles respectively. We can rearrange this equation to $P_f = P_i(n_f/n_i)$, and since there are 2 moles of gas in the products for every 3 moles in the reactants, we can say that $P_f = (2/3)P_i$.

21. **B** From Dalton's law, the partial pressures of nitrogen and water vapor must add up to the total pressure in the container. The partial pressure of water vapor in a closed container will be equal to its vapor pressure, so the partial pressure of nitrogen is

796 mmHg – 55 mmHg = 741 mmHg

22. **E** According to the ideal gas laws, at constant temperature and pressure, the volume of a gas is directly proportional to the number of moles.

We increased the number of grams by a factor of 5 ([0.16][5] = [0.80]). That's the same as increasing the number of moles by a factor of 5. So we must have increased the volume by a factor of 5. Therefore, (5)(1 L) = 5 L.

23. **C** From the ideal gas laws, for a gas sample at constant pressure

$\dfrac{V_1}{T_1} = \dfrac{V_2}{T_2}$

Solving for V_2 we get $V_2 = V_1 \dfrac{T_2}{T_1}$

So V_1 is multiplied by a factor of $\dfrac{T_2}{T_1}$

Remember to convert Celsius to Kelvin, $\dfrac{127°C + 273}{27°C + 273} = \dfrac{400 \text{ K}}{300 \text{ K}} = \dfrac{4}{3}$

24. **A** We can find the number of moles of gas from $PV = nRT$.

Remember to convert 26°C to 299 K.

Then solve for n.

$$n = \frac{PV}{RT} = \frac{(2.0)(6.0)}{(0.08)(299)} \text{ mol}$$

Now, remember:

$$MW = \frac{\text{grams}}{\text{moles}} = \frac{(10 \text{ g})}{\left(\dfrac{(2.0)(6.0)}{(0.08)(299)} \text{ mol}\right)} = \frac{(10)(0.08)(299)}{(2.0)(6.0)} \text{ g/mol}$$

25. **E** All three assumptions are included in kinetic molecular theory.

PROBLEMS

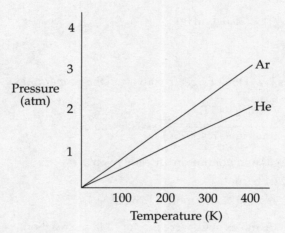

1. (a) Read the graph, and add the two pressures.

$P_{\text{Total}} = P_{\text{He}} + P_{\text{Ar}}$

$P_{\text{Total}} = (1 \text{ atm}) + (1.5 \text{ atm}) = 2.5 \text{ atm}$

(b) Read the pressure (1 atm) at 200 K, and use the ideal gas equation.

$$n = \frac{PV}{RT} = \frac{(1.0 \text{ atm})(2.0 \text{ L})}{(0.082 \text{ L-atm/mol-K})(200 \text{ K})} = 0.12 \text{ moles}$$

(c) Use the definition of a mole.

Molecules = (moles)(6.02×10^{23})

Molecules (atoms) of helium = $(0.12)(6.02 \times 10^{23}) = 7.2 \times 10^{22}$

(d) Use Graham's law.

$$\frac{v_1}{v_2} = \sqrt{\frac{MW_2}{MW_1}}$$

$$\frac{v_{He}}{v_{Ar}} = \sqrt{\frac{MW_{Ar}}{MW_{He}}} = \sqrt{\frac{(40)}{(4)}} = 3.2 \text{ to } 1$$

(e) Use the relationship

$$\frac{P_1 V_1}{T_1} = \frac{P_2 V_2}{T_2}$$

With T constant

$P_1 V_1 = P_2 V_2$

$(1.5 \text{ atm})(2.0 \text{ L}) = P_2 (1.0 \text{ L})$

$P_2 = 3.0 \text{ atm}$

2. (a) Use Dalton's law.

$P_{Total} = P_{Oxygen} + P_{Water}$

$(755 \text{ mmHg}) = (P_{Oxygen}) + (30.0 \text{ mmHg})$

$P_{Oxygen} = 725 \text{ mmHg}$

(b) Use the ideal gas law. Don't forget to convert to the proper units.

$$n = \frac{PV}{RT} = \frac{\left(\frac{725}{760} \text{ atm}\right)(1.45 \text{ L})}{(0.082 \text{ L-atm / mol-K})(302 \text{ K})} = 0.056 \text{ moles}$$

(c) At STP, moles of gas and volume are directly related.

Volume = (moles)(22.4 L/mol)

Volume of O_2 = (0.056 mol)(22.4 L/mol) = 1.25 L

(d) We know that 0.056 moles of O_2 were produced in the reaction.

From the balanced equation, we know that for every 3 moles of O_2 produced, 2 moles of $KClO_3$ are consumed. So there are $\frac{2}{3}$ as many moles of $KClO_3$ as O_2.

Moles of $KClO_3$ = $\left(\frac{2}{3}\right)$(moles of O_2)

Moles of $KClO_3$ = $\left(\frac{2}{3}\right)$(0.056 mol) = 0.037 moles

Grams = (moles)(MW)

Grams of $KClO_3$ = (0.037 mol)(122 g/mol) = 4.51 g

ESSAYS

3. (a) The partial pressures depend on the number of moles of gas present. Because the number of moles of the two gases are the same, the partial pressures are the same.

(b) O_2 has the greater density. Density is mass per unit volume. Both gases have the same number of moles in the same volume, but oxygen has heavier molecules, so it has greater density.

(c) Concentration is moles per volume. Both gases have the same number of moles in the same volume, so their concentrations are the same.

(d) According to kinetic-molecular theory, the average kinetic energy of a gas depends only on the temperature. Both gases are at the same temperature, so they have the same average kinetic energy.

(e) H_2O will deviate most from ideal behavior. Ideal behavior for gas molecules assumes that there will be no intermolecular interactions.

H_2O is polar, and O_2 is not. H_2O undergoes hydrogen bonding while O_2 does not. So H_2O has stronger intermolecular interactions, which will cause it to deviate more from ideal behavior.

(f) H_2O has the greater average molecular speed. From Graham's law, if two gases are at the same temperature, the one with the smaller molecular weight (MW of H_2O = 18, MW of O_2 = 32) will have the greater average molecular speed.

4. (a) H_2O deviates most drastically from ideal behavior due to very strong hydrogen bonding interactions. A truly ideal gas has no interactions with surrounding molecules. These hydrogen bonding interactions are why H_2O is a liquid at room temperature and the others are gases.

(b) As per Graham's law, the species with the lowest molecular weight would have the greatest average molecular speed. H_2O, MW = 18 g/mol, is the lightest, and hence would be the fastest.

(c) By Dalton's law of partial pressures, if SF_6 was added until the pressure was 4 times the pressure with just (an equal mixture of) H_2O and CO_2, then there must be 3 times as much SF_6 as H_2O and CO_2 combined. This means that, in the end, there 6 times more SF_6 than H_2O. For example, imagine starting with 1 atm each of CO_2 and H_2O. In this case the initial pressure is 2 atm, and to get to 8 atm one must add 6 atm of SF_6 which is 6 times that of the H_2O present.

(d) All three gases will have identical partial pressures. The equation for partial pressure of any component a is $P_a = (P_{total})(X_a)$. Since we know that each has the same mol fraction the total pressure doesn't matter, and each will have the same partial pressure, relative to one another.

(e) SF_6 will have the fewest collisions per unit time. Since the kinetic theory assumes that the gas molecules are moving randomly, the one with the most collisions with the walls of the containers will be the one moving the fastest. On the other hand, the slowest molecule will have the fewest collisions. The slowest molecule will be the most massive one, as per Graham's law, which in this case is SF_6.

7

Phase Changes

HOW OFTEN DO PHASE CHANGES APPEAR ON THE EXAM?

In the multiple-choice section, this topic appears in about 4 out of 75 questions. In the free-response section, this topic appears occasionally.

THE PHASE DIAGRAM

You should know how to work with the phase diagram below.

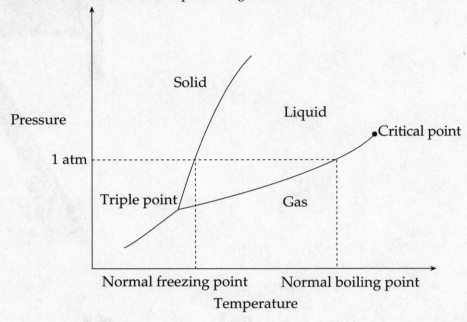

At high pressure and low temperature, a substance is in a solid phase. At low pressure and high temperature, a substance is in a gas phase. Liquid phase is in between these two.

The lines that separate the phases correspond to points at which the substance can exist simultaneously in both phases at equilibrium. The normal freezing point is the temperature at which the solid-liquid phase equilibrium line crosses the 1 atmosphere pressure line. The normal boiling point is the temperature at which the liquid-gas equilibrium line crosses the 1 atmosphere pressure line.

The **triple point** is the temperature and pressure at which all three phases can exist simultaneously in equilibrium.

The **critical point** is the temperature beyond which the molecules of a substance have too much kinetic energy to stick together and form a liquid.

PHASE CHANGES

NAMING THE PHASE CHANGES

Solid to *liquid*	—	**Melting**
Liquid to *solid*	—	**Freezing**
Liquid to *gas*	—	**Vaporization**
Gas to *liquid*	—	**Condensation**
Solid to *gas*	—	**Sublimation**
Gas to *solid*	—	**Deposition**

Phase changes occur because of changes in temperature and/or pressure. You should be able to read the phase diagram and see how a change in pressure or temperature will affect the phase of a substance.

Some particles in a liquid or solid will have enough energy to break away from the surface and become gaseous. The pressure exerted by these molecules as they escape from the surface is called the **vapor pressure**. When the liquid or solid phase of a substance is in equilibrium with the gas phase, the pressure of the gas will be equal to the vapor pressure of the substance. As temperature increases, the vapor pressure of a liquid will increase. When the vapor pressure of a liquid increases to the point where it is equal to the surrounding atmospheric pressure, the liquid boils.

HEAT OF FUSION

The **heat of fusion** is the energy that must be put into a solid to melt it. This energy is needed to overcome the forces holding the solid together. Alternatively, the heat of fusion is the heat given off by a substance when it freezes. The intermolecular forces within a solid are more stable and, therefore, have lower energy than the forces within a liquid, so energy is released in the freezing process.

HEAT OF VAPORIZATION

The **heat of vaporization** is the energy that must be put into a liquid to turn it into a gas. This energy is needed to overcome the forces holding the liquid together. Alternatively, the heat of vaporization is the heat given off by a substance when it condenses. Intermolecular forces become stronger when a gas condenses; the gas becomes a liquid, which is more stable, and energy is released.

As heat is added to a substance in equilibrium, the temperature of the substance can increase or the substance can change phases, but both changes cannot occur simultaneously.

THE PHASE DIAGRAM FOR WATER

You should recognize how the phase diagram for water differs from the phase diagram for most other substances.

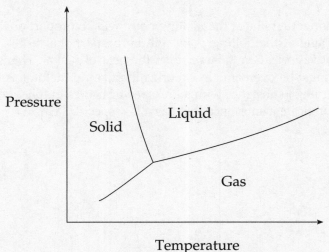

In the phase diagram for substances other than water, the solid-liquid equilibrium line slopes upward. In the phase diagram for water, the solid-liquid equilibrium line slopes *downward*. What this means is that when pressure is increased, a normal substance will change from liquid to solid, but water will change from solid to liquid.

Water has this odd property because its hydrogen bonds form a lattice structure when it freezes. This forces the molecules to remain farther apart in ice than in water, making the solid phase less dense than the liquid phase. That's why ice floats on water, and this natural phenomenon preserves most marine life.

THE HEATING CURVE

You should be familiar with the heating curve below, which shows how the temperature of a substance changes as it is heated. In the diagram shown, the substance starts as a solid. As heat is added, the kinetic energy of the substance increases and the temperature increases. When the temperature reaches the melting point, the line flattens out. At this point, during the phase change from solid to liquid, all added heat goes toward overcoming the intermolecular forces in the solid; therefore, there is no increase in kinetic energy during the phase change, and the temperature stays the same. The same thing occurs at the boiling point (during the phase change from liquid to gas).

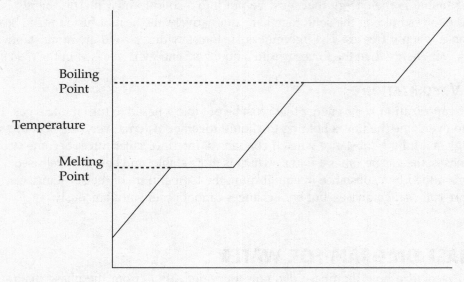

The length of the horizontal line at the melting point will be proportional to the heat of fusion, and the length of the line at the boiling point will be proportional to the heat of vaporization. Notice that the heat of vaporization is larger than the heat of fusion. That's because many more intermolecular forces must be overcome in vaporization than in melting. Also, because it's easier to change the temperature of substance with a low specific heat, a graph whose temperature curve has a steep slope will indicate a substance that has a low specific heat.

CHAPTER 7 QUESTIONS

MULTIPLE-CHOICE QUESTIONS

Questions 1–4 refer to the phase diagram below.

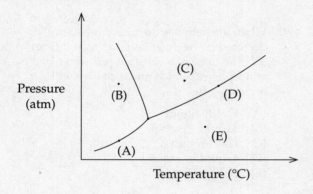

1. At this point, the substance represented by the phase diagram will be solely in the solid phase at equilibrium.

2. This point represents a boiling point of the substance.

3. At this point, the substance represented by the phase diagram will be undergoing sublimation.

4. At this point, the substance represented by the phase diagram will be solely in the liquid phase at equilibrium.

5. When a substance undergoes a phase change from liquid to solid, which of the following will occur?

    (A) Energy will be released by the substance because intermolecular forces are being weakened.
    (B) Energy will be released by the substance because intermolecular forces are being strengthened.
    (C) Energy will be absorbed by the substance because intermolecular forces are being weakened.
    (D) Energy will be absorbed by the substance because intermolecular forces are being strengthened.
    (E) The energy of the substance will not be changed.

6. During which of the following phase changes must heat be added to overcome intermolecular forces?

    I. Vaporization
    II. Sublimation
    III. Deposition

    (A) I only
    (B) II only
    (C) I and II only
    (D) I and III only
    (E) I and III only

Questions 7–10 are based on the phase diagram below.

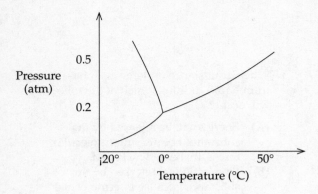

7. As pressure on the substance depicted in the diagram is increased at constant temperature, which of the following phase changes CANNOT occur?

    I.   Condensation
    II. Melting
    III. Freezing

    (A) I only
    (B) II only
    (C) III only
    (D) I and II only
    (E) I and III only

8. At a temperature of 50°C and a pressure of 0.2 atmospheres, the substance depicted in the diagram is

    (A) in the gas phase.
    (B) in the liquid phase.
    (C) in the solid phase.
    (D) at its triple point.
    (E) at its critical point.

9. Which of the following lists the three phases of the substance shown in the diagram in order of increasing density at –5°C?

    (A) Solid, gas, liquid
    (B) Solid, liquid, gas
    (C) Gas, liquid, solid
    (D) Gas, solid, liquid
    (E) Liquid, solid, gas

10. When the temperature of the substance depicted in the diagram is decreased from 10°C to –10°C at a constant pressure of 0.3 atmospheres, which phase change will occur?

    (A) Gas to liquid
    (B) Liquid to solid
    (C) Gas to solid
    (D) Liquid to gas
    (E) Solid to liquid

11. Which of the following processes can occur when the temperature of a substance is increased at constant pressure?

    I.   Sublimation
    II. Melting
    III. Boiling

    (A) I only
    (B) II only
    (C) I and II only
    (D) II and III only
    (E) I, II, and III

12. The temperature above which gas molecules become too energetic to form a true liquid, no matter what the pressure, is called the

    (A) melting point.
    (B) critical point.
    (C) boiling point.
    (D) triple point.
    (E) freezing point.

Questions 13–15 are based on the information given below.

In an experiment, a solid 1 molar sample of Substance A was gradually heated by a source of constant energy for several hours and the temperature was measured periodically. At the end of the heating period, Substance A had been converted to the gas phase. The heating curve produced by this experiment is shown below.

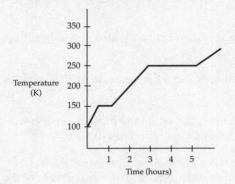

13. During the course of the experiment, there was a period of time when the the solid phase of Substance A was in equilibrium with the liquid phase. At what temperature did this occur?

(A) Between 100 K and 150 K
(B) At 150 K
(C) Between 150 K and 250 K
(D) At 250 K
(E) Between 250 K and 350 K

14. Based on the data given in the heating curve, which of the following statements is NOT true regarding Substance A?

(A) The boiling point of Substance A is 250 K.
(B) The freezing point of Substance A is 150 K.
(C) The heat of vaporization of Substance A is greater than the heat of fusion.
(D) Substance A is a liquid at room temperature.
(E) The intermolecular forces exhibited by Substance A are weaker than those of water.

15. During the course of the experiment, Substance A was gradually heated from 100 K to 350 K. When the temperature reached 250 K, the energy absorbed by Substance A

(A) was used to change from liquid to gas phase.
(B) was used to change from gas to liquid phase.
(C) was used to change from solid to liquid phase.
(D) was used to change from liquid to solid phase.
(E) was reduced to zero.

ESSAYS

1.

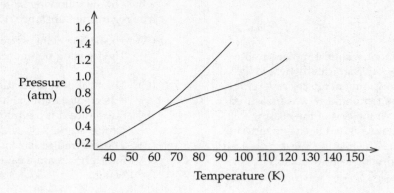

The phase diagram for a substance is shown above. Use the diagram and your knowledge of phase changes to answer the following questions.

(a)   When the substance is at a pressure of 0.8 atmospheres and a temperature of 50 K, what is its phase?

(b)   Describe the change in phase that the substance undergoes when the pressure is decreased from 1.2 atmospheres to 0.6 atmospheres at a constant temperature of 110 K.

(c)   A constant source of heat was applied to the substance at a constant pressure of 0.8 atmospheres. The substance was initially at a temperature of 50 K. The temperature of the substance increased at a constant rate until 70 K was reached. At this point, the temperature remained constant for a period of time, then continued to climb at a constant rate. Explain what has happened.

(d)   What is the normal boiling point for this substance?

2.

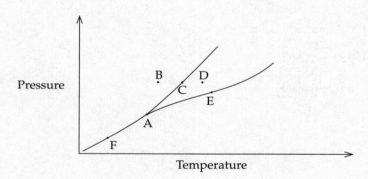

The phase diagram for a substance is shown above. Use the diagram and your knowledge of phase changes to answer the following questions.

(a)   Describe the phase change that the substance undergoes as the temperature is increased from points B to C to D at constant pressure.

(b)   Which of the phases are in equilibrium at point F? Give a name for one of the phase changes between these two phases.

(c)   Could this phase diagram represent water? Explain why or why not.

(d)   What is the name given to point A, and what is the situation particular to this point?

CHAPTER 7 ANSWERS AND EXPLANATIONS

MULTIPLE-CHOICE QUESTIONS

1. **B** Point (B) is in a region of high pressure and low temperature. That corresponds to the solid phase.

2. **D** Point (D) is on the phase change line between liquid and gas. The temperature at point (D) is the boiling point of the substance at the pressure at point (D).

3. **A** Point (A) is on the phase change line between solid and gas. Sublimation is the phase change from solid directly to gas.

4. **C** Point (C) is in the liquid region, which is the middle region between solid and gas.

5. **B** The intermolecular forces in the solid phase of any substance are stronger than the intermolecular forces in the liquid phase. Whenever intermolecular forces are strengthened, energy is given off.

6. **C** Intermolecular forces must be overcome to convert a liquid to a gas (vaporization). Likewise, in sublimation (the conversion of a solid to a gas), intermolecular forces must be overcome. However, in deposition (as a gas changes to a solid), intermolecular forces are strengthened.

7. **C** This diagram has the downward slope between the solid and liquid phases that is typical of water, so you've got to be careful in dealing with solid-liquid phase changes. The possible phase changes for increasing pressure and constant temperature are given in the diagram below.

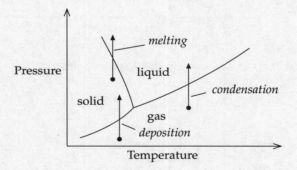

This substance will melt under increased pressure instead of freezing, so while choices (I) and (II) can occur, choice (III) cannot.

8. **A** At 50°C and 0.2 atm the substance is a gas.

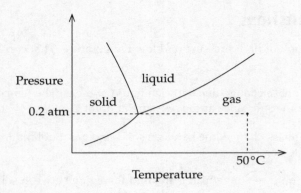

9. **D** As pressure is increased, density will increase; so increasing density is shown by the arrow in the diagram below.

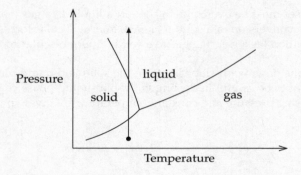

10. **B** The phase change will occur as shown in the diagram below.

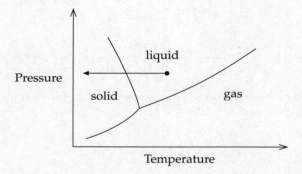

11. **E** All three phase changes listed can occur when temperature is increased at constant pressure. This is shown in the diagram below.

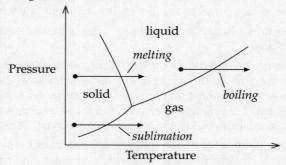

12. **B** At temperatures above the critical point, gases can become dense, but liquids do not form.

13. **B** The first flat part of the heating curve corresponds to the phase change from solid to liquid. During the phase change, an equilibrium exists between the two phases.

14. **D** Choice (D) is the only choice that's not true. Room temperature is about 298 K. At 298 K, Substance *A* is in the gas phase, not the liquid phase. As for the other answers, Choices (A) and (B) accurately give the boiling and freezing points. Choice (C) is accurate because the vaporization line is much longer than the fusion line. Heat of vaporization is usually much higher than heat of fusion. Choice (E) is accurate because the boiling and freezing points of Substance *A* are lower than those of water, so Substance *A* must be held together by weaker intermolecular forces.

15. **A** At 250 K, Substance *A* boiled. That means that the temperature remained constant while the energy absorbed was used to break up the forces holding the liquid together and convert the substance to a gas.

ESSAYS

1. (a) The phase is solid, which corresponds to high pressure and low temperature.

   (b) The substance starts as a liquid and changes to a gas (vaporizes) when the pressure has dropped to about 0.9 atm.

   (c) At 70 K, the substance melted. While the substance was melting, the heat from the constant source went toward overcoming the strong intermolecular forces of the solid instead of increasing the temperature. The heat that must be put into a solid to melt it is called the heat of fusion. When the heat of fusion was overcome and the substance was entirely in the liquid phase, the temperature of the liquid began to increase at a constant rate.

   (d) The normal boiling point is the boiling point at a pressure of 1 atmosphere. We can find the normal boiling point by looking for the spot at which the gas-liquid equilibrium line crosses the 1 atm line. This happens at 115 K.

2. (a) At point B, the substance is a solid.

   At point C, the substance is in equilibrium between solid and liquid phase; that is, it's melting.

   At point D, the substance is entirely in liquid phase.

   (b) At point F, the substance is in equilibrium between the gas and solid phases.

   When a substance changes from solid to gas phase, it's called sublimation.

   When a substance changes from gas to solid phase, it's called deposition.

   (c) The diagram CANNOT represent water.

   For water, the slope of the solid-liquid equilibrium line must be negative, indicating that the liquid phase of water is denser than the solid phase.

   (d) Point A is called the triple point.

   At the triple point, all three phases can exist simultaneously in equilibrium.

8

Thermodynamics

HOW OFTEN DOES THERMODYNAMICS APPEAR ON THE EXAM?

In the multiple-choice section, this topic appears in about 5 out of 75 questions. In the free-response section, this topic appears every year.

THE FIRST AND SECOND LAWS OF THERMODYNAMICS

The first law of thermodynamics says that the energy of the universe is constant. Energy can be neither created nor destroyed, so while energy can be *converted* in a chemical process, the total energy remains constant.

The second law of thermodynamics says that if a process is spontaneous in one direction, then it can't be spontaneous in the reverse direction, and that the entropy of the universe always increases during spontaneous reactions.

STATE FUNCTIONS

Enthalpy change (ΔH), entropy change (ΔS), and free-energy change (ΔG) are **state functions.** That means they all depend only on the change between the initial and final states of a system, not on the process by which the change occurs. For a chemical reaction, this means that the thermodynamic state functions are independent of reaction pathway; for instance, the addition of a catalyst to a reaction will have no effect on the overall energy or entropy change of the reaction.

STANDARD STATE CONDITIONS

When the values of thermodynamic quantities are given on the test, they are almost always given for standard state conditions. A thermodynamic quantity under standard state conditions is indicated by the little superscript circle, so the following is true under standard state conditions:

$$\Delta H = \Delta H^\circ$$

$$\Delta S = \Delta S^\circ$$

$$\Delta G = \Delta G^\circ$$

Standard State Conditions

- All gases are at 1 atmosphere pressure.
- All liquids are pure.
- All solids are pure.
- All solutions are at 1-molar (1 M) concentration.
- The energy of formation of an element in its normal state is defined as zero.
- The temperature used for standard state values is almost invariably room temperature: 25°C (298 K). Standard state values can be calculated for other temperatures, however.

ENTHALPY

ENTHALPY CHANGE, ΔH

The enthalpy of a substance is a measure of the energy that is released or absorbed by the substance when bonds are broken and formed during a reaction.

> **The Basic Rules of Enthalpy**
> When bonds are *formed*, energy is *released*.
> When bonds are *broken*, energy is *absorbed*.

The change in enthalpy, ΔH, that takes place over the course of a reaction can be calculated by subtracting the enthalpy of the reactants from the enthalpy of the products.

> **Enthalpy Change**
> $$\Delta H = H_{products} - H_{reactants}$$

If the products have stronger bonds than the reactants, then the products have lower enthalpy than the reactants and are more stable; in this case, energy is released by the reaction, or the reaction is **exothermic.**

If the products have weaker bonds than the reactants, then the products have higher enthalpy than the reactants and are less stable; in this case, energy is absorbed by the reaction, or the reaction is **endothermic**.

All substances like to be in the lowest possible energy state, which gives them the greatest stability. This means that, in general, exothermic processes are more likely to occur spontaneously than endothermic processes.

HEAT OF FORMATION, ΔH_f°

Heat of formation is the change in energy that takes place when one mole of a compound is formed from its component pure elements under standard state conditions. Heat of formation is almost always calculated at a temperature of 25°C (298 K).

Remember: ΔH_f for a pure element is defined as zero.

- If ΔH_f° for a compound is negative, energy is released when the compound is formed from pure elements, and the product is *more* stable than its constituent elements. That is, the process is exothermic.

- If ΔH_f° for a compound is positive, energy is absorbed when the compound is formed from pure elements, and the product is *less* stable than its constituent elements. That is, the process is endothermic.

If the ΔH_f° values of the products and reactants are known, ΔH for a reaction can be calculated.

$$\Delta H^\circ = \Sigma \, \Delta H_f^\circ \text{ products} - \Sigma \Delta H_f^\circ \text{ reactants}$$

Let's find ΔH° for the reaction below.

$$2 \, CH_3OH(g) + 3 \, O_2(g) \rightarrow 2 \, CO_2(g) + 4 \, H_2O(g)$$

Compound	ΔH (kJ/mol)
$CH_3OH(g)$	−201
$O_2(g)$	0
$CO_2(g)$	−394
$H_2O(g)$	−242

$\Delta H^\circ = \Sigma \, \Delta H_f^\circ \text{ products} - \Sigma \, \Delta H_f^\circ \text{ reactants}$

$\Delta H^\circ = [(2)(\Delta H_f^\circ \, CO_2) + (4)(\Delta H_f^\circ \, H_2O)] - [(2)(\Delta H_f^\circ \, CH_3OH) + (3)(\Delta H_f^\circ \, O_2)]$

$\Delta H^\circ = [(2)(-394 \text{ kJ}) + (4)(-242 \text{ kJ})] - [(2)(-201 \text{ kJ}) + (3)(0 \text{ kJ})]$

$\Delta H^\circ = (-1{,}756 \text{ kJ}) - (-402 \text{ kJ})$

$\Delta H^\circ = -1{,}354 \text{ kJ}$

BOND ENERGY

Bond energy is the energy required to break a bond. Because the breaking of a bond is an endothermic process, bond energy is always a positive number. When a bond is formed, energy equal to the bond energy is released.

$$\Delta H° = \Sigma \text{ Bond energies of bonds broken} - \Sigma \text{ Bond energies of bonds formed}$$

The bonds broken will be the reactant bonds, and the bonds formed will be the product bonds. Let's find ΔH for the reaction below.

$$2 H_2(g) + O_2(g) \rightarrow 2 H_2O(g)$$

Bond	Bond Energy (kJ/mol)
H–H	436
O=O	499
O–H	463

$\Delta H° = \Sigma$ Bond energies of bonds broken $- \Sigma$ Bond energies of bonds formed
$\Delta H° = [(2)(H–H) + (1)(O=O)] - [(4)(O–H)]$
$\Delta H° = [(2)(436 \text{ kJ}) + (1)(499 \text{ kJ})] - [(4)(463 \text{ kJ})]$
$\Delta H° = (1{,}371 \text{ kJ}) - (1{,}852 \text{ kJ})$
$\Delta H° = -481 \text{ kJ}$

HESS'S LAW

Hess's law states that if a reaction can be described as a series of steps, then ΔH for the overall reaction is simply the sum of the ΔH values for all the steps.

For example, let's say you wanted to calculate the enthalpy change for the following reaction:

$$C_2H_2(g) + H_2O(l) \rightarrow C_2H_5OH(l) \qquad\qquad \Delta H° = ?$$

And let's say that you know the enthalpy changes for the following two reactions:

$$C_2H_2(g) + 3 O_2(g) \rightarrow 2 CO_2(g) + 2 H_2O(l) \qquad \Delta H° = -1{,}411 \text{ kJ}$$

$$2 CO_2(g) + 3 H_2O(l) \rightarrow C_2H_5OH(l) + 3 O_2(g) \qquad \Delta H° = +1{,}368 \text{ kJ}$$

If you add the two reactions whose enthalpy changes you know (as though they were simultaneous equations) and cancel the substances that appear on both sides, you'll get the reaction that you're looking for. This means that you can add the enthalpy changes for the reactions that you know to get the enthalpy change that you're looking for.

$$-1{,}411 \text{ kJ} + 1{,}368 \text{ kJ} = -43 \text{ kJ}$$

Heat Capacity and Specific Heat

Heat capacity, C_p, is a measure of how much the temperature of an object is raised when it absorbs heat.

Heat Capacity
$$C_P = \frac{\Delta H}{\Delta T}$$ C_P = heat capacity ΔH = heat added (J or cal) ΔT = temperature change (K or °C)

An object with a large heat capacity can absorb a lot of heat without undergoing much of a change in temperature, whereas an object with a small heat capacity shows a large increase in temperature even if only a small amount of heat is absorbed.

Specific heat is the amount of heat required to raise the temperature of one gram of a substance one degree Celsius.

Specific Heat
$$q = mc\Delta T$$ q = heat added (J or cal) m = mass of the substance (g or kg) c = specific heat ΔT = temperature change (K or °C)

ENTROPY

The entropy, S, of a system is a measure of the randomness or disorder of the system; the greater the disorder of a system, the greater its entropy. Because zero entropy is defined as a solid crystal at 0 K, and because 0 K has never been reached experimentally, all substances that we encounter will have some positive value for entropy. Standard entropies, S, are calculated at 25°C (298 K).

You should be familiar with several simple rules concerning entropies.

- Liquids have higher entropy values than solids.

- Gases have higher entropy values than liquids.

- Particles in solution have higher entropy values than solids.

- Two moles of a substance have higher entropy value than one mole.

The standard entropy change, $\Delta S°$, that has taken place at the completion of a reaction is the difference between the standard entropies of the products and the standard entropies of the reactants.

$$\Delta S° = \Sigma \ S°_{products} - \Sigma \ S°_{reactants}$$

GIBBS FREE ENERGY

The **Gibbs free energy**, or simply free energy, G, of a process is a measure of the spontaneity of the process.

For a given reaction

- if ΔG is negative, the reaction is spontaneous
- if ΔG is positive, the reaction is not spontaneous
- if $\Delta G = 0$, the reaction is at equilibrium

FREE ENERGY CHANGE, ΔG

The standard free energy change, ΔG, for a reaction can be calculated from the standard free energies of formation, ΔG°_f, of its products and reactants in the same way that ΔS° was calculated.

$$\Delta G^\circ = \Sigma \; \Delta G^\circ_{f \text{ products}} - \Sigma \; \Delta G^\circ_{f \text{ reactants}}$$

ΔG, ΔH, AND ΔS

In general, nature likes to move toward two different and seemingly contradictory states—low energy and high disorder, so spontaneous processes must result in decreasing enthalpy or increasing entropy or both.

There is an important equation that relates spontaneity (ΔG), enthalpy (ΔH), and entropy (ΔS) to one another.

$$\Delta G^o = \Delta H^\circ - T\Delta S^\circ$$
$$T = \text{absolute temperature (K)}$$

The chart below shows how different values of enthalpy and entropy affect spontaneity.

ΔH	ΔS	T	ΔG	
–	+	Low	–	Always spontaneous
		High	–	
+	–	Low	+	Never spontaneous
		High	+	
+	+	Low	+	Not spontaneous at low temperature
		High	–	Spontaneous at high temperature
–	–	Low	–	Spontaneous at low temperature
		High	+	Not spontaneous at high temperature

You should note that at low temperature, enthalpy is dominant, while at high temperature, entropy is dominant.

ΔG and $\Delta G°$

The standard free energy change, $\Delta G°$, gives the spontaneity of a reaction when all the concentrations of reactants and products are in their standard state concentrations (at 1 molar). The free energy change, and thus the spontaneity, of a reaction will be different from the standard free energy change if the initial concentrations of reactants and products are not 1 molar.

The standard free energy change, $\Delta G°$, can be related to ΔG for other conditions by the following equations:

$$\Delta G = \Delta G° + RT \ln Q$$
$$\text{or}$$
$$\Delta G = \Delta G° + 2.303 RT \log Q$$

$\Delta G°$ = standard free energy change (J)

ΔG = free energy change under given initial conditions (J)

R = the gas constant, 8.31 J/mol-K

T = absolute temperature (K)

Q = the reaction quotient for the given initial conditions

STANDARD FREE ENERGY CHANGE AND THE EQUILIBRIUM CONSTANT

Let's look at the equation relating ΔG and $\Delta G°$.

$$\Delta G = \Delta G° + RT \ln Q$$

At equilibrium, $\Delta G = 0$ and $Q = K$.

Knowing this, we can derive an equation relating the standard free energy change, $\Delta G°$, and the equilibrium constant, K.

$$\Delta G° = -RT \ln K$$
$$\text{or}$$
$$\Delta G° = -2.303 RT \log K$$

$\Delta G°$ = the gas constant, 8.31 J/mol-K

T = absolute temperature (K)

K = the equilibrium constant

Notice that if $\Delta G°$ is negative, K must be greater than 1, and products will be favored at equilibrium. Alternatively, if $\Delta G°$ is positive, K must be less than 1, and reactants will be favored at equilibrium.

ENERGY DIAGRAMS

EXOTHERMIC AND ENDOTHERMIC REACTIONS

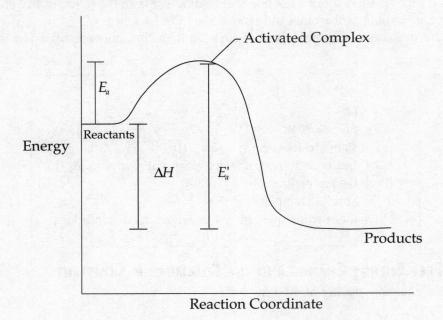

EXOTHERMIC REACTION

The diagram above shows the energy change that takes place during an exothermic reaction. The reactants start with a certain amount of energy (read the graph from left to right). For the reaction to proceed, the reactants must have enough energy to reach the transition state, where they are part of an activated complex. This is the highest point on the graph above. The amount of energy needed to reach this point is called the **activation energy, E_a**. At this point, all reactant bonds have been broken, but no product bonds have been formed, so this is the point in the reaction with the highest energy and lowest stability. The energy needed for the reverse reaction is shown as line E_a'.

Moving to the right past the activated complex, product bonds start to form, and we eventually reach the energy level of the products.

This diagram represents an exothermic reaction, so the products are at a lower energy level than the reactants and ΔH is negative.

The diagram below shows an endothermic reaction.

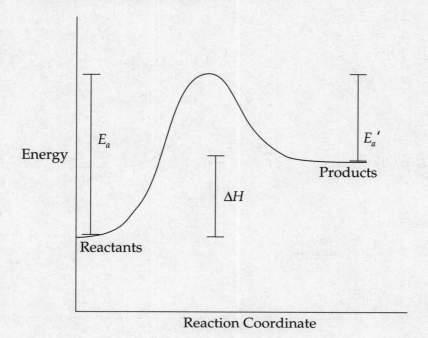

ENDOTHERMIC REACTION

In this diagram, the energy of the products is greater than the energy of the reactants, so ΔH is positive.

Reaction diagrams can be read in both directions, so the reverse reaction for an exothermic reaction is endothermic and vice versa.

CATALYSTS AND ENERGY DIAGRAMS

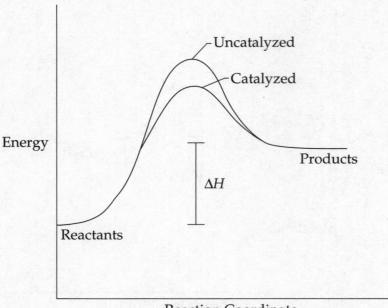

A catalyst speeds up a reaction by providing the reactants with an alternate pathway that has a lower activation energy, as shown in the diagram above.

Notice that the only difference between the catalyzed reaction and the uncatalyzed reaction is that the energy of the activated complex is lower for the catalyzed reaction. A catalyst lowers the activation energy, but it has no effect on the energy of the reactants, the energy of the products, or ΔH for the reaction.

Also note that a catalyst lowers the activation energy for both the forward and the reverse reaction, so it has no effect on the equilibrium conditions.

CHAPTER 8 QUESTIONS

MULTIPLE-CHOICE QUESTIONS

Questions 1–4

    (A) Free energy change (ΔG)
    (B) Entropy change (ΔS)
    (C) Heat of vaporization
    (D) Heat of fusion
    (E) Heat capacity

1. If this has a negative value for a process, then the process occurs spontaneously.

2. This is a measure of how the disorder of a system is changing.

3. This is the energy given off when a substance condenses.

4. This is the energy taken in by a substance when it melts.

5. $2\,Al(s) + 3\,Cl_2(g) \rightarrow 2\,AlCl_3(s)$

The reaction above is not spontaneous under standard conditions, but becomes spontaneous as the temperature decreases toward absolute zero. Which of the following is true at standard conditions?

    (A) ΔS and ΔH are both negative.
    (B) ΔS and ΔH are both positive.
    (C) ΔS is negative, and ΔH is positive.
    (D) ΔS is positive, and ΔH is negative.
    (E) ΔS and ΔH are both equal to zero.

6.

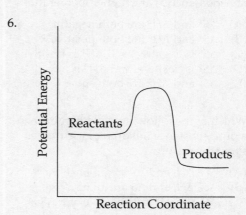

Which of the following is true of the reaction shown in the diagram above?

    (A) The reaction is endothermic because the reactants are at a higher energy level than the products.
    (B) The reaction is endothermic because the reactants are at a lower energy level than the products.
    (C) The reaction is exothermic because the reactants are at a higher energy level than the products.
    (D) The reaction is exothermic because the reactants are at a lower energy level than the products.
    (E) The reaction is endothermic because the reactants are at the same energy level as the products.

7. $$2 H_2(g) + O_2(g) \rightarrow 2 H_2O(g)$$

Based on the information given in the table below, what is $\Delta H°$ for the above reaction?

Bond	Average bond energy (kJ/mol)
H–H	500
O=O	500
O–H	500

(A) –2,000 kJ
(B) –1,500 kJ
(C)   –500 kJ
(D) +1,000 kJ
(E) +2,000 kJ

8. Which of the following is true of a reaction that is spontaneous at 298 K but becomes nonspontaneous at a higher temperature?

(A) $\Delta S°$ and $\Delta H°$ are both negative.
(B) $\Delta S°$ and $\Delta H°$ are both positive.
(C) $\Delta S°$ is negative, and $\Delta H°$ is positive.
(D) $\Delta S°$ is positive, and $\Delta H°$ is negative.
(E) $\Delta S°$ and $\Delta H°$ are both equal to zero.

9. Which of the following will be true when a pure substance in liquid phase freezes spontaneously?

(A) ΔG, ΔH, and ΔS are all positive.
(B) ΔG, ΔH, and ΔS are all negative.
(C) ΔG and ΔH are negative, but ΔS is positive.
(D) ΔG and ΔS are negative, but ΔH is positive.
(E) ΔS and ΔH are negative, but ΔG is positive.

10.

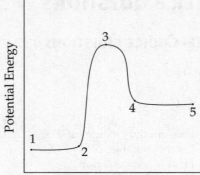

Which point on the graph shown above corresponds to activated complex or transition state?

(A) 1
(B) 2
(C) 3
(D) 4
(E) 5

11. $$C(s) + O_2(g) \rightarrow CO_2(g) \qquad \Delta H° = -390 \text{ kJ/mol}$$

$$H_2(g) + \frac{1}{2} O_2(g) \rightarrow H_2O(l) \quad \Delta H° = -290 \text{ kJ/mol}$$

$$2 C(s) + H_2(g) \rightarrow C_2H_2(g) \qquad \Delta H° = +230 \text{ kJ/mol}$$

Based on the information given above, what is ΔH for the following reaction?

$$C_2H_2(g) + \frac{5}{2} O_2(g) \rightarrow 2 CO_2(g) + H_2O(l)$$

(A) –1,300 kJ
(B) –1,070 kJ
(C)   –840 kJ
(D)   –780 kJ
(E)   –680 kJ

12. If an endothermic reaction is spontaneous at 298 K, which of the following must be true for the reaction?

    I. ΔG is greater than zero.
    II. ΔH is greater than zero.
    III. ΔS is greater than zero.

(A) I only
(B) II only
(C) I and II only
(D) II and III only
(E) I, II, and III

13. The addition of a catalyst will have which of the following effects on a chemical reaction?

    I. The enthalpy change will decrease.
    II. The entropy change will decrease.
    III. The activation energy will decrease.

(A) I only
(B) II only
(C) III only
(D) I and II only
(E) II and III only

14. $C(s) + 2\,H_2(g) \rightarrow CH_4(g)$ $\Delta H = x$

$C(s) + O_2(g) \rightarrow CO_2(g)$ $\Delta H = y$

$H_2(g) + \dfrac{1}{2}\,O_2(g) \rightarrow H_2O(l)$ $\Delta H = z$

Based on the information given above, what is $\Delta H°$ for the following reaction?

$CH_4(g) + 2\,O_2(g) \rightarrow CO_2(g) + 2\,H_2O(l)$

(A) $x + y + z$
(B) $x + y - z$
(C) $z + y - 2x$
(D) $2z + y - x$
(E) $2z + y - 2x$

15. For which of the following processes will ΔS be positive?

    I. $NaCl(s) \rightarrow Na^+(aq) + Cl^-(aq)$
    II. $2\,H_2(g) + O_2(g) \rightarrow 2\,H_2O(g)$
    III. $CaCO_3(s) \rightarrow CaO(s) + CO_2(g)$

(A) I only
(B) II only
(C) I and II only
(D) I and III only
(E) I, II, and III

16. In which of the following reactions is entropy increasing?

(A) $2\,SO_2(g) + O_2(g) \rightarrow 2\,SO_3(g)$
(B) $CO(g) + H_2O(g) \rightarrow H_2(g) + CO_2(g)$
(C) $H_2(g) + Cl_2(g) \rightarrow 2\,HCl(g)$
(D) $2\,NO_2(g) \rightarrow 2\,NO(g) + O_2(g)$
(E) $2\,H_2S(g) + 3\,O_2(g) \rightarrow 2\,H_2O(g) + 2\,SO_2(g)$

17. When pure sodium is placed in an atmosphere of chlorine gas, the following spontaneous reaction occurs.

$$2\,Na(s) + Cl_2(g) \rightarrow 2\,NaCl(s)$$

Which of the following statements is true about the reaction?

    I. $\Delta S > 0$
    II. $\Delta H < 0$
    III. $\Delta G > 0$

(A) I only
(B) II only
(C) I and II only
(D) II and III only
(E) I, II, and III

18. $H_2(g) + F_2(g) \rightarrow 2\ HF(g)$

Gaseous hydrogen and fluorine combine in the reaction above to form hydrogen fluoride with an enthalpy change of –540 kJ. What is the value of the heat of formation of $HF(g)$?

(A) –1,080 kJ/mol
(B) –540 kJ/mol
(C) –270 kJ/mol
(D) 270 kJ/mol
(E) 540 kJ/mol

19. $H_2O(s) \rightarrow H_2O(l)$

Which of the following is true of the reaction shown above at room temperature?

I. ΔG is greater than zero.
II. ΔH is greater than zero.
III. ΔS is greater than zero.

(A) II only
(B) III only
(C) I and II only
(D) I and III only
(E) II and III only

20. $2\ S(s) + 3\ O_2(g) \rightarrow 2\ SO_3(g)$
$\Delta H = +800$ kJ/mol

$2\ SO_3(g) \rightarrow 2\ SO_2(g) + O_2(g)$
$\Delta H = -200$ kJ/mol

Based on the information given above, what is ΔH for the following reaction?

$$S(s) + O_2(g) \rightarrow SO_2(g)$$

(A) 300 kJ
(B) 500 kJ
(C) 600 kJ
(D) 1,000 kJ
(E) 1,200 kJ

PROBLEMS

1.

Substance	Absolute Entropy, S (J/mol-K)	Molecular Weight
$C_6H_{12}O_6$ (s)	212.13	180
O_2 (g)	205	32
CO_2 (g)	213.6	44
H_2O (l)	69.9	18

Energy is released when glucose is oxidized in the following reaction, which is a metabolism reaction that takes place in the body.

$$C_6H_{12}O_6(s) + 6\,O_2(g) \rightarrow 6\,CO_2(g) + 6\,H_2O(l)$$

The standard enthalpy change, $\Delta H°$, for the reaction is $-2{,}801$ kJ at 298 K.

(a)   Calculate the standard entropy change, $\Delta S°$, for the oxidation of glucose.

(b)   Calculate the standard free energy change, $\Delta G°$, for the reaction at 298 K.

(c)   What is the value of K_{eq} for the reaction?

(d)   How much energy is given off by the oxidation of 1.00 gram of glucose?

2.

Bond	Average Bond Dissociation Energy (kJ/mol)
C–H	415
O=O	495
C=O	799
O–H	463

$$CH_4(g) + 2\,O_2(g) \rightarrow CO_2(g) + 2\,H_2O(g)$$

The standard free energy change, $\Delta G°$, for the reaction above is -801 kJ at 298 K.

(a)   Use the table of bond dissociation energies to find $\Delta H°$ for the reaction above.

(b)   What is the value of K_{eq} for the reaction?

(c)   What is the value of $\Delta S°$ for the reaction at 298 K?

(d)   Give an explanation for the size of the entropy change found in (c).

3.
$$N_2(g) + 3\,H_2(g) \rightleftharpoons 2\,NH_3(g)$$

The heat of formation, ΔH°_f, of $NH_3(g)$ is -46.2 kJ/mol. The free energy of formation, ΔG°_f, of $NH_3(g)$ is -16.7 kJ/mol.

(a)　What are the values of ΔH° and ΔG° for the reaction?

(b)　What is the value of the entropy change, ΔS°, for the reaction above at 298 K?

(c)　As the temperature is increased, what is the effect on ΔG for the reaction? How does this affect the spontaneity of the reaction?

(d)　At what temperature can N_2, H_2, and NH_3 gases be maintained together in equilibrium, each with a partial pressure of 1 atm?

ESSAYS

4.
$$2\,H_2(g) + O_2(g) \rightarrow 2\,H_2O(l)$$

The reaction above proceeds spontaneously from standard conditions at 298 K.

(a)　Predict the sign of the entropy change, ΔS°, for the reaction. Explain.

(b)　How would the value of ΔS° for the reaction change if the product of the reaction was $H_2O(g)$?

(c)　What is the sign of ΔG° at 298 K? Explain.

(d)　What is the sign of ΔH° at 298 K? Explain.

5.
$$CaO(s) + CO_2(g) \rightarrow CaCO_3(s)$$

The reaction above is spontaneous at 298 K, and the heat of reaction, ΔH°, is -178 kJ.

(a)　Predict the sign of the entropy change, ΔS°, for the reaction. Explain.

(b)　What is the sign of ΔG° at 298 K? Explain.

(c)　What change, if any, occurs to the value of ΔG° as the temperature is increased from 298 K?

(d)　As the reaction takes place in a closed container, what changes will occur in the concentration of CO_2 and the temperature?

6.
$$H_2O(l) \rightleftharpoons H_2O(g)$$

At 298 K, the value of the equilibrium constant, K, for the reaction above is 0.036.

(a)　What is the sign of ΔS° for the reaction above at 298 K?

(b)　What is the sign of ΔH° for the reaction above at 298 K?

(c)　What is the sign of ΔG° for the reaction above at 298 K?

(d)　At approximately what temperature will ΔG for the reaction be equal to zero?

CHAPTER 8 ANSWERS AND EXPLANATIONS

MULTIPLE-CHOICE QUESTIONS

1. **A** A negative value for ΔG means that the process is spontaneous. A positive value for ΔG means that the process is nonspontaneous.

2. **B** Entropy is a measure of the disorder of a system. A positive value for ΔS means that a system has become more disordered. A negative value for ΔS means that a system has become more orderly.

3. **C** The heat of vaporization is the heat given off when a substance condenses. It is also the heat that must be put into a substance to make it vaporize.

4. **D** The heat of fusion is the heat that must be put into a substance to melt it. It is also the heat given off when a substance freezes.

    By the way, choice (E), heat capacity, is a measure of how much heat must be added to a given object to raise its temperature 1°C.

5. **A** Remember that $\Delta G = \Delta H - T\Delta S$.

    If the reaction is spontaneous only when the temperature is very low, then ΔG is negative only when T is very small. This can happen only when ΔH is negative (which favors spontaneity) and ΔS is negative (which favors nonspontaneity). A very small value for T will eliminate the influence of ΔS.

6. **C** In an exothermic reaction, energy is given off as the products are created because the products have less potential energy than the reactants.

7. **C** The bond energy is the energy that must be put into a bond to break it. First let's figure out how much energy must be put into the reactants to break their bonds.

    To break 2 moles of H–H bonds, it takes (2)(500) kJ = 1,000 kJ

    To break 1 mole of O=O bonds, it takes 500 kJ.

    So to break up the reactants, it takes +1,500 kJ.

    Energy is given off when a bond is formed; that's the negative of the bond energy. Now let's see how much energy is given off when 2 moles of H_2O are formed.

    2 moles of H_2O molecules contain 4 moles of O–H bonds, so (4)(–500) kJ = –2,000 kJ are given off.

    So the value of $\Delta H°$ for the reaction is

    (–2,000 kJ, the energy given off) + (1,500 kJ, the energy put in) = –500 kJ.

8. **A** Remember that $\Delta G = \Delta H - T \Delta S$. If the reaction is spontaneous at standard temperature but becomes nonspontaneous at higher temperatures, then ΔG is negative only at lower temperatures. This can happen only when ΔH is negative (which favors spontaneity) and ΔS is negative (which favors nonspontaneity). As the value of T increases, the influence of ΔS increases, eventually making the reaction nonspontaneous.

9. **B** The process is spontaneous, so ΔG must be negative. The intermolecular forces become stronger and the substance moves to a lower energy level when it freezes, so ΔH must be negative. The substance becomes more orderly when it freezes, so ΔS must be negative.

10. **C** Point 3 represents the activated complex, which is the point of highest energy. This point is the transition state between the reactants and the products.

11. **A** The equations given on top give the heats of formation of all the reactants and products (remember: the heat of formation of O_2, an element in its most stable form, is zero).

    $\Delta H°$ for a reaction = ($\Delta H°$ for the products) – ($\Delta H°$ for the reactants).

    First, the products:

    From CO_2, we get (2)(–390 kJ) = –780 kJ

    From H_2O, we get –290 kJ

    So $\Delta H°$ for the products = (–780 kJ) + (–290 kJ) = –1,070 kJ

    Now the reactants:

    From C_2H_2, we get +230 kJ. The heat of formation of O_2 is defined as zero, so that's it for the reactants.

    $\Delta H°$ for the reaction = (–1,070 kJ) – (+230 kJ) = –1,300 kJ.

12. **D** The reaction is spontaneous, so ΔG must be less than zero, so (I) is not true. The reaction is endothermic, so ΔH must be greater than zero; therefore, (II) is true.

    The only way that an endothermic reaction can be spontaneous is if the entropy is increasing, so ΔS is greater than zero, and (III) is true.

13. **C** The addition of a catalyst speeds up a reaction by lowering the activation energy. A catalyst has no effect on the entropy or enthalpy change of a reaction.

14. **D** The equations given above the question give the heats of formation of all the reactants and products (remember: the heat of formation of O_2, an element in its most stable form, is zero).

    $\Delta H°$ for a reaction = ($\Delta H°$ for the products) – ($\Delta H°$ for the reactants).

    First, the products:

    From $2 H_2O$, we get $2z$

    From CO_2, we get y

    So $\Delta H°$ for the products = $2z + y$

    Now the reactants:

    From CH_4, we get x. The heat of formation of O_2 is defined to be zero, so that's it for the reactants.

    $\Delta H°$ for the reaction = $(2z + y) – (x) = 2z + y – x$.

15. **D** In (I), NaCl goes from a solid to aqueous particles. Aqueous particles are more disorderly than a solid, so ΔS will be positive.

    In (II), we go from 3 moles of gas to 2 moles of gas. Fewer moles and less gas means less entropy, so ΔS will be negative.

    In (III), we go from a solid to a solid and a gas. More moles and the production of more gas means increasing entropy, so ΔS will be positive.

16. **D** Choice (D) is the only reaction where the number of moles of gas is increasing, going from 2 moles of gas on the reactant side to 3 moles of gas on the product side. In all the other choices, the number of moles of gas either decreases or remains constant.

17. **B** In the reaction, a solid combines with a gas to produce fewer moles of solid, so the entropy change is negative and (I) is not true. The reaction is spontaneous, so the free energy change is negative and (III) is not true. For a reaction with decreasing entropy to be spontaneous, it must be exothermic. The enthalpy change is negative for an exothermic reaction and (II) is true.

18. **C** The reaction that forms 2 moles of HF(g) from its constituent elements has an enthalpy change of –540 kJ. The heat of formation is given by the reaction that forms 1 mole from these elements, so you can just divide –540 kJ by 2 to get –270 kJ.

19. **E** Ice melts spontaneously at room temperature, so ΔG is less than zero; therefore, (I) is not true.

    In this reaction, ice melts, so heat is absorbed to break up the intermolecular forces, and ΔH is greater than zero; therefore, (II) is true.

    Liquid water is less orderly than ice, so the entropy change when ice melts is positive; therefore, ΔS is greater than zero, and (III) is true.

20. **A** You can use Hess's law. Add the two reactions together, and cancel things that appear on both sides.

    $$2\,S(s) + 3\,O_2(g) \rightarrow 2\,SO_3(g) \qquad \Delta H = +800 \text{ kJ}$$
    $$2\,SO_3(g) \rightarrow 2\,SO_2(g) + O_2(g) \qquad \Delta H = -200 \text{ kJ}$$

    $$2\,S(s) + 3\,O_2(g) + 2\,SO_3(g) \rightarrow 2\,SO_2(g) + 2\,SO_3(g) + O_2(g)$$

    This reduces to

    $$2\,S(s) + 2\,O_2(g) \rightarrow 2\,SO_2(g) \qquad \Delta H = +600 \text{ kJ}$$

    Now we can cut everything in half to get the equation we want.

    $$S(s) + O_2(g) \rightarrow SO_2(g) \qquad \Delta H = +300 \text{ kJ}$$

PROBLEMS

1. (a) Use the entropy values in the table.

$$\Delta S^\circ = \sum S^\circ_{products} - \sum S^\circ_{reactants}$$

$\Delta S^\circ = [(6)(213.6) + (6)(69.9)] - [(212.13) + (6)(205)]$ J/K

$\Delta S^\circ = 259$ J/K

(b) Use the equation below. Remember that enthalpy values are given in kJ and entropy values are given in J.

$$\Delta G^\circ = \Delta H^\circ - T\,\Delta S^\circ$$

$\Delta G^\circ = (-2{,}801$ kJ$) - (298)(0.259$ kJ$) = -2{,}880$ kJ

(c) Use the equation below. Remember that the gas constant is given in terms of J.

$$\log K = \frac{\Delta G^\circ}{-2.303RT}$$

$$\log K = \frac{(-2{,}880{,}000)}{(-2.303)(8.31)(298)} = 505$$
$K = 10^{505}$

(d) The enthalpy change of the reaction, H°, is a measure of the energy given off by 1 mole of glucose.

$$\text{Moles} = \frac{\text{grams}}{\text{MW}}$$

$$\text{Moles of glucose} = \frac{(1.00\text{ g})}{(180\text{ g / mol})} = 0.00556\text{ moles}$$

$(0.00556$ mol$)(2{,}801$ kJ/mol$) = 15.6$ kJ

2. (a) Use the relationship below.

$$\Delta H^\circ = \sum \text{Energies of the bonds broken} - \sum \text{Energies of the bonds formed}$$

$\Delta H^\circ = [(4)(415) + (2)(495)] - [(2)(799) + (4)(463)]$ kJ

$\Delta H^\circ = -800$ kJ

(b) Use the following equation. Remember that the gas constant is given in terms of J.

$$\log K = \frac{\Delta G^\circ}{-2.303RT}$$

$$\log K = \frac{(-801{,}000)}{(-2.303)(8.31)(298)} = 140$$

$K = 10^{140}$

(c) Use $\Delta G° = \Delta H° - T\,\Delta S°$

Remember that enthalpy values are given in kJ and entropy values are given in J.

$$\Delta S° = \frac{\Delta H - \Delta G}{T} = \frac{(-800 \text{ kJ}) - (-801 \text{ kJ})}{(298 \text{ K})}$$
$$\Delta S° = 0.003 \text{ kJ/K} = 3 \text{ J/K}$$

(d) $\Delta S°$ is very small, which means that the entropy change for the process is very small. This makes sense because the number of moles remains constant, the number of moles of gas remains constant, and the complexity of the molecules remains about the same.

3. (a) By definition, $\Delta H°_f$ and $\Delta G°_f$ for $N_2(g)$ and $H_2(g)$ are equal to zero.

$$\Delta H° = \sum \Delta H°_f \text{products} - \sum \Delta H°_f \text{reactants}$$

$$\Delta H° = [(2 \text{ mol})(-46.2 \text{ kJ/mol})] - 0 = -92.4 \text{ kJ}$$

$$\Delta G° = \sum \Delta G°_f \text{products} - \sum \Delta G°_f \text{reactants}$$

$$\Delta G° = [(2 \text{ mol})(-16.7 \text{ kJ/mol})] - 0 = -33.4 \text{ kJ}$$

(b) Use $\Delta G° = \Delta H° - T\,\Delta S°$

Remember that enthalpy values are given in kJ and entropy values are given in J.

$$\Delta S° = \frac{\Delta H - \Delta G}{T} = \frac{(-92.4 \text{ kJ}) - (-33.4 \text{ kJ})}{(298 \text{ K})}$$
$$\Delta S° = -0.198 \text{ kJ/K} = -198 \text{ J/K}$$

(c) Use $\Delta G = \Delta H° - T\,\Delta S°$

From (b), $\Delta S°$ is negative, so increasing the temperature increases the value of $\Delta G°$, making the reaction less spontaneous.

(d) Use $\Delta G = \Delta H° - T\,\Delta S°$

At equilibrium, $\Delta G = 0$
$$T = \frac{\Delta H°}{\Delta S°} = \frac{(-92,400 \text{ J})}{(-198 \text{ J / K})} = 467 \text{ K}$$

ESSAYS

4. (a) $\Delta S°$ is negative because the products are less random than the reactants. That's because gas is converted into liquid in the reaction.

   (b) The value of $\Delta S°$ would increase, becoming less negative because $H_2O(g)$ is more random than water but remaining negative because the entropy would still decrease from reactants to products.

   (c) $\Delta G°$ is negative because the reaction proceeds spontaneously.

   (d) $\Delta H°$ must be negative at 298 K. For a reaction to occur spontaneously from standard conditions, either $\Delta S°$ must be positive or $\Delta H°$ must be negative. This reaction is spontaneous although $\Delta S°$ is negative, so $\Delta H°$ must be negative.

5. (a) $\Delta S°$ is negative because the products are less random than the reactants. That's because two moles of reactants are converted to one mole of products and gas is converted into solid in the reaction.

   (b) $\Delta G°$ is negative because the reaction proceeds spontaneously.

   (c) Use $\Delta G° = \Delta H° - T \Delta S°$

   ΔG will become less negative because as temperature is increased, the entropy change of a reaction becomes more important in determining its spontaneity. The entropy change for this reaction is negative, which discourages spontaneity, so increasing temperature will make the reaction less spontaneous, thus making $\Delta G°$ less negative.

   (d) The concentration of CO_2 will decrease as the reaction proceeds in the forward direction and the reactants are consumed. The temperature will increase as heat is given off by the exothermic reaction.

6. (a) $\Delta S°$ is positive because the product is more random than the reactant. That's because liquid is converted into gas in the reaction.

   (b) $\Delta H°$ is positive because $H_2O(g)$ is less stable than water. Energy must be put into water to overcome intermolecular forces and create water vapor.

   (c) Use $\Delta G° = -2.203RT \log K$

   If K is less than 1, $\log K$ will be negative, making $\Delta G°$ positive.

   (d) ΔG will be equal to zero at 373 K or 100°C. ΔG will be equal to zero at equilibrium. The point at which water and $H_2O(g)$ are in equilibrium is the boiling point.

Solutions

HOW OFTEN DO SOLUTIONS APPEAR ON THE EXAM?

In the multiple-choice section, this topic appears in about 9 out of 75 questions. In the free-response section, this topic appears almost every year.

CONCENTRATION MEASUREMENTS

MOLARITY

Molarity (*M*) expresses the concentration of a solution in terms of volume. It is the most widely used unit of concentration, turning up in calculations involving equilibrium, acids and bases, and electrochemistry, among others.

When you see a chemical symbol in brackets on the test, that means they are talking about molarity. For instance, "[Na^+]" is the same as "the molar concentration (molarity) of sodium ions."

$$\text{Molarity } (M) = \frac{\text{moles of solute}}{\text{liters of solution}}$$

MOLALITY

Molality (*m*) expresses concentration in terms of the mass of a solvent. It is the unit of concentration used for determining the effect of most colligative properties, where the number of moles of solute is more important than the nature of the solute.

$$\text{Molality } (m) = \frac{\text{moles of solute}}{\text{kilograms of solvent}}$$

Molarity and molality differ in two ways: Molarity tells you about moles of solute per *volume* of the *entire solution* (that is, the solute and the solvent), whereas molality tells you about moles of solute per *mass* of the *solvent*. Keeping in mind that one liter of water weighs one kilogram, and that for a dilute solution, the amount of solution is about the same as the amount of solvent, you should be able to see that for dilute aqueous solutions, molarity and molality are basically the same.

MOLE FRACTION

Mole fraction (X_s) gives the fraction of moles of a given substance (S) out of the total moles present in a sample. It is used in determining how the vapor pressure of a solution is lowered by the addition of a solute.

$$\text{Mole Fraction } (X_s) = \frac{\text{moles of substance S}}{\text{total number of moles in solution}}$$

SOLUTES AND SOLVENTS

There is a basic rule for remembering which solutes will dissolve in which solvents.

Like dissolves like

That means that polar or ionic solutes (like salt) will dissolve in polar solvents (like water). That also means that nonpolar solutes (like organic compounds) are best dissolved in nonpolar solvents. When an ionic substance dissolves, it breaks up into ions. That's **dissociation.** Free ions in a solution are called electrolytes because they can conduct electricity.

The **van't Hoff factor (*i*)** tells how many ions one unit of a substance will dissociate into in solution. For instance

- $C_6H_{12}O_6$ does not dissociate, so $i = 1$
- NaCl dissociates into Na^+ and Cl^-, so $i = 2$
- HNO_3 dissociates into H^+ and NO_3^-, so $i = 2$
- $CaCl_2$ dissociates into Ca^{2+}, Cl^-, and Cl^-, so $i = 3$

COLLIGATIVE PROPERTIES

Colligative properties are properties of a solution that depend on the number of solute particles in the solution. For colligative properties, the identity of the particles is not important.

BOILING-POINT ELEVATION

When a solute is added to a solution, the boiling point of the solution increases.

Boiling-Point Elevation
$\Delta T = ik_b m$
i = the van't Hoff factor, the number of particles into which the added solute dissociates
k_b = the boiling-point elevation constant for the solvent
m = molality

FREEZING-POINT DEPRESSION

When solute is added to a solution, the freezing point of the solution decreases.

Freezing-Point Depression
$\Delta T = ik_f m$
i = the van't Hoff factor, the number of particles into which the added solute dissociates
k_f = the freezing-point depression constant for the solvent
m = molality

VAPOR PRESSURE LOWERING (RAOULT'S LAW)

When a solute is added to a solution, the vapor pressure of the solution will decrease. You may want to note that a direct result of the lowering of vapor pressure of a solution is the raising of its boiling point.

Vapor Pressure Lowering, Raoult's Law
$P = XP^\circ$
P = vapor pressure of the solution
P° = vapor pressure of the pure solvent
X = the mole fraction of the solvent

OSMOTIC PRESSURE

When a pure solvent and a solution are separated by a membrane that allows only solvent to pass through, the solvent will try to pass through the membrane to dilute the solution. The pressure that must be applied to stop this process is called the **osmotic pressure**. The greater the concentration of solute in the solution, the greater the osmotic pressure. The equation for osmotic pressure takes a form that is similar to the ideal gas equation, as shown below.

Osmotic Pressure
$$\lambda = \frac{nRT}{V}i = MRTi$$
λ = osmotic pressure (atm)
n = moles of solute
R = the gas constant, 0.0821 (L-atm)/(mol-K)
T = absolute temperature (K)
V = volume of the solution (L)
i = the van't Hoff factor, the number of particles into which the added solute dissociates
M = molarity of the solution ($M = \frac{n}{V}$)

DENSITY

Density is the measure of mass per unit volume. Density can be used to describe liquids, solids, or gases. Because density relates mass and volume, it is useful if you need to convert between molarity, which deals with volume, and molality, which deals with mass.

Density of a Solution
$$D = \frac{m}{V}$$
m = mass of the solution
V = volume of the solution

SOLUBILITY

Roughly speaking, a salt can be considered "soluble" if more than 1 gram of the salt can be dissolved in 100 milliliters of water. Soluble salts are usually assumed to dissociate completely in aqueous solution. Most, but not all, solids become more soluble in a liquid as the temperature is increased.

SOLUBILITY PRODUCT (K_{sp})

Salts that are "slightly soluble" and "insoluble" still dissociate in solution to some extent. The solubility product (K_{sp}) is a measure of the extent of a salt's dissociation in solution. The K_{sp} is one of the forms of the equilibrium expression, which we'll discuss in Chapter 10. The greater the value of the solubility product for a salt, the more soluble the salt.

Solubility Product

For the reaction

$$A_aB_b(s) \rightleftharpoons a\ A^{b+}(aq) + b\ B^{a-}(aq)$$

The solubility expression is

$$K_{sp} = [A^{b+}]^a[B^{a-}]^b$$

Here are some examples:

$CaF_2(s) \Leftrightarrow Ca^{2+}(aq) + 2\ F^-(aq)$ $K_{sp} = [Ca^{2+}][F^-]^2$

$Ag_2CrO_4(s) \Leftrightarrow 2\ Ag^+(aq) + CrO_4^{2-}(aq)$ $K_{sp} = [Ag^+]^2[CrO_4^{2-}]$

$CuI(s) \Leftrightarrow Cu^+(aq) + I^-(aq)$ $K_{sp} = [Cu^+][I^-]$

THE COMMON ION EFFECT

Let's look at the solubility expression for AgCl.

$$K_{sp} = [Ag^+][Cl^-] = 1.6 \times 10^{-10}$$

If we throw a block of solid AgCl into a beaker of water, we can tell from the K_{sp} what the concentrations of Ag^+ and Cl^- will be at equilibrium. For every unit of AgCl that dissociates, we get one Ag^+ and one Cl^-, so we can solve the equation above as follows:

$$[Ag^+][Cl^-] = 1.6 \times 10^{-10}$$

$$(x)(x) = 1.6 \times 10^{-10}$$

$$x^2 = 1.6 \times 10^{-10}$$

$$x = [Ag^+] = [Cl^-] = 1.3 \times 10^{-5}\ M$$

So there are very small amounts of Ag^+ and Cl^- in the solution.

Let's say we add 0.10 mole of NaCl to 1 liter of the AgCl solution. NaCl dissociates completely, so that's the same thing as adding 1 mole of Na^+ ions and 1 mole of Cl^- ions to the solution. The Na^+ ions will not affect the AgCl equilibrium, so we can ignore them; but the Cl^- ions must be taken into account. That's because of the **common ion effect.**

The common ion effect says that the newly added Cl⁻ ions will affect the AgCl equilibrium, although the newly added Cl⁻ ions did not come from AgCl.

Let's look at the solubility expression again. Now we have 0.10 mole of Cl⁻ ions in 1 liter of the solution, so [Cl⁻] = 0.10 M.

$$[Ag^+][Cl^-] = 1.6 \times 10^{-10}$$

$$[Ag^+](0.10\ M) = 1.6 \times 10^{-10}$$

$$[Ag^+] = \frac{\left(1.6 \times 10^{-10}\right)}{(0.10)} M$$

$$[Ag^+] = 1.6 \times 10^{-9}\ M$$

Now the number of Ag⁺ ions in the solution has decreased drastically because of the Cl⁻ ions introduced to the solution by NaCl. So when solutions of AgCl and NaCl, which share a common Cl⁻ ion, are mixed, the more soluble salt (NaCl) can cause the less soluble salt (AgCl) to precipitate. In general, when two salt solutions that share a common ion are mixed, the salt with the lower value for K_{sp} will precipitate first.

SOLUBILITY RULES

You should have a good working knowledge of the solubilities of common salts. This is especially useful for the part in Section II in which you are asked to predict the outcome of chemical reactions.

Cations

- **Alkali Metals:** Li⁺, Na⁺, K⁺, Rb⁺, Cs⁺
  All salts of the alkali metals are **soluble.**

- **Ammonium:** NH_4^+
  All ammonium salts are **soluble.**

- **Alkaline Earths and Transition Metals**
  The solubility of these elements varies depending on the identity of the anion.

Anions

The following are mostly soluble:

- **Nitrate:** NO_3^-
  All nitrate salts are **soluble.**

- **Chlorate:** ClO_3^-
  All chlorate salts are **soluble.**

- **Perchlorate:** ClO_4^-
  All perchlorate salts are **soluble.**

- **Acetate:** $C_2H_3O_2^-$
  All acetate salts are **soluble.**

- **Chloride, Bromide, Iodide:** Cl^-, Br^-, I^-
  Salts containing Cl^-, Br^-, and I^- are **soluble**
  EXCEPT for those containing Ag^+, Pb^{2+}, and Hg_2^{2+}.

- **Sulfate:** SO_4^{2-}
  Sulfate salts are **soluble**
  EXCEPT for those containing Ag^+, Pb^{2+}, Hg_2^{2+}, Ca^{2+}, Sr^{2+}, and Ba^{2+}.

The following are mostly insoluble:

- **Hydroxide:** OH^-
  Hydroxide salts are **insoluble**
  EXCEPT for those containing alkali metals, which are soluble
  AND those containing Ca^{2+}, Sr^{2+}, and Ba^{2+}, which fall in the gray area of moderate solubility.

- **Carbonate:** CO_3^{2-}
  Carbonate salts are **insoluble**
  EXCEPT for those containing alkali metals and ammonium, which are soluble.

- **Phosphate:** PO_4^{3-}
  Phosphate salts are **insoluble**
  EXCEPT for those containing alkali metals and ammonium, which are soluble.

- **Sulfite:** SO_3^{2-}
  Sulfite salts are **insoluble**
  EXCEPT for those containing alkali metals and ammonium, which are soluble.

- **Chromate:** CrO_4^{2-}
  Chromate salts are **insoluble**
  EXCEPT for those containing alkali metals and ammonium, which are soluble.

- **Sulfide:** S^{2-}
  Sulfide salts are **insoluble**
  EXCEPT for those containing alkali metals, the alkaline earths, and ammonium, which are soluble.

SOLUBILITY OF GASES

The lower the temperature and the higher the pressure of the gas, the more soluble the gas will be. Think of what happens when you open a bottle of warm seltzer. The gas suddenly escapes from the warm liquid when you release the pressure.

CHAPTER 9 QUESTIONS

MULTIPLE-CHOICE QUESTIONS

Questions 1–4

    (A) Molarity (M)
    (B) Molality (m)
    (C) Density
    (D) pH
    (E) pOH

1. Has the units moles/kg.

2. This is the negative logarithm of the hydrogen ion concentration.

3. Can have the units grams/liter.

4. Has the units moles per liter.

5. Which of the following is (are) colligative properties?

    I. Freezing-point depression
    II. Vapor pressure lowering
    III. Boiling-point elevation

    (A) I only
    (B) I and II only
    (C) I and III only
    (D) II and III only
    (E) I, II, and III

6. Which of the following aqueous solutions has the highest boiling point?

    (A) 0.5 m NaCl
    (B) 0.5 m KBr
    (C) 0.5 m $CaCl_2$
    (D) 0.5 m $C_6H_{12}O_6$
    (E) 0.5 m $NaNO_3$

7. When sodium chloride is added to a saturated aqueous solution of silver chloride, which of the following precipitates would be expected to appear?

    (A) Sodium
    (B) Silver
    (C) Chlorine
    (D) Sodium chloride
    (E) Silver chloride

8. A substance is dissolved in water, forming a 0.50-molar solution. If 4.0 liters of solution contains 240 grams of the substance, what is the molecular mass of the substance?

    (A) 60 grams/mole
    (B) 120 grams/mole
    (C) 240 grams/mole
    (D) 480 grams/mole
    (E) 640 grams/mole

9. The solubility product, K_{sp}, of AgCl is 1.8×10^{-10}. Which of the following expressions is equal to the solubility of AgCl?

(A) $\left(1.8 \times 10^{-10}\right)^2$ molar

(B) $\dfrac{1.8 \times 10^{-10}}{2}$ molar

(C) 1.8×10^{-10} molar

(D) $(2)\left(1.8 \times 10^{-10}\right)$ molar

(E) $\sqrt{1.8 \times 10^{-10}}$ molar

10. A 0.1-molar solution of which of the following acids will be the best conductor of electricity?

(A) $HC_2H_3O_2$
(B) H_2CO_3
(C) H_2S
(D) HF
(E) HNO_3

11. When 31.0 grams of a nonionic substance is dissolved in 2.00 kg of water, the observed freezing-point depression of the solution is 0.93°C. If k_f for water is $1.86°C/m$, which of the following expressions is equal to the molar mass of the substance?

(A) $\dfrac{(31.0)(0.93)(2.00)}{(1.86)}$ g/mol

(B) $\dfrac{(31.0)(1.86)}{(0.93)(2.00)}$ g/mol

(C) $\dfrac{(1.86)(2.00)}{(31.0)(0.93)}$ g/mol

(D) $\dfrac{(0.93)}{(31.0)(1.86)(2.00)}$ g/mol

(E) $(31.0)(0.93)(1.86)(2.00)$ g/mol

12. What is the boiling point of a 2 m solution of NaCl in water? (The boiling point elevation constant, k_b, for water is 0.5°C/m.)

(A) 100°C
(B) 101°C
(C) 102°C
(D) 103°C
(E) 104°C

13. When an aqueous solution of potassium chloride is compared with water, the salt solution will have

(A) a higher boiling point, a lower freezing point, and a lower vapor pressure.
(B) a higher boiling point, a higher freezing point, and a lower vapor pressure.
(C) a higher boiling point, a higher freezing point, and a higher vapor pressure.
(D) a lower boiling point, a lower freezing point, and a lower vapor pressure.
(E) a lower boiling point, a higher freezing point, and a higher vapor pressure.

14. If 46 grams of $MgBr_2$ (molar mass 184 grams) are dissolved in water to form 0.50 liters of solution, what is the concentration of bromine ions in the solution?

(A) 0.25-molar
(B) 0.50-molar
(C) 1.0-molar
(D) 2.0-molar
(E) 4.0-molar

15. A solution contains equal masses of glucose (molecular mass 180) and toluene (molecular mass 90). What is the mole fraction of glucose in the solution?

(A) $\dfrac{1}{4}$

(B) $\dfrac{1}{3}$

(C) $\dfrac{1}{2}$

(D) $\dfrac{2}{3}$

(E) $\dfrac{3}{4}$

16. When benzene and toluene are mixed together, they form an ideal solution. If benzene has a higher vapor pressure than toluene, then the vapor pressure of a solution that contains an equal number of moles of benzene and toluene will be

    (A) higher than the vapor pressure of benzene.
    (B) equal to the vapor pressure of benzene.
    (C) lower than the vapor pressure of benzene and higher than the vapor pressure of toluene.
    (D) equal to the vapor pressure of toluene.
    (E) lower than the vapor pressure of toluene.

17. How many moles of Na_2SO_4 must be added to 500 milliliters of water to create a solution that has a 2-molar concentration of the Na^+ ion? (Assume the volume of the solution does not change).

    (A) 0.5 moles
    (B) 1 mole
    (C) 2 moles
    (D) 4 moles
    (E) 5 moles

18. Given that a solution of NaCl (molar mass 58.5 g/mole) in water (molar mass 18 g/mole) has a molality of 0.5 m, which of the following can be determined?

    I. The mass of the NaCl in the solution
    II. The total mass of the solution
    III. The mole fraction of the NaCl in the solution

    (A) I only
    (B) III only
    (C) I and II only
    (D) II and III only
    (E) I, II, and III

19. How many liters of water must be added to 4 liters of a 6-molar HNO_3 solution to create a solution that is 2-molar?

    (A) 2 liters
    (B) 4 liters
    (C) 6 liters
    (D) 8 liters
    (E) 12 liters

20. Which of the following expressions is equal to the K_{sp} of Ag_2CO_3?

    (A) $K_{sp} = [Ag^+][CO_3^{2-}]$
    (B) $K_{sp} = [Ag^+][CO_3^{2-}]^2$
    (C) $K_{sp} = [Ag^+]^2[CO_3^{2-}]$
    (D) $K_{sp} = [Ag^+]^2[CO_3^{2-}]^2$
    (E) $K_{sp} = [Ag^+]^2[CO_3^{2-}]^3$

21. If the solubility of BaF_2 is equal to x, which of the following expressions is equal to the solubility product, K_{sp}, for BaF_2?

    (A) x^2
    (B) $2x^2$
    (C) x^3
    (D) $2x^3$
    (E) $4x^3$

22. A beaker contains 50.0 ml of a 0.20 M Na_2SO_4 solution. If 50.0 ml of a 0.10 M solution of $Ba(NO_3)_2$ is added to the beaker, what will be the final concentration of sulfate ions in the solution?

    (A) 0.20 M
    (B) 0.10 M
    (C) 0.050 M
    (D) 0.025 M
    (E) 0.012 M

23. The solubility of strontium fluoride in water is 1×10^{-3} M at room temperature. What is the value of the solubility product for SrF_2?

    (A) 2×10^{-3}
    (B) 4×10^{-6}
    (C) 2×10^{-6}
    (D) 4×10^{-9}
    (E) 2×10^{-9}

24. The bottler of a carbonated beverage dissolves carbon dioxide in water by placing carbon dioxide in contact with water at a pressure of 1 atm at room temperature. The best way to increase the amount of dissolved CO_2 would be to

(A) increase the temperature and increase the pressure of CO_2.
(B) decrease the temperature and decrease the pressure of CO_2.
(C) decrease the temperature and increase the pressure of CO_2.
(D) increase the temperature without changing the pressure of CO_2.
(E) increase the pressure of CO_2 without changing the temperature.

25. When 300. ml of a 0.60 M NaCl solution is combined with 200. ml of a 0.40 M $MgCl_2$ solution, what will be the molar concentration of Cl^- ions in the solution?

(A) 0.20 M
(B) 0.34 M
(C) 0.68 M
(D) 0.80 M
(E) 1.0 M

26. Silver hydroxide will be LEAST soluble in a solution with a pH of

(A) 3
(B) 5
(C) 7
(D) 9
(E) 11

27. Copper (II) chloride will be LEAST soluble in a 0.02-molar solution of which of the following compounds?

(A) NaCl
(B) $CuNO_3$
(C) $CaCl_2$
(D) $NaCO_3$
(E) KI

28. A student added 0.10 mol of NaBr and 0.20 mol of $BaBr_2$ to 2 liters of water to create an aqueous solution. What is the minimum number of moles of $Ag(C_2H_3O_2)$ that the student must add to the solution to precipitate out all of the Br^- ions as AgBr?

(A) 0.20
(B) 0.30
(C) 0.40
(D) 0.50
(E) 1.00

29. A student added 1 liter of a 1.0 M KCl solution to 1 liter of a 1.0 M $Pb(NO_3)_2$ solution. A lead chloride precipitate formed, and nearly all of the lead ions disappeared from the solution. Which of the following lists the ions remaining in the solution in order of decreasing concentration?

(A) $[NO_3^-] > [K^+] > [Pb^{2+}]$
(B) $[NO_3^-] > [Pb^{2+}] > [K^+]$
(C) $[K^+] > [Pb^{2+}] > [NO_3^-]$
(D) $[K^+] > [NO_3^-] > [Pb^{2+}]$
(E) $[Pb^{2+}] > [NO_3^-] > [K^+]$

30. The solubility of PbS in water is 3×10^{-14} molar. What is the solubility product constant, K_{sp}, for PbS?

(A) 2×10^{-7}
(B) 9×10^{-7}
(C) 3×10^{-14}
(D) 3×10^{-28}
(E) 9×10^{-28}

PROBLEMS

1. The molecular weight and formula of a hydrocarbon are to be determined through the use of the freezing-point depression method. The hydrocarbon is known to be 86 percent carbon and 14 percent hydrogen by mass. In the experiment, 3.72 grams of the unknown hydrocarbon were placed into 50.0 grams of liquid benzene, C_6H_6. The freezing point of the solution was measured to be 0.06°C. The normal freezing point of benzene is 5.50°C, and the freezing-point depression constant for benzene is 5.12°C/m.

   (a)   What is the molecular weight of the compound?

   (b)   What is the molecular formula of the hydrocarbon?

   (c)   What is the mole fraction of benzene in the solution?

   (d)   If the density of the solution is 875 grams per liter, what is the molarity of the solution?

2. The value of the solubility product, K_{sp}, for calcium hydroxide, $Ca(OH)_2$, is 5.5×10^{-6}, at 25°C.

   (a)   Write the K_{sp} expression for calcium hydroxide.

   (b)   What is the mass of $Ca(OH)_2$ in 500 ml of a saturated solution at 25°C?

   (c)   What is the pH of the solution in (b)?

   (d)   If 1.0 mole of OH^- is added to the solution in (b), what will be the resulting Ca^{2+} concentration? Assume that the volume of the solution does not change.

ESSAYS

3. Explain the following statements in terms of the chemical properties of the substances involved.

   (a)   A 1-molal aqueous solution of sodium chloride has a lower freezing point than a 1-molal aqueous solution of ethanol.

   (b)   NaCl is a strong electrolyte, whereas $PbCl_2$ is a weak electrolyte.

   (c)   Propanol is soluble in water, but propane is not.

   (d)   In a dilute aqueous solution, molarity and molality will have the same value.

4. For sodium chloride, the solution process with water is endothermic.

   (a)   Describe the change in entropy when sodium chloride dissociates into aqueous particles.

   (b)   Two saturated aqueous NaCl solutions, one at 20°C and one at 50°C, are compared. Which one will have higher concentration? Justify your answer.

   (c)   The solubility product of $Ce_2(SO_4)_3$ decreases as temperature increases. Is the solution process for this salt endothermic or exothermic? Justify your answer.

   (d)   When equal molar quantities of HF and HCl are added to separate containers filled with the same amount of water, the HCl solution will freeze at a lower temperature. Explain.

CHAPTER 9 ANSWERS AND EXPLANATIONS

MULTIPLE-CHOICE QUESTIONS

1. **B** Molality is the measure of moles of solute per kilograms of solvent.

2. **D** $pH = -\log[H^+]$

3. **C** Density is a measure of mass per unit volume (e.g., grams per liter).

4. **A** Molarity is the measure of moles of solute per liter of solution.

5. **E** All of the choices are colligative properties, which means they depend only on the number of particles in solution, not on the identity of those particles.

6. **C** Boiling-point elevation is a colligative property. That is, it depends only on the number of particles in solution, not on the specific particles.

   Remember the formula: $\Delta T = kmx$.

   All of the solutions have the same molality, so the one with the greatest boiling-point elevation will be the one that breaks up into the most ions in solution. $CaCl_2$ breaks up into 3 ions, $C_6H_{12}O_6$ doesn't break up into ions, and the other three break up into 2 ions.

7. **E** Sodium chloride is much more soluble than silver chloride. Because of the common ion effect, the chloride ions introduced into the solution by sodium chloride will disrupt the silver chloride equilibrium, causing silver chloride to precipitate from the solution.

8. **B** First find the number of moles.

   Moles = (molarity)(volume)

   Moles of substance = (0.50 M)(4.0 L) = 2 moles

   $$Moles = \frac{grams}{MW}$$

   $$So\ MW = \frac{240\ g}{2\ mol} = 120\ g/mol$$

9. **E** The solubility of a substance is equal to its maximum concentration in solution.

   For every AgCl in solution, we get one Ag^+ and one Cl^-, so the solubility of AgCl—let's call it x—will be the same as $[Ag^+]$, which is the same as $[Cl^-]$.

   So for AgCl, $K_{sp} = [Ag^+][Cl^-] = 1.8 \times 10^{-10} = x^2$.

   $$x = \sqrt{1.8 \times 10^{-10}}$$

10. **E** The best conductor of electricity (also called the strongest electrolyte) will be the solution that contains the most charged particles. HNO_3 is the only strong acid listed in the answer choices, so it is the only choice where the acid has dissociated completely in solution into H^+ and NO_3^- ions. So a 0.1-molar HNO_3 solution will contain the most charged particles and, therefore, be the best conductor of electricity.

11. **B** We can get the molality from the freezing-point depression with the expression
$\Delta T = k_f mx$. Because the substance is nonionic, it will not dissociate, and x will be equal to 1, so we can leave it out of the calculation.

$$m = \frac{\Delta T}{k_f}$$

We know the molality and the mass of the solvent, so we can calculate the number of moles of solute.

$$\text{Moles} = (\text{molality})(\text{kilograms of solvent}) = \frac{\Delta T}{k_f}(\text{kg})$$

Now we use one of our stoichiometry relationships.

$$\text{Moles} = \frac{\text{grams}}{\text{MW}}$$

$$\text{MW} = \frac{\text{grams}}{\text{moles}} = \frac{(\text{grams})}{\left(\frac{(\Delta T)(\text{kg})}{k_f}\right)} = \frac{(\text{grams})(k_f)}{(\Delta T)(\text{kg})} = \frac{(31.0\text{g})(1.86\text{ K-kg/mol})}{(0.93\text{ K})(2.00\text{kg})} = \frac{(31.0)(1.86)}{(0.93)(2.00)}\text{grams/mol}$$

12. **C** $\Delta T = k_b mx$.

Each NaCl dissociates into two particles, so $x = 2$.
$$\Delta T = (0.5°C/m)(2\ m)(2) = 2°C$$
So the boiling point of the solution is 102°C.

13. **A** Particles in solution tend to interfere with phase changes, so the boiling point is raised, the freezing point is lowered, and the vapor pressure is lowered.

14. **C** First we'll find the molarity of the $MgBr_2$ solution.

$$\text{Moles} = \frac{\text{grams}}{\text{MW}}$$

$$\text{Moles of } MgBr_2 \text{ added} = \frac{(46\text{g})}{(184\text{g / mol})} = 0.25 \text{ moles}$$

$$\text{Molarity} = \frac{\text{moles}}{\text{liters}} = \frac{(0.25 \text{ mol})}{(0.50 \text{ L})} = 0.50\text{-molar}$$

For every $MgBr_2$ in solution, 2 Br^- ions are produced, so a 0.50-molar $MgBr_2$ solution will have twice the concentration of Br^- ions, so the bromine ion concentration is 1.0-molar.

15. **B** Let's say the solution contains 180 grams of glucose and 180 grams of toluene. That's 1 mole of glucose and 2 moles of toluene. So that's 1 mole of glucose out of a total of 3 moles, for a mole fraction of $\frac{1}{3}$.

16. **C** From Raoult's law, the vapor pressure of an ideal solution depends on the mole fractions of the components of the solution. The vapor pressure of a solution with equal amounts of benzene and toluene will look like as follows.

$$(P_{\text{solution}}) = (\frac{1}{2})(P_{\text{benzene}}) + (\frac{1}{2})(P_{\text{toluene}})$$

That's just the average of the two vapor pressures.

17. **A** Let's find out how many moles of Na^+ we have to add.

Moles = (molarity)(volume)

Moles of Na^+ = (2 M)(0.5 L) = 1 mole

Because we get 2 moles of Na^+ ions for every mole of Na_2SO_4 we add, we need to add only 0.5 moles of Na_2SO_4.

18. **B** We can't determine (I) and (II) because we don't know how much solution we have. We can figure out (III) because molality tells us the number of moles of NaCl found in 1 kilogram of water. We can figure out how many moles of water there are in 1 kilogram. So if we have a ratio of moles of NaCl to moles of water, we can figure out the mole fraction of NaCl.

19. **D** The number of moles of HNO_3 remains constant.

Moles = (molarity)(volume)

Moles of HNO_3 = (6 M)(4 L) = (2 M)(x)

x = 12 liters, but that's not the answer.

To get a 2-molar solution we need 12 liters, but the solution already has 4 liters, so we need to add 8 liters of water. That's the answer.

20. **C** K_{sp} is just the equilibrium constant without a denominator.

When Ag_2CO_3 dissociates, we get the following reaction:

$$Ag_2CO_3(s) \rightleftharpoons 2\ Ag^+ + CO_3^{2-}$$

In the equilibrium expression, coefficients become exponents, so we get

$$K_{sp} = [Ag^+]^2[CO_3^{2-}]$$

21. **E** For BaF_2, $K_{sp} = [Ba^{2+}][F^-]^2$.

For every BaF_2 that dissolves, we get one Ba^{2+} and two F^-.

So if the solubility of BaF_2 is x, then $[Ba^{2+}] = x$, and $[F^-] = 2x$

So $K_{sp} = (x)(2x)^2 = (x)(4x^2) = 4x^3$

22. **C** The Ba^{2+} ions and the SO_4^- ions will combine and precipitate out of the solution, so let's find out how many of each we have.

Moles = (molarity)(volume)

Moles of SO_4^- = (0.20 M)(0.050 L) = 0.010 mole

Moles of Ba^{2+} = (0.10 M)(0.050 L) = 0.0050 mole

To find the number of moles of SO_4^- left in the solution, subtract the moles of Ba^{2+} from the moles of SO_4^-.

0.010 mole – 0.0050 mole = 0.0050 mole

Now use the formula for molarity to find the concentration of SO_4^- ions. Don't forget to add the volumes of the two solutions.

$$Molarity = \frac{moles}{liters} = \frac{(0.0050\ mol)}{(0.10 L)} = 0.050\ M$$

23. **D** Use the formula for K_{sp}. For every SrF_2 in solution, there will be one strontium ion and two fluoride ions, so $[Sr^{2+}]$ will be 1×10^{-3} M and $[F^-]$ will be 2×10^{-3} M.

$$K_{sp} = [Sr^{2+}][F^-]^2$$
$$K_{sp} = (1 \times 10^{-3}\ M)(2 \times 10^{-3}\ M)^2$$
$$K_{sp} = 4 \times 10^{-9}$$

24. **C** The lower the temperature, the more soluble a gas will be in water. The greater the pressure of the gas, the more soluble it will be.

25. **C** The Cl^- ions from the two salts will both be present in the solution, so we need to find the number of moles of Cl^- contributed by each salt.

Moles = (molarity)(volume)

Each NaCl produces 1 Cl^-

Moles of Cl^- from NaCl = $(0.60\ M)(0.300\ L) = 0.18$ mole

Each $MgCl_2$ produces 2 Cl^-

Moles of Cl^- from $MgCl_2$ = $(2)(0.40\ M)(0.200\ L) = 0.16$ mole

To find the number of moles of Cl^- in the solution, add the two together.

0.18 mole + 0.16 mole = 0.34 mole

Now use the formula for molarity to find the concentration of Cl^- ions. Don't forget to add the volumes of the two solutions.

$$\text{Molarity} = \frac{\text{moles}}{\text{liters}} = \frac{(0.34\,\text{mol})}{(0.500\,\text{L})} = 0.68\ \text{M}$$

26. **E** Silver hydroxide will be least soluble in the solution with the highest hydroxide concentration. That would be the solution with the highest pH.

27. **C** According to the common ion effect, ions already present in a solution will affect the solubility of compounds that also produce those ions. So a solution containing Cu^+ ions or Cl^- ions will inhibit the solubility of CuCl.

A 0.02-molar solution of NaCl will have a 0.02-molar concentration of Cl^- ions and a 0.02-molar solution of $CuNO_3$ will have a 0.02-molar concentration of Cu^+ ions, so choices (A) and (B) will affect the solubility of CuCl to the same extent.

The correct answer is choice (C), $CaCl_2$, because a 0.02-molar solution of $CaCl_2$ will have a 0.04-molar concentration of Cl^- ions, so this solution will do the most to inhibit the solubility of CuCl.

Choices (D) and (E) have no effect on the solubility of CuCl.

28. **D** The solution contains 0.50 moles of Br^- ions, 0.10 from NaBr and 0.40 from $BaBr_2$ (each $BaBr_2$ provides 2 Br^- ions). Each Ag^+ ion will remove 1 Br^- ion, so the student needs to add 0.50 moles of $Ag(C_2H_3O_2)$.

29. **A** At the start, the concentrations of the ions are as follows:

$[K^+] = 1\ M$

$[Cl^-] = 1\ M$

$[Pb^{2+}] = 1\ M$

$[NO_3^-] = 2\ M$

After $PbCl_2$ forms, the concentrations are as follows:

$[K^+] = 1\ M$

$[Cl^-] = 0.5\ M$

$[Pb^{2+}] = 0\ M$

$[NO_3^-] = 2\ M$

So from greatest to least

$[NO_3^-] > [K^+] > [Pb^{2+}]$

30. **E** The solubility of a substance is equal to its maximum concentration in solution. For every PbS in solution, we get one Pb^{2+} and one S^{2-}, so the concentration of PbS, $3 \times 10^{-14}\ M$, will be the same as the concentrations of Pb^{2+} and S^{2-}.

$K_{sp} = [Pb^{2+}][S^{2-}]$

$K_{sp} = (3 \times 10^{-14}\ M)(3 \times 10^{-14}\ M) = 9 \times 10^{-28}$

PROBLEMS

1. (a) First we'll find the molality of the solution. The freezing point depression, ΔT, is
5.50°C − 0.06°C = 5.44°C.

$\Delta T = km$

Solve for m

$$m = \frac{\Delta T}{k} = \frac{(5.44\,°C)}{(5.12\,°C\,/\,m)} = 1.06m$$

From the molality of the solution, we can find the number of moles of unknown hydrocarbon.

$$\text{Molality} = \frac{\text{moles of solute}}{\text{kg of solvent}}$$

Solve for moles.

Moles = (molality)(kg of solvent)

Moles of hydrocarbon = (1.06 m)(0.050 kg) = 0.053 moles

Now we can find the molecular weight of the hydrocarbon.

$$MW = \frac{\text{grams}}{\text{moles}} = \frac{(3.72\ g)}{(0.053\ mol)} = 70.2\ g/mol$$

(b) You can use the percent by mass and the molecular weight.

For carbon

(86%)(70 g/mol) = 60 g/mol

Carbon has an atomic weight of 12, so there must be $\frac{60}{12}$ = 5 moles of carbon in 1 mole of the hydrocarbon.

For hydrogen

(14%)(70 g/mol) = 10 g/mol

Hydrogen has an atomic weight of 1, so there must be $\frac{10}{1}$ = 10 moles of hydrogen in 1 mole of the hydrocarbon.

So the molecular formula for the hydrocarbon is C_5H_{10}.

(c) We know that there are 0.053 moles of hydrocarbon. We need to find the number of moles of benzene.

$$Moles = \frac{grams}{MW}$$

$$Moles\ of\ benzene = \frac{(50.00\ g)}{(78\ g/mol)} = 0.64\ mol$$

Total moles = 0.64 mol + 0.053 mol = 0.69 mol

$$Mole\ fraction\ of\ benzene = \frac{0.64\ mol}{0.69\ mol} = 0.93$$

(d) Remember the definition of molarity.

$$Molarity = \frac{moles\ of\ solute}{liters\ of\ solution}$$

We know that the moles of solute is 0.053. We need to find the liters of solution.

The weight of the solution is

50.00 g + 3.72 g = 53.72 g

$$Density = \frac{grams}{liters}$$

Solve for liters.

$$Liters\ of\ solution = \frac{grams}{density} = \frac{(53.72\ g)}{(875\ g/L)} = 0.0614\ L$$

$$Molarity = \frac{(0.053\ mol)}{(0.0614\ L)} = 0.863\ M$$

2. (a) The solubility product is the same as the equilibrium expression, but because the reactant is a solid, there is no denominator.

$K_{sp} = [Ca^{2+}][OH^-]^2$

(b) Use the solubility product.

$K_{sp} = [Ca^{2+}][OH^-]^2$

$5.5 \times 10^{-6} = (x)(2x)^2 = 4x^3$

$x = 0.01\ M$ for Ca^{2+}

One mole of calcium hydroxide produces 1 mole of Ca^{2+}, so the concentration of $Ca(OH)_2$ must be 0.01 M.

Moles = (molarity)(volume)

Moles of $Ca(OH)_2$ = (0.01 M)(0.500 L) = 0.005 moles

Grams = (moles)(MW)

Grams of $Ca(OH)_2$ = (0.005 mol)(74 g/mol) = 0.37 g

(c) We can find $[OH^-]$ from (b).

If $[Ca^{2+}]$ = 0.01 M, then $[OH^-]$ must be twice that, so $[OH^-]$ = 0.02 M

pOH = $-\log[OH^-]$ = 1.7

pH = 14 − pOH = 14 − 1.7 = 12.3

(d) Find the new $[OH^-]$. The hydroxide already present is small enough to ignore, so we'll use only the hydroxide just added.

$$Molarity = \frac{moles}{liters}$$

$$[OH^-] = \frac{(1.0 \text{ mol})}{(0.500 \text{ L})} = 2.0 \ M$$

Now use the K_{sp} expression.

$K_{sp} = [Ca^{2+}][OH^-]^2$

$5.5 \times 10^{-6} = [Ca^{2+}](2.0 \ M)^2$

$[Ca^{2+}] = 1.4 \times 10^{-6} \ M$

ESSAYS

3. (a) Freezing-point depression is a colligative property, which means that it depends on the number of particles in solution, not their identity.

Sodium chloride dissociates into Na^+ and Cl^-, so every unit of sodium chloride produces two particles in solution. Ethanol does not dissociate, so sodium chloride will put twice as many particles in solution as ethanol.

(b) An electrolyte is a substance that ionizes in solution, thus causing the solution to conduct electricity.

Both of the salts dissociate into ions, but $PbCl_2$ is almost insoluble, so it will produce very few ions in solution, while NaCl is extremely soluble and produces many ions.

(c) Water is best at dissolving polar substances.

Propanol (C_3H_7OH) has a hydroxide group, which makes it polar, and thus soluble in water. Propane (C_3H_8) is nonpolar and is best dissolved in nonpolar solvents.

(d) Remember the definitions, and remember that a dilute solution has very little solute.

$$Molarity = \frac{moles \ of \ solute}{liters \ of \ solution}$$

$$Molality = \frac{moles \ of \ solute}{kilograms \ of \ solvent}$$

For water, 1 liter weighs 1 kilogram, so for a dilute solution this distinction disappears.

If there is very little solute, the mass and volume of the solution will be indistinguishable from the mass and volume of the solvent.

4. (a) Entropy increases when a salt dissociates because aqueous particles have more randomness than a solid.

(b) Most salt solution processes are endothermic, and endothermic processes are favored by an increase in temperature, therefore increasing temperature will increase the solubility of most salts.

(c) $Ce_2(SO_4)_3$ becomes less soluble as temperature increases, so the solution process for this salt must be exothermic.

(d) Freezing-point depression is a colligative property, which means that it depends on the number of particles in solution, not their identity.

HCl is a strong acid, which means that it dissociates completely. This means that 1 mole of HCl in solution will produce 2 moles of particles. HF is a weak acid, which means that it dissociates very little. This means that 1 mole of HF in solution will remain at about 1 mole of particles in solution.

Therefore, the HCl solution will have more particles than the HF solution.

10

Equilibrium

HOW OFTEN DOES EQUILIBRIUM APPEAR ON THE EXAM?

In the multiple-choice section, this topic appears in about 4 out of 75 questions. In the free-response section, this topic appears every year.

THE EQUILIBRIUM CONSTANT, K_{eq}

Most chemical processes are reversible. That is, reactants react to form products, but those products can also react to form reactants.

A reaction is at equilibrium when the rate of the forward reaction is equal to the rate of the reverse reaction.

The relationship between the concentrations of reactants and products in a reaction at equilibrium is given by the equilibrium expression, also called the **law of mass action.**

The Equilibrium Expression

For the reaction

$$aA + bB \rightleftharpoons cC + dD$$

$$K_{eq} = \frac{[C]^c [D]^d}{[A]^a [B]^b}$$

1. $[A]$, $[B]$, $[C]$, and $[D]$ are molar concentrations or partial pressures at equilibrium.
2. Products are in the numerator, and reactants are in the denominator.
3. Coefficients in the balanced equation become exponents in the equilibrium expression.
4. Solids and pure liquids are not included in the equilibrium expression—only aqueous reactants and products are included.
5. Units are not given for K_{eq}.

Let's look at a few examples:

1. $HC_2H_3O_2(aq) \rightleftharpoons H^+(aq) + C_2H_3O_2^-(aq)$

$$K_{eq} = K_a = \frac{[H^+][C_2H_3O_2^-]}{[HC_2H_3O_2]}$$

This reaction shows the dissociation of acetic acid in water. All of the reactants and products are aqueous particles, so they are all included in the equilibrium expression. None of the reactants or products have coefficients, so there are no exponents in the equilibrium expression. This is the standard form of K_a, the acid dissociation constant.

2. $2H_2S(g) + 3O_2(g) \rightleftharpoons 2H_2O(g) + 2SO_2(g)$

$$K_{eq} = K_c = \frac{[H_2O]^2[SO_2]^2}{[H_2S]^2[O_2]^3}$$

$$K_{eq} = K_p = \frac{P^2_{H_2O} P^2_{SO_2}}{P^2_{H_2S} P^3_{O_2}}$$

All of the reactants and products in this reaction are gases, so K_{eq} can be expressed in terms of concentration (K_c, moles/liter or molarity) or in terms of partial pressure (K_p, atmospheres). In the next section, we'll see how these two different ways of looking at the same equilibrium situation are related. All of the reactants and products are included here, and the coefficients in the reaction become exponents in the equilibrium expression.

3. $CaF_2(s) \rightleftharpoons Ca^{2+}(aq) + 2 F^-(aq)$

$$K_{eq} = K_{sp} = [Ca^{2+}][F^-]^2$$

This reaction shows the dissociation of a slightly soluble salt. There is no denominator in this equilibrium expression because the reactant is a solid. Solids are left out of the equilibrium expression because the concentration of a solid is constant. There must be some solid present for equilibrium to exist, but you do not need to include it in your calculations. This form of K_{eq} is called the solubility product, K_{sp}, which we already saw in Chapter 9.

4. $NH_3(aq) + H_2O(l) \rightleftharpoons NH_4^+(aq) + OH^-(aq)$

$$K_{eq} = K_b = \frac{\left[NH_4^+\right]\left[OH^-\right]}{\left[NH_3\right]}$$

This is the acid-base reaction between ammonia and water. We can leave water out of the equilibrium expression because it is a pure liquid. By pure liquid, we mean that the concentration of water is so large (about 50 molar) that nothing that happens in the reaction is going to change it significantly, so we can consider it to be constant. This is the standard form for K_b, the base dissociation constant.

Here is a roundup of the equilibrium constants you need to be familiar with for the test.

- K_c is the constant for molar concentrations.

- K_p is the constant for partial pressures.

- K_{sp} is the solubility product, which has no denominator because the reactants are solids.

- K_a is the acid dissociation constant for weak acids.

- K_b is the base dissociation constant for weak bases.

- K_w describes the ionization of water ($K_w = 1 \times 10^{-14}$).

The equilibrium constant has a lot of aliases, but they all take the same form and tell you the same thing. The **equilibrium constant** tells you the relative amounts of products and reactants at equilibrium.

A large value for K_{eq} means that products are favored over reactants at equilibrium, while a small value for K_{eq} means that reactants are favored over products at equilibrium.

K_{eq} AND GASES

As we saw in the example above, the equilibrium constant for a gas phase reaction can be written in terms of molar concentrations, K_c, or partial pressures, K_p. These two forms of K can be related by the following equation, which is derived from the ideal gas law.

$$K_p = K_c \left(RT\right)^{\Delta n}$$

K_p = partial pressure constant (using atmospheres as units)
K_c = molar concentration constant (using molarities as units)
R = the ideal gas constant, 0.0821 (L-atm)/(mol-K)
T = absolute temperature (K)
Δn = (moles of product gas − moles of reactant gas)

THE REACTION QUOTIENT, Q

The reaction quotient is determined in exactly the same way as the equilibrium constant, but initial conditions are used in place of equilibrium conditions. The reaction quotient can be used to predict the direction in which a reaction will proceed from a given set of initial conditions.

The Reaction Quotient
For the reaction $$aA + bB \rightleftharpoons cC + dD$$ $$Q = \frac{[C]^c [D]^d}{[A]^a [B]^b}$$ $[A]$, $[B]$, $[C]$, and $[D]$ are initial molar concentrations or partial pressures. • If Q is less than the calculated K for the reaction, the reaction proceeds forward, generating products. • If Q is greater than K, the reaction proceeds backward, generating reactants. • If $Q = K$, the reaction is already at equilibrium.

K_{eq} AND MULTISTEP PROCESSES

There is a simple relationship between the equilibrium constants for the steps of a multistep reaction and the equilibrium constant for the overall reaction.

If two reactions can be added together to create a third reaction, then the K_{eq} for the two reactions can be multiplied together to get the K_{eq} for the third reaction.

If	$A + B \rightleftharpoons C$	$K_{eq} = K_1$
and	$C \rightleftharpoons D + E$	$K_{eq} = K_2$
then	$A + B \rightleftharpoons D + E$	$K_{eq} = K_1 K_2$

LE CHÂTELIER'S LAW

Le Châtelier's law says that whenever a stress is placed on a situation at equilibrium, the equilibrium will shift to relieve that stress.

Let's use the **Haber process,** which is used in the industrial preparation of ammonia, as an example.

$$N_2(g) + 3\,H_2(g) \rightleftharpoons 2\,NH_3(g) \qquad\qquad \Delta H° = -92.6 \text{ kJ}$$

CONCENTRATION

- When the concentration of a reactant or product is increased, the reaction will proceed in the direction that will use up the added substance.

If N_2 or H_2 is added, the reaction proceeds in the forward direction. If NH_3 is added, the reaction proceeds in the reverse direction.

- When the concentration of a reactant or product is decreased, the reaction will proceed in the direction that will produce more of the substance that has been removed.

If N_2 or H_2 is removed, the reaction will proceed in the reverse direction. If NH_3 is removed, the reaction will proceed in the forward direction.

VOLUME

- When the volume in which a reaction takes place is increased, the reaction will proceed in the direction that produces more moles of gas.

When the volume for the Haber process is increased, the reaction proceeds in the reverse direction because the reactants have more moles of gas (4) than the products (2).

- When the volume in which a reaction takes place is decreased, the reaction will proceed in the direction that produces fewer moles of gas.

When the volume for the Haber process is decreased, the reaction proceeds in the forward direction because the products have fewer moles of gas (2) than the reactants (4).

- If there is no gas involved in the reaction, or if the reactants and products have the same number of moles of gas, then volume changes have no effect on the equilibrium.

TEMPERATURE

- When temperature is increased, the reaction will proceed in the endothermic direction.

When the temperature for the Haber process is increased, the reaction proceeds in the reverse direction because the reverse reaction is endothermic ($\Delta H°$ is positive).

- When temperature is decreased, the reaction will proceed in the exothermic direction.

When the temperature for the Haber process is decreased, the reaction proceeds in the forward direction because the forward reaction is exothermic ($\Delta H°$ is negative).

PRESSURE

- When pressure is increased, the reaction will proceed toward the side with the fewest molecules of gas.

- When pressure is decreased, the reaction will proceed toward the side with the greatest number of molecules of gas.

When an inert gas, like Argon, is added to the system, although the total pressure goes up, the partial pressures of the gases involved in the reaction will remain the same. Therefore, there is no effect on the equilibrium position, because the reaction quotient, Q, is determined by the individual partial pressures of the gases involved in the reaction.

CHAPTER 10 QUESTIONS

MULTIPLE-CHOICE QUESTIONS

Questions 1–4

(A) K_c
(B) K_p
(C) K_a
(D) K_w
(E) K_{sp}

1. This equilibrium constant uses partial pressures of gases as units.

2. This equilibrium constant always has a value of 1×10^{-14} at 25°C.

3. This equilibrium constant is used for the dissociation of an acid.

4. The equilibrium expression for this equilibrium constant does not contain a denominator.

5. For a particular salt, the solution process is endothermic. As the temperature at which the salt is dissolved increases, which of the following will occur?

(A) K_{sp} will increase, and the salt will become more soluble.
(B) K_{sp} will decrease, and the salt will become more soluble.
(C) K_{sp} will increase, and the salt will become less soluble.
(D) K_{sp} will decrease, and the salt will become less soluble.
(E) K_{sp} will not change, and the salt will become more soluble.

6. $2\,HI(g) + Cl_2(g) \rightleftharpoons 2\,HCl(g) + I_2(g) + energy$

A gaseous reaction occurs and comes to equilibrium as shown above. Which of the following changes to the system will serve to increase the number of moles of I_2 present at equilibrium?

(A) Increasing the volume at constant temperature
(B) Decreasing the volume at constant temperature
(C) Adding a mole of inert gas at constant volume
(D) Increasing the temperature at constant volume
(E) Decreasing the temperature at constant volume

7. A sealed isothermal container initially contained 2 moles of CO gas and 3 moles of H_2 gas. The following reversible reaction occurred:

$$CO(g) + 2 H_2(g) \rightleftharpoons CH_3OH(g)$$

At equilibrium, there was 1 mole of CH_3OH in the container. What was the total number of moles of gas present in the container at equilibrium?

(A) 1
(B) 2
(C) 3
(D) 4
(E) 5

8. $4 NH_3(g) + 3 O_2(g) \rightleftharpoons$

$$2 N_2(g) + 6 H_2O(g) + energy$$

Which of the following changes to the system at equilibrium shown above would cause the concentration of H_2O to increase?

(A) The volume of the system was decreased at constant temperature.
(B) The temperature of the system was increased at constant volume.
(C) NH_3 was removed from the system.
(D) N_2 was removed from the system.
(E) O_2 was removed from the system.

9. A sample of solid potassium nitrate is placed in water. The solid potassium nitrate comes to equilibrium with its dissolved ions by the endothermic process shown below.

$$KNO_3(s) + energy \rightleftharpoons K^+(aq) + NO_3^-(aq)$$

Which of the following changes to the system would increase the concentration of K^+ ions at equilibrium?

(A) The volume of the solution is increased.
(B) The volume of the solution is decreased.
(C) Additional solid KNO_3 is added to the solution.
(D) The temperature of the solution is increased.
(E) The temperature of the solution is decreased.

10. For which of the following gaseous equilibria do K_p and K_c differ the most?

(A) $2H_2(g) + O_2(g) \rightleftharpoons 2H_2O(g)$
(B) $NH_3BH_3(g) \rightleftharpoons NH_2BH_2(g) + H_2(g)$
(C) $NO(g) + O_3(g) \rightleftharpoons NO_2(g) + O_2(g)$
(D) $B_3N_3H_6(g) + 3H_2(g) \rightleftharpoons B_3N_3H_{12}(g)$
(E) $BH_3(g) + 3HCl(g) \rightleftharpoons BCl_3(g) + 3H_2(g)$

11. A 1M solution of $SbCl_5$ in organic solvent shows no noticeable reactivity at room temperature. The sample is heated to 200°C and the following equilibrium reaction is found to occur:

$$SbCl_5 \rightleftharpoons SbCl_3 + Cl_2$$

The value of K_{eq} at 200°C was measured to be 10^{-6}. The reaction was then heated to 350°C, and the equilibrium concentration of Cl_2 was found to be 0.1 M, which of the following best approximated the value of K_{eq} at 350°C?

(A) 1.5
(B) 1.0
(C) 0.1
(D) 0.01
(E) 0.001

12. $2 \, NOBr(g) \rightleftharpoons 2 \, NO(g) + Br_2(g)$

The reaction above came to equilibrium at a temperature of 100°C. At equilibrium the partial pressure due to NOBr was 4 atmospheres, the partial pressure due to NO was 4 atmospheres, and the partial pressure due to Br_2 was 2 atmospheres. What is the equilibrium constant, K_p, for this reaction at 100°C?

(A) $\dfrac{1}{4}$

(B) $\dfrac{1}{2}$

(C) 1

(D) 2

(E) 4

13. $HCrO_4^- + Ca^{2+} \leftrightarrow H^+ + CaCrO_4$

If the acid dissociation constant for $HCrO_4^-$ is K_a and the solubility product for $CaCrO_4$ is K_{sp}, which of the following gives the equilibrium expression for the reaction above?

(A) $K_a K_{sp}$

(B) $\dfrac{K_a}{K_{sp}}$

(C) $\dfrac{K_{sp}}{K_a}$

(D) $\dfrac{1}{K_{sp} K_a}$

(E) $\dfrac{K_a K_{sp}}{2}$

14. $Br_2(g) + I_2(g) \leftrightarrow 2 \, IBr(g)$

At 150°C, the equilibrium constant, K_c, for the reaction shown above has a value of 300. This reaction was allowed to reach equilibrium in a sealed container and the partial pressure due to $IBr(g)$ was found to be 3 atm. Which of the following could be the partial pressures due to $Br_2(g)$ and $I_2(g)$ in the container?

	$Br_2(g)$	$I_2(g)$
(A)	0.1 atm	0.3 atm
(B)	0.3 atm	1 atm
(C)	1 atm	1 atm
(D)	1 atm	3 atm
(E)	3 atm	3 atm

15. $H_2(g) + CO_2(g) \leftrightarrow H_2O(g) + CO(g)$

Initially, a sealed vessel contained only $H_2(g)$ with a partial pressure of 6 atm and $CO_2(g)$ with a partial pressure of 4 atm. The reaction above was allowed to come to equilibrium at a temperature of 700 K. At equilibrium, the partial pressure due to $CO(g)$ was found to be 2 atm. What is the value of the equilibrium constant K_p, for the reaction?

(A) $\dfrac{1}{24}$

(B) $\dfrac{1}{6}$

(C) $\dfrac{1}{4}$

(D) $\dfrac{1}{3}$

(E) $\dfrac{1}{2}$

PROBLEMS

1.
$$BaF_2(s) \rightleftharpoons Ba^{2+}(aq) + 2\ F^-(aq)$$

The value of the solubility product, K_{sp}, for the reaction above is 1.0×10^{-6} at 25°C.

(a)   Write the K_{sp} expression for BaF_2.

(b)   What is the concentration of F^- ions in a saturated solution of BaF_2 at 25°C?

(c)   500 milliliters of a 0.0060-molar NaF solution is added to 400 ml of a 0.0060-molar $Ba(NO_3)_2$ solution. Will there be a precipitate?

(d)   What is the value of ΔG for the dissociation of BaF_2 at 25°C?

2.
$$H_2CO_3 \rightleftharpoons H^+ + HCO_3^- \qquad K_1 = 4.3 \times 10^{-7}$$
$$HCO_3^- \rightleftharpoons H^+ + CO_3^{2-} \qquad K_2 = 5.6 \times 10^{-11}$$

The acid dissociation constants for the reactions above are given at 25°C.

(a)   What is the pH of a 0.050-molar solution of H_2CO_3 at 25°C?

(b)   What is the concentration of CO_3^{2-} ions in the solution in (a)?

(c)   How would the addition of each of the following substances affect the pH of the solution in (a)?

(i)   HCl

(ii) $NaHCO_3$

(iii)   NaOH

(iv)   NaCl

(d)   What is the value of K_{eq} for the following reaction?
$$H_2CO_3 \rightleftharpoons 2\ H^+ + CO_3^{2-}$$

3.
$$N_2(g) + 3 H_2(g) \rightleftharpoons 2 NH_3(g) \qquad \Delta H = -92.4 \text{ kJ}$$

When the reaction above took place at a temperature of 570 K, the following equilibrium concentrations were measured:

$[NH_3] = 0.20 \text{ mol/L}$

$[N_2] = 0.50 \text{ mol/L}$

$[H_2] = 0.20 \text{ mol/L}$

(a)   Write the expression for K_c and calculate its value.

(b)   What is the value of K_p for the reaction?

(c)   Describe how the concentration of H_2 will be affected by each of the following changes to the system at equilibrium:

   (i)   The temperature is increased.

   (ii)   The volume of the reaction chamber is increased.

   (iii) N_2 gas is added to the reaction chamber.

   (iv)   Helium gas is added to the reaction chamber.

4.
$$CaCO_3(s) \rightleftharpoons Ca^{2+}(aq) + CO_3^{2-}(aq) \qquad K_{sp} = 2.8 \times 10^{-9}$$
$$CaSO_4(s) \rightleftharpoons Ca^{2+}(aq) + SO_4^{2-}(aq) \qquad K_{sp} = 9.1 \times 10^{-6}$$

The values for the solubility products for the two reactions above are given at 25°C.

(a)   What is the concentration of CO_3^{2-} ions in a saturated 1.00 liter solution of $CaCO_3$ at 25°C?

(b)   Excess $CaSO_4(s)$ is placed in the solution in (a). Assume that the volume of the solution does not change.

   (i)   What is the concentration of the SO_4^{2-} ion?

   (ii)   What is the concentration of the CO_3^{2-} ion?

(c)   A 0.20 mole sample of $CaCl_2$ is placed in the solution in (b). Assume that the volume of the solution does not change.

   (i)   What is the concentration of the Ca^{2+} ion?

   (ii)   What is the concentration of the SO_4^{2-} ion?

   (iii)   What is the concentration of the CO_3^{2-} ion?

CHAPTER 10 ANSWERS AND EXPLANATIONS

MULTIPLE-CHOICE QUESTIONS

1. **B** K_p is used for gaseous reactions, and the units used are partial pressures.

2. **D** K_w is the dissociation constant for water.

   At 25°C, $K_w = [H^+][OH^-] = 1 \times 10^{-14}$.

3. **C** K_a is known as the acid dissociation constant.

4. **E** K_{sp} is the solubility product. It always has a solid as the reactant. Because the reactant is always in the denominator and solids are ignored in the equilibrium expression, K_{sp} never has a denominator.

5. **A** From Le Châtelier's law, the equilibrium will shift to counteract any stress that is placed on it. Increasing temperature favors the endothermic direction of a reaction because the endothermic reaction absorbs the added heat. So the salt becomes more soluble, increasing the number of dissociated particles, thus increasing the value of K_{sp}.

6. **E** According to Le Châtelier's law, the equilibrium will shift to counteract any stress that is placed on it. If the temperature is decreased, the equilibrium will shift toward the side that produces energy or heat. That's the product side where I_2 is produced.

   Choices (A) and (B) are wrong because there are equal numbers of moles of gas (3 moles) on each side, so changing the volume will not affect the equilibrium. Choice (C) is wrong because the addition of a substance that does not affect the reaction will not affect the equilibrium conditions.

7. **C** Because the equation is balanced, the following will occur:

   If 1 mole of CH_3OH was created, then 1 mole of CO was consumed and 1 mole of CO remains; and if 1 mole of CH_3OH was created, then 2 moles of H_2 were consumed and 1 mole of H_2 remains. So at equilibrium, there are

   (1 mol CH_3OH) + (1 mol CO) + (1 mol H_2) = 3 moles of gas

8. **D** According to Le Châtelier's law, equilibrium will shift to relieve any stress placed on a system. If N_2 is removed, the equilibrium will shift to the right to produce more N_2, with the result that more H_2O will also be produced.

   If the volume is decreased (A), the equilibrium will shift toward the left, where there are fewer moles of gas. If the temperature is increased (B), the equilibrium will shift to the left. That's the endothermic reaction, which absorbs the added energy of the temperature increase. If NH_3 (C) or O_2 (E) is removed, the equilibrium will shift to the left to replace the substance removed.

9. **D** According to Le Châtelier's law, equilibrium will shift to relieve any stress placed on a system. If the temperature is increased, the equilibrium will shift to favor the endothermic reaction because it absorbs the added energy. In this case, the equilibrium will be shifted to the right, increasing the concentration of both K^+ and NO_3^- ions.

   Changing the volume of the solution, (A) and (B), will change the *number* of K^+ ions in solution, but not the *concentration* of K^+ ions. Because solids are not considered in the equilibrium expression, adding more solid KNO_3 to the solution (C) will not change the equilibrium. Decreasing the temperature (E) will favor the exothermic reaction, driving the equilibrium toward the left and decreasing the concentration of K^+ ions.

10. **D** K_p and K_c will differ the most in situations where the moles of gas change drastically from the reactants to the products. This relation is given in the equation

$$K_p = K_c(RT)^{\Delta n}$$

The exponential Δn is the moles of gas in the products minus the moles of gas in the reactants. The biggest change in number of moles of gas is found in the equation in choice D.

11. **D** The information about K at 200°C is useless information and can be ignored. Since all the Cl_2 found in solution must have come from $SbCl_5$, we know that at equilibrium $[Cl_2] = [SbCl_3] = 0.1$ M, and $[SbCl_5] = (1.0 - 0.1)$ M $= .99$ M. We can then say that $K = (0.1)(0.1)/.99 = .0101$ which is most closely approximated by choice D.

12. **D** $K_p = \dfrac{[NO]^2[Br_2]}{[NOBr]^2} = \dfrac{(4)^2(2)}{(4)^2} = 2$

13. **B** We can think of the reaction given in the question as the sum of two other reactions.

$$HCrO_4^- \leftrightarrow H^+ + CrO_4^{2-} \quad K_{eq} = K_a$$

$$Ca^{2+} + CrO_4^{2-} \leftrightarrow CaCrO_4 \quad K_{eq} = \dfrac{1}{K_{sp}}$$

Notice that we are using the reverse reaction for the solvation of $CaCrO_4$, so the reactants and products are reversed and we must take the reciprocal of the solubility product.

When reactions can be added to get another reaction, their equilibrium constants can be multiplied to get the equilibrium constant of the resulting reaction.

So $K_{eq} = (K_a)(\dfrac{1}{K_{sp}}) = \dfrac{K_a}{K_{sp}}$

14. **A** The equilibrium expression for the reaction is as follows:

$$\dfrac{P_{IBr}^{\;2}}{P_{Br_2}P_{I_2}} = 300$$

When all of the values are plugged into the expression, (A) is the only choice that works.

$$\dfrac{(3)^2}{(0.1)(0.3)} = \dfrac{9}{0.03} = 300$$

15. **E** Use a table to see how the partial pressures change. Based on the balanced equation, we know that if 2 atm of CO(g) were formed, then 2 atm of $H_2O(g)$ must also have formed. We also know that the reactants must have lost 2 atm each.

	$H_2(g)$	$CO_2(g)$	$H_2O(g)$	CO(g)
Before	6 atm	4 atm	0	0
Change	−2	−2	+2	+2
At Equilibrium	4 atm	2 atm	2 atm	2 atm

Now plug the numbers into the equilibrium expression.

$$K_{eq} = \dfrac{P_{H_2O}P_{CO}}{P_{H_2}P_{CO_2}} = \dfrac{(2)(2)}{(4)(2)} = \dfrac{1}{2}$$

PROBLEMS

1. (a) $K_{sp} = [Ba^{2+}][F^-]^2$

   (b) Use the K_{sp} expression.

   $K_{sp} = [Ba^{2+}][F^-]^2$

   Two F^-s are produced for every Ba^{2+}, so $[F^-]$ will be twice as large as $[Ba^{2+}]$.

   Let $x = [F^-]$

   $$1.0 \times 10^{-6} = \left(\frac{x}{2}\right)(x)^2 = \frac{x^3}{2}$$

   $x = [F^-] = 0.01\ M$

   (c) First we need to find the concentrations of the Ba^{2+} and F^- ions.

   Moles = (molarity)(volume)

   Moles of Ba^{2+} = (0.0060 M)(0.400 L) = 0.0024 mol

   Moles of F^- = (0.0060 M)(0.500 L) = 0.0030 mol

   Remember to add the two volumes: (0.400 L) + (0.500 L) = 0.900 L

   $$\text{Molarity} = \frac{\text{moles}}{\text{liters}}$$

   $$\left[Ba^{2+}\right] = \frac{(0.0024\ \text{mol})}{(0.900\ \text{L})} = 0.0027\,M$$

   $$\left[F^-\right] = \frac{(0.0030\ \text{mol})}{(0.900\ \text{L})} = 0.0033\,M$$

   Now test the solubility expression using the initial values to find the reaction quotient.

   $Q = [Ba^{2+}][F^-]^2$

   $Q = (0.0027)(0.0033)^2 = 2.9 \times 10^{-8}$

   Q is less than K_{sp}, so no precipitate forms.

   (d) Use the standard free energy expression.

   $\Delta G° = -2.303RT \log K$

   $\Delta G° = (-2.303)(8.31\,\text{J/mol} - \text{K})(298\ \text{K})(\log 1.0 \times 10^{-6}) = 34,000\ \text{J/mol}$

   The positive value of $\Delta G°$ means that the reaction is not spontaneous under standard conditions.

2. (a) Use the equilibrium expression.

$$K_1 = \frac{\left[H^+\right]\left[HCO_3^-\right]}{\left[H_2CO_3\right]}$$

$$\left[H^+\right] = \left[HCO_3^-\right] = x$$

$$\left[H_2CO_3\right] = (0.050 \ M - x)$$

Assume that x is small enough that we can use $[H_2CO_3] = (0.050 \ M)$.

$$4.3 \times 10^{-7} = \frac{x^2}{(0.050)}$$

$$x = \left[H^+\right] = 1.5 \times 10^{-4}$$

$$pH = -\log\left[H^+\right] = -\log(1.5 \times 10^{-4}) = 3.8$$

(b) Use the equilibrium expression.

$$K_2 = \frac{\left[H^+\right]\left[CO_3^{2-}\right]}{\left[HCO_3^-\right]}$$

From (a) we know that $[H^+] = \left[HCO_3^{\pm}\right] = 1.5 \times 10^{-4}$.

$$5.6 \times 10^{-11} = \frac{(1.5 \times 10^{-4})\left[CO_3^{2-}\right]}{(1.5 \times 10^{-4})} = \left[CO_3^{2-}\right]$$

$$\left[CO_3^{2-}\right] = 5.6 \times 10^{-11} \ M$$

(c) (i) Adding HCl will increase [H+], lowering the pH.

(ii) From Le Châtelier's law, you can see that adding NaHCO$_3$ will cause the first equilibrium to shift to the left to try to use up the excess HCO$_3^-$. This will cause a decrease in [H+], raising the pH.

You may notice that adding NaHCO$_3$ will also cause the second equilibrium to shift toward the right, which should increase [H+], but because K_2 is much smaller than K_1, this shift is insignificant.

(iii) Adding NaOH will neutralize hydrogen ions, decreasing [H+] and raising the pH.

(iv) Adding NaCl will have no effect on the pH.

(d) The reaction in (d) is just the sum of the two reactions given. When two reactions can be added to give a third reaction, the equilibrium constants for those reactions can be multiplied to give K_{eq} for the third reaction.

$$K_{eq} = (K_1)(K_2) = (4.3 \times 10^{-7})(5.6 \times 10^{-11}) = 2.4 \times 10^{-17}$$

3. (a) $K_c = \dfrac{[NH_3]^2}{[N_2][H_2]^3}$

$K_c = \dfrac{(0.20)^2}{(0.50)(0.20)^3} = 10$

(b) Use the formula that relates the two constants.

$K_p = K_c(RT)^{\Delta n}$

Δn is the change in the number of moles of gas from reactants to products. So $n = -2$.

$K_p = (10)[(0.082)(570)]^{-2} = (10)(46.7)^{-2} = 4.7 \times 10^{-3}$

(c) (i) An increase in temperature favors the endothermic direction. In this case, that's the reverse reaction, so the concentration of H_2 will increase.

(ii) An increase in volume favors the direction that produces more moles of gas. In this case, that's the reverse direction, so the concentration of H_2 will increase.

(iii) According to Le Châtelier's law, increasing the concentration of the reactants forces the reaction to proceed in the direction that will use up the added reactants. In this case, adding the reactant N_2 will shift the reaction to the right and decrease the concentration of H_2.

(iv) The addition of He, a gas that takes no part in the reaction, will have no effect on the concentration of H_2.

4. (a) Use the solubility product.

$$K_{sp} = \left[Ca^{2+}\right]\left[CO_3^{2-}\right]$$

$$\left[Ca^{2+}\right] = \left[CO_3^{2-}\right] = x$$

$$2.8 \times 10^{-9} = x^2$$

$$x = \left[CO_3^{2-}\right] = 5.3 \times 10^{-5} M$$

(b) Use the solubility product.

(i) $K_{sp} = \left[Ca^{2+}\right]\left[SO_4^{2-}\right]$

$$\left[Ca^{2+}\right] = \left[SO_4^{2-}\right] = x$$

$$K_{sp} = 9.1 \times 10^{-6} = x^2$$

$$x = \left[SO_4^{2-}\right] = 3.0 \times 10^{-3} M$$

(ii) $K_{sp} = \left[Ca^{2+}\right]\left[CO_3^{2-}\right]$

Now use the value of $[Ca^{2+}]$ that you found in (b)(i).

$$\left[Ca^{2+}\right] = x = 3.0 \times 10^{-3} M$$

$$K_{sp} = 2.8 \times 10^{-9} = \left(3.0 \times 10^{-3}\right)\left[CO_3^{2-}\right]$$

$$\left[CO_3^{2-}\right] = 9.3 \times 10^{-7} M$$

(c) (i) The $CaCl_2$ dissociates completely, so the solution can be assumed to contain 0.2 moles of Ca^{2+} ions. We can ignore the ions from $CaCO_3$ and $CaSO_4$ because there are so few of them.

$$\text{Molarity} = \frac{\text{moles}}{\text{volume}}$$

$$\left[Ca^{2+}\right] = \frac{(0.20 \text{mol})}{(1 \text{ L})} = 0.20 M$$

(ii) Use K_{sp} again with the new value of $[Ca^{2+}]$.

$$K_{sp} = \left[Ca^{2+}\right]\left[SO_4^{2-}\right]$$

$$9.1 \times 10^{-6} = (0.20)\left[SO_4^{2-}\right]$$

$$\left[SO_4^{2-}\right] = 4.6 \times 10^{-5} M$$

(iii) Use K_{sp} again with the new value of $[Ca^{2+}]$.

$$K_{sp} = \left[Ca^{2+}\right]\left[CO_3^{2-}\right]$$

$$2.8 \times 10^{-9} = (0.20)\left[CO_3^{2-}\right]$$

$$\left[CO_3^{2-}\right] = 1.4 \times 10^{-8} M$$

Acids and Bases

HOW OFTEN DO ACIDS AND BASES APPEAR ON THE EXAM?

In the multiple-choice section, this topic appears in about 10 out of 75 questions. In the free-response section, this topic appears every year.

DEFINITIONS

ARRHENIUS

S. A. Arrhenius defined an acid as a substance that ionizes in water and produces hydrogen ions (H^+ ions). For instance, HCl is an acid.

$$HCl \rightarrow H^+ + Cl^-$$

He defined a base as a substance that ionizes in water and produces hydroxide ions (OH^- ions). For instance, NaOH is a base.

$$NaOH \rightarrow Na^+ + OH^-$$

BRØNSTED-LOWRY

J. N. Brønsted and T. M. Lowry defined an acid as a substance that is capable of donating a proton, which is the same as donating an H^+ ion, and they defined a base as a substance that is capable of accepting a proton.

Look at the reversible reaction below.

$$HC_2H_3O_2 + H_2O \leftrightarrow C_2H_3O_2^- + H_3O^+$$

According to Brønsted-Lowry

$HC_2H_3O_2$ and H_3O^+ are acids.

$C_2H_3O_2^-$ and H_2O are bases.

Now look at this reversible reaction.

$$NH_3 + H_2O \leftrightarrow NH_4^+ + OH^-$$

According to Brønsted-Lowry

NH_3 and OH^- are bases.

H_2O and NH_4^+ are acids.

So in each case, the species with the H^+ ion is the acid, and the same species without the H^+ ion is the base; the two species are called a **conjugate pair.** The following are the acid-base conjugate pairs in the reactions above:

$HC_2H_3O_2$ and $C_2H_3O_2^-$

NH_4^+ and NH_3

H_3O^+ and H_2O

H_2O and OH^-

Notice that water can act either as an acid or a base.

LEWIS

G. N. Lewis focused on electrons, and his definitions are the most broad of the acid-base definitions. Lewis defined a base as an electron pair donor and an acid as an electron pair acceptor; according to Lewis's rule, all of the Brønsted-Lowry bases above are also Lewis bases, and all of the Brønsted-Lowry acids are Lewis acids.

The following reaction is exclusively a Lewis acid-base reaction:

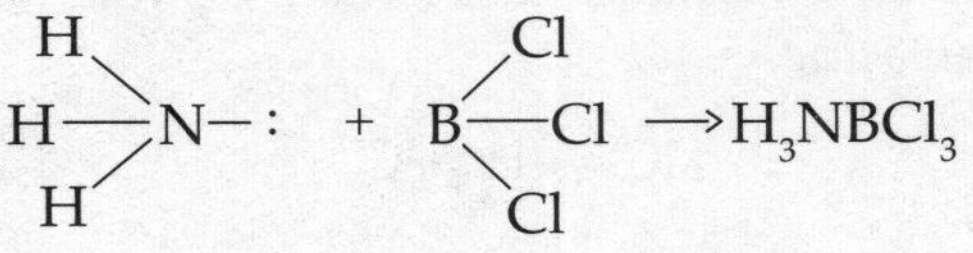

NH_3 is the Lewis base, donating its electron pair, and BCl_3 is the Lewis acid, accepting the electron pair.

pH

Many of the concentration measurements in acid-base problems are given to us in terms of pH and pOH.

$$p \text{ (anything)} = -\log \text{ (anything)}$$

$$pH = -\log [H^+]$$
$$pOH = -\log [OH^-]$$
$$pK_a = -\log K_a$$
$$pK_b = -\log K_b$$

In a solution

- when $[H^+] = [OH^-]$, the solution is neutral, and pH = 7

- when $[H^+]$ is greater than $[OH^-]$, the solution is acidic, and pH is less than 7

- when $[H^+]$ is less than $[OH^-]$, the solution is basic, and pH is greater than 7

It is important to remember that *increasing* pH means *decreasing* $[H^+]$, which means that there are fewer H^+ ions floating around and the solution is *less acidic*. Alternatively, *decreasing* pH means *increasing* $[H^+]$, which means that there are more H^+ ions floating around and the solution is *more acidic*.

WEAK ACIDS

When a weak acid is placed in water, a small fraction of its molecules will dissociate into hydrogen ions (H^+) and conjugate base ions (B^-). Most of the acid molecules will remain in solution as undissociated aqueous particles.

The dissociation constants, K_a and K_b, are measures of the strengths of weak acids and bases. K_a and K_b are just the equilibrium constants specific to acids and bases.

Acid Dissociation Constant
$$K_a = \frac{\left[H^+\right]\left[A^-\right]}{[HA]}$$
$[H^+]$ = molar concentration of hydrogen ions (M) $[A^-]$ = molar concentration of conjugate base ions (M) $[HA]$= molar concentration of undissociated acid molecules (M)

Base Dissociation Constant
$$K_b = \frac{\left[HA^+\right]\left[OH^-\right]}{[B]}$$
$[HA^+]$ = protonated base ions (M) $[OH^-]$ = molar concentration of hydroxide ions (M) $[B]$ = unprotonated base molecules (M)

The greater the value of K_a, the greater the extent of the dissociation of the acid and the stronger the acid. The same thing goes for K_b.

If you know the K_a for an acid and the concentration of the acid, you can find the pH. For instance, let's look at 0.20-molar solution of $HC_2H_3O_2$, with $K_a = 1.8 \times 10^{-5}$.

First we set up the K_a equation, plugging in values.

$$HC_2H_3O_2 \rightarrow H^+ + C_2H_3O_2^-$$

$$K_a = \frac{[H^+][C_2H_3O_2^-]}{[HC_2H_3O_2]}$$

Therefore, the ICE (Initial, Change, Equilibrium) table for the above problem is as follows:

	$[HC_2H_3O_2]$	$[H^+]$	$[C_2H_3O_2^-]$
Initial	0.20	0.0	0.0
Change	$-x$	$+x$	$+x$
Equilibrium	$0.20 - x$	x	x

Because every acid molecule that dissociates produces one H^+ and one $C_2H_3O_2^-$,

$$[H^+] = [C_2H_3O_2^-] = x$$

and because, strictly speaking, the molecules that dissociate should be subtracted from the initial concentration of $HC_2H_3O_2$, $[HC_2H_3O_2]$ should be ($0.20\ M - x$). In practice, however, x is almost always insignificant compared with the initial concentration of acid, so we just use the initial concentration in the calculation.

$$[HC_2H_3O_2] = 0.20\ M$$

Now we can plug our values and variable into the K_a expression.

$$1.8 \times 10^{-5} = \frac{x^2}{0.20}$$

Solve for x.

$$x = [H^+] = 1.9 \times 10^{-3}$$

Now that we know $[H^+]$, we can calculate the pH.

$$pH = -\log [H^+] = -\log (1.9 \times 10^{-3}) = 2.7$$

This is the basic approach to solving many of the weak acid/base problems that will be on the test.

STRONG ACIDS

Strong acids dissociate completely in water, so the reaction goes to completion and they never reach equilibrium with their conjugate bases. Because there is no equilibrium, there is no equilibrium constant, so there is no dissociation constant for strong acids or bases.

Important Strong Acids
HCl, HBr, HI, HNO_3, $HClO_4$, H_2SO_4
Important Strong Bases
LiOH, NaOH, KOH, $Ba(OH)_2$, $Sr(OH)_2$

Because the dissociation of a strong acid goes to completion, there is no tendency for the reverse reaction to occur, which means that the conjugate base of a strong acid must be extremely weak.

Oxoacids are acids that contain oxygen. The greater the number of oxygen atoms attached to the central atom in an oxoacid, the stronger the acid. For instance, $HClO_4$ is stronger than $HClO_3$, which is stronger than $HClO_2$. That's because increasing the number of oxygen atoms that are attached to the central atom weakens the attraction that the central atom has for the H^+ ion.

It's much easier to find the pH of a strong acid solution than it is to find the pH of a weak acid solution. That's because strong acids dissociate completely, so the final concentration of H^+ ions will be the same as the initial concentration of the strong acid.

Let's look at a 0.010-molar solution of HCl.

HCl dissociates completely, so $[H^+] = 0.010\ M$

$pH = -\log [H^+] = -\log (0.010) = -\log (10^{-2}) = 2$

So you can always find the pH of a strong acid solution directly from its concentration.

K_w

Water comes to equilibrium with its ions according to the following reaction:

$$H_2O(l) \rightleftharpoons H^+(aq) + OH^-(aq) \qquad K_w = 1 \times 10^{-14} \text{ at } 25°C.$$
$$K_w = 1 \times 10^{-14} = [H^+][OH^-]$$
$$pH + pOH = 14$$

The common ion effect tells us that the hydrogen ion and hydroxide ion concentrations for any acid or base solution must be consistent with the equilibrium for the ionization of water. That is, no matter where the H^+ and OH^- ions came from, when you multiply $[H^+]$ and $[OH^-]$, you must get 1×10^{-14}. So for any aqueous solution, if you know the value of $[H^+]$, you can find out the value of $[OH^-]$ and vice versa.

The acid and base dissociation constants for conjugates must also be consistent with the equilibrium for the ionization of water.

$$K_w = 1 \times 10^{-14} = K_a K_b$$
$$pK_a + pK_b = 14$$

So if you know K_a for a weak acid, you can find K_b for its conjugate base and vice versa.

ACID AND BASE SALTS

A salt solution can be acidic or basic, depending on the identities of the anion and cation in the salt.

- *If a salt is composed of the conjugates of a strong base and a strong acid, its solution will be neutral.*
  For example, let's look at NaCl.
  Na^+ is the conjugate acid of NaOH (a strong base), and Cl^- is the conjugate base of HCl (a strong acid).

  $$NaCl(s) \rightarrow Na^+(aq) + Cl^-(aq) \qquad\qquad pH = 7$$

  That's because neither ion will react in water because, as you may recall, the conjugates of strong acids and bases are very weak and unreactive.

- *If a salt is composed of the conjugates of a weak base and a strong acid, its solution will be acidic.*
  For example, look at NH_4Cl.
  NH_4^+ is the conjugate acid of NH_3 (a weak base), and Cl^- is the conjugate base of HCl (a strong acid).

  $$NH_4Cl(s) \rightarrow NH_4^+(aq) + Cl^-(aq) \qquad pH < 7$$

  When NH_4Cl ionizes in water, the Cl^- ions won't react at all, but the NH_4^+ will dissociate to some extent to produce some H^+ ions, as shown in the reaction below, making the solution acidic.

  $$NH^+ + H_2O \rightarrow NH_3 + H_3O^+$$

- *If a salt is composed of the conjugates of a strong base and a weak acid, its solution will be basic.*
  For instance, look at $NaC_2H_3O_2$.
  Na^+ is the conjugate acid of NaOH (a strong base), and $C_2H_3O_2^-$ is the conjugate base of $HC_2H_3O_2$ (a weak acid).

  $$NaC_2H_3O_2(s) \rightarrow Na^+(aq) + C_2H_3O_2^-(aq) \qquad pH > 7$$

  When $NaC_2H_3O_2$ ionizes in water, the Na^+ ions don't react at all, but the $C_2H_3O_2^-$ ions will react with water according to the following reaction:

  $$C_2H_3O_2^-(aq) + H_2O(l) \rightarrow HC_2H_3O_2(aq) + OH^-(aq)$$

  This is an acid-base reaction, which produces some OH^- ions and increases the pH.

- *If a salt is composed of the conjugates of a weak base and a weak acid, the pH of its solution will depend on the relative strengths of the conjugate acid and base of the specific ions in the salt.*

Calculations with salts aren't much different from regular acid-base calculations, except that you may need to convert between K_a and K_b and between pH and pOH.

Remember:

$$K_a K_b = 1 \times 10^{-14}, \text{ and pH + pOH = 14}$$

Let's find the pH of a 0.10-molar solution of $NaC_2H_3O_2$. K_a for $HC_2H_3O_2$ is 1.8×10^{-5}. The solution will be basic, so let's find K_b for $C_2H_3O_2^-$.

$$K_b = \frac{\left(1.0 \times 10^{-14}\right)}{K_a} = \frac{\left(1.0 \times 10^{-14}\right)}{\left(1.8 \times 10^{-5}\right)} = 5.6 \times 10^{-10}$$

Therefore, the ICE table for the above equation is as follows:

	$[HC_2H_3O_2]$	$[OH^-]$	$[C_2H_3O^{2-}]$
Initial	0.10	0.0	0.0
Change	$-x$	$+x$	$+x$
Equilibrium	$0.10 - x$	x	x

Now we can use the base dissociation expression.

$$K_b = \frac{\left[HC_2H_3O_2\right]\left[OH^-\right]}{\left[C_2H_3O_2^-\right]}$$

Every $C_2H_3O_2^-$ polyatomic ion that reacts with a water molecule produces one $HC_2H_3O_2$ and one OH^-, so

$$[HC_2H_3O_2] = [OH^-] = x$$

Again, we assume that x is insignificant compared with the initial concentration of $C_2H_3O_2^-$; 0.10 M, so

$$[C_2H_3O_2^-] = (0.10\ M - x) = 0.10\ M$$

Now we can plug our values and variable into the K_b expression.

$$5.6 \times 10^{-10} = \frac{x^2}{0.10}$$

Solve for x.

$$x = [OH^-] = 7.5 \times 10^{-6}$$

Once we know the value of $[OH^-]$, we can calculate the pOH.

$$\text{pOH} = -\log[OH^-] = -\log(7.5 \times 10^{-6}) = 5.1$$

Now we use pOH to find pH.

$$\text{pH} = 14 - \text{pOH} = 14 - 5.1 = 8.9$$

BUFFERS

A **buffer** is a solution with a very stable pH. You can add acid or base to a buffer solution without greatly affecting the pH of the solution. The pH of a buffer will also remain unchanged if the solution is diluted with water or if water is lost through evaporation.

A buffer is created by placing a large amount of a weak acid or base into a solution along with its conjugate, in the form of salt. A weak acid and its conjugate base can remain in solution together without neutralizing each other. This is called the **common ion effect**.

When both the acid and the conjugate base are together in the solution, any hydrogen ions that are added will be neutralized by the base, while any hydroxide ions that are added will be neutralized by the acid, without this having much of an effect on the solution's pH.

When dealing with buffers, it is useful to rearrange the equilibrium constant to create the **Henderson-Hasselbalch equation**.

The Henderson-Hasselbalch Equation

$$pH = pK_a + \log \frac{\left[A^-\right]}{[HA]}$$

[HA] = molar concentration of undissociated weak acid (M)

[A$^-$] = molar concentration of conjugate base (M)

$$pOH = pK_b + \log \frac{\left[HB^+\right]}{[B]}$$

[B] = molar concentration of weak base (M)

[HB$^+$] = molar concentration of conjugate acid (M)

Let's say we have a buffer solution with concentrations of 0.20 M $HC_2H_3O_2$ and 0.50 M $C_2H_3O_2^-$. The acid dissociation constant for $HC_2H_3O_2$ is 1.8×10^{-5}. Let's find the pH of the solution.

We can just plug the values we have into the Henderson-Hasselbalch equation for acids.

$$pH = pK_a + \log \frac{\left[C_2H_3O_2^-\right]}{[HC_2H_3O]}$$

$$pH = -\log (1.8 \times 10^{-5}) + \log \frac{(0.50M)}{(0.20)}$$

$$pH = -\log (1.8 \times 10^{-5}) + \log (2.5)$$

$$pH = (4.7) + (0.40) = 5.1$$

Now let's see what happens when [$HC_2H_3O_2$] and [$C_2H_3O_2^-$] are both equal to 0.20 M.

$$pH = pK_a + \log \frac{\left[C_2H_3O_2^-\right]}{[HC_2H_3O]}$$

$$pH = -\log (1.8 \times 10^{-5}) + \log \frac{(0.20M)}{(0.20M)}$$

$$pH = -\log (1.8 \times 10^{-5}) + \log (1)$$

$$pH = (4.7) + (0) = 4.7$$

Notice that when the concentrations of acid and conjugate base in a solution are the same, $pH = pK_a$ (and $pOH = pK_b$). When you choose an acid for a buffer solution, it is best to pick an acid with a pK_a that is close to the desired pH. That way you can have almost equal amounts of acid and conjugate base in the solution, which will make the buffer as flexible as possible in neutralizing both added H^+ and OH^-.

POLYPROTIC ACIDS AND AMPHOTERIC SUBSTANCES

Some acids, such as H_2SO_4 and H_3PO_4, can give up more than one hydrogen ion in solution. These are called **polyprotic** acids.

Polyprotic acids are always more willing to give up their first protons than later protons. For example, H_3PO_4 gives up an H^+ ion (proton) more easily than does $H_2PO_4^{\pm}$, so H_3PO_4 is a stronger acid. In the same way, $H_2PO_4^{\pm}$ is a stronger acid than $HPO_4^{2\pm}$.

Substances that can act as either acids or bases are called **amphoteric** substances.
For instance

- $H_2PO_4^-$ can act as an acid, giving up a proton to become $HPO_4^{2\pm}$, or it can act as a base, accepting a proton to become H_3PO_4

- HSO_4^- can act as an acid, giving up a proton to become SO_4^{2-}, or it can act as a base, accepting a proton to become H_2SO_4

- H_2O can act as an acid, giving up a proton to become OH^-, or it can act as a base, accepting a proton to become H_3O^+

ANHYDRIDES

An **acid anhydride** is a substance that combines with water to form an acid. Generally, oxides of nonmetals are acid anhydrides.

$$CO_2 + H_2O \rightarrow H_2CO_3$$

$$SO_3 + H_2O \rightarrow H^+ + HSO_4^{\pm}$$

A **basic anhydride** is a substance that combines with water to form a base. Generally, oxides of metals are basic anhydrides.

$$CaO + H_2O \rightarrow Ca(OH)_2$$

$$Na_2O + H_2O \rightarrow 2\ Na^+ + 2\ OH^-$$

TITRATION

When an acid and a base are mixed, a **neutralization reaction** occurs. Neutralization reactions can be written in the following form:

$$\text{Acid} + \text{Base} \rightarrow \text{Water} + \text{Salt}$$

Neutralization reactions are generally performed by titration, where a base of known concentration is slowly added to an acid (or vice versa). The progress of a neutralization reaction can be shown in a titration curve. The diagram below shows the titration of a strong acid by a strong base.

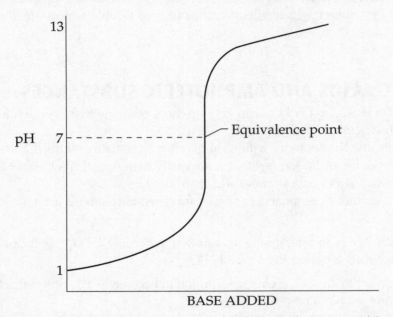

In the diagram above, the pH increases slowly but steadily from the beginning of the titration until just before the equivalence point. The **equivalence point** is the point in the titration when exactly enough base has been added to neutralize all the acid that was initially present. Just before the equivalence point, the pH increases sharply as the last of the acid is neutralized. The equivalence point of a titration can be recognized through the use of an indicator. An indicator is a substance that changes color over a specific pH range. When choosing an indicator, it's important to make sure the equivalence point falls within the pH range for the color change.

For this titration, the pH at the equivalence point is exactly 7 because the titration of a strong acid by a strong base produces a neutral salt solution.

The following diagram shows the titration of a weak acid by a strong base:

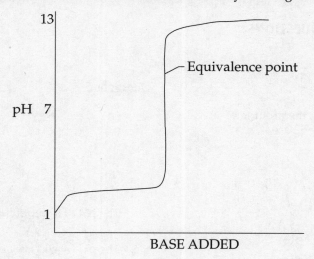

In this diagram, the pH increases more quickly at first, then levels out into a buffer region. At the center of the buffer region is the **half-equivalence point.** At this point, enough base has been added to convert exactly half of the acid into conjugate base; here the concentration of acid is equal to the concentration of conjugate base (pH = pK_a). The curve remains fairly flat until just before the equivalence point, when the pH increases sharply. For this titration, the pH at the equivalence point is greater than 7 because the titration of a weak acid by a strong base produces a basic salt solution.

The following diagram shows the titration curve of a polyprotic acid:

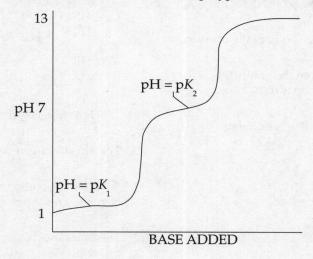

For a polyprotic acid, the titration curve will have as many bumps as there are hydrogen ions to give up. The curve above has two bumps, so it represents the titration of a diprotic acid.

CHAPTER 11 QUESTIONS

MULTIPLE-CHOICE QUESTIONS

Questions 1–4

The diagram below shows the titration of a weak monoprotic acid by a strong base.

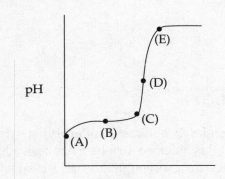

Base Added

1. At this point in the titration, the pH of the solution is equal to the pK_a of the acid.

2. This is the equivalence point of the titration.

3. Of the points shown on the graph, this is the point when the solution is most basic.

4. At this point the solution is buffered.

Questions 5–8

(A) 1
(B) 3
(C) 7
(D) 11
(E) 13

5. The pH of a solution with a pOH of 11

6. The pH of a 0.1-molar solution of HCl

7. The pH of a 0.001-molar solution of HNO_3

8. The pH of a 0.1-molar solution of NaOH

Questions 9–12

(A) HNO_3
(B) HCN
(C) H_2CO_3
(D) HF
(E) H_2O

9. This is a strong electrolyte.

10. This substance can act as a Lewis base.

11. A 0.1-molar solution of this substance will have the lowest pH of the substances listed.

12. This substance is formed when carbon dioxide is dissolved in water.

13. What is the pH of a 0.01-molar solution of NaOH?

    (A) 1
    (B) 2
    (C) 8
    (D) 10
    (E) 12

14. What is the volume of 0.05-molar HCl that is required to neutralize 50 ml of a 0.10-molar $Mg(OH)_2$ solution?

    (A) 100 ml
    (B) 200 ml
    (C) 300 ml
    (D) 400 ml
    (E) 500 ml

15. Which of the following best describes the pH of a 0.01-molar solution of HBrO? (For HBrO, $K_a = 2 \times 10^{-9}$)

    (A) Less than or equal to 2
    (B) Between 2 and 7
    (C) 7
    (D) Between 7 and 11
    (E) Greater than or equal to 11

16. A 0.5-molar solution of which of the following salts will have the lowest pH?

    (A) KCl
    (B) $NaC_2H_3O_2$
    (C) NaI
    (D) KNO_3
    (E) NH_4Cl

17. Which of the following salts will produce a solution with a pH of greater than 7 when placed in distilled water?

    (A) NaCN
    (B) KCl
    (C) $NaNO_3$
    (D) NH_4NO_3
    (E) KI

18. A laboratory technician wishes to create a buffered solution with a pH of 5. Which of the following acids would be the best choice for the buffer?

    (A) $H_2C_2O_4$, $K_a = 5.9 \times 10^{-2}$
    (B) H_3AsO_4, $K_a = 5.6 \times 10^{-3}$
    (C) $H_2C_2H_3O_2$, $K_a = 1.8 \times 10^{-5}$
    (D) HOCl, $K_a = 3.0 \times 10^{-8}$
    (E) HCN, $K_a = 4.9 \times 10^{-10}$

19. Which of the following species is amphoteric?

    (A) H^+
    (B) CO_3^{2-}
    (C) HCO_3^-
    (D) H_2CO_3
    (E) H_2

20. How many liters of distilled water must be added to 1 liter of an aqueous solution of HCl with a pH of 1 to create a solution with a pH of 2?

    (A) 0.1 L
    (B) 0.9 L
    (C) 2 L
    (D) 9 L
    (E) 100 L

21. A 1-molar solution of a very weak monoprotic acid has a pH of 5. What is the value of K_a for the acid?

    (A) $K_a = 1 \times 10^{-10}$
    (B) $K_a = 1 \times 10^{-7}$
    (C) $K_a = 1 \times 10^{-5}$
    (D) $K_a = 1 \times 10^{-2}$
    (E) $K_a = 1 \times 10^{-1}$

22. The value of K_a for $HSO_4^{\pm}$ is 1×10^{-2}. What is the value of K_b for SO_4^{2-}?

    (A) $K_b = 1 \times 10^{-12}$

    (B) $K_b = 1 \times 10^{-8}$

    (C) $K_b = 1 \times 10^{-2}$

    (D) $K_b = 1 \times 10^2$

    (E) $K_b = 1 \times 10^5$

23. How much 0.1-molar NaOH solution must be added to 100 milliliters of a 0.2-molar H_2SO_3 solution to neutralize all of the hydrogen ions in H_2SO_3?

    (A) 100 ml
    (B) 200 ml
    (C) 300 ml
    (D) 400 ml
    (E) 500 ml

24. The concentrations of which of the following species will be increased when HCl is added to a solution of $HC_2H_3O_2$ in water?

    I. H^+
    II. $C_2H_3O_2^-$
    III. $HC_2H_3O_2$

    (A) I only
    (B) I and II only
    (C) I and III only
    (D) II and III only
    (E) I, II, and III

25. Which of the following species is amphoteric?

    (A) HNO_3
    (B) $HC_2H_3O_2$
    (C) HSO_4^-
    (D) H_3PO_4
    (E) ClO_4^-

26. If 0.630 grams of HNO_3 (molecular weight 63.0) are placed in 1 liter of distilled water at 25°C, what will be the pH of the solution? (Assume that the volume of the solution is unchanged by the addition of the HNO_3.)

    (A) 0.01
    (B) 0.1
    (C) 1
    (D) 2
    (E) 3

27. Which of the following procedures will produce a buffered solution?

    I. Equal volumes of 0.5 M NaOH and 1 M HCl solutions are mixed.
    II. Equal volumes of 0.5 M NaOH and 1 M $HC_2H_3O_2$ solutions are mixed.
    III. Equal volumes of 1 M $NaC_2H_3O_2$ and 1 M $HC_2H_3O_2$ solutions are mixed.

    (A) I only
    (B) III only
    (C) I and II only
    (D) II and III only
    (E) I, II, and III

28. Which of the following compounds will most likely act as a *Lewis* acid?

(A) $AlBr_3$
(B) CH_4
(C) NH_3
(D) PCl_3
(E) Xe

29. The following titration curve shows the titration of a weak base with a strong acid:

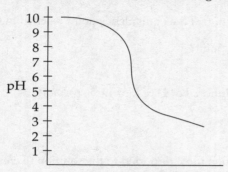

Mol acid titrated

Which of the following values most accurately approximates the pK_b of the weak base?

(A) 9.8
(B) 8.4
(C) 7
(D) 3.3
(E) 4.2

30. Which of the following expressions is equal to the hydrogen ion concentration of a 1-molar solution of a very weak monoprotic acid, HA, with an ionization constant K_a?

(A) K_a

(B) K_a^2

(C) $2K_a$

(D) $2K_a^2$

(E) $\sqrt{K_a}$

PROBLEMS

1. A beaker contains 100 milliliters of a solution of hypochlorous acid, HOCl, of unknown concentration.

   (a) The solution was titrated with 0.100 molar NaOH solution, and the equivalence point was reached when 40.0 milliliters of NaOH solution was added. What was the original concentration of the HOCl solution?

   (b) If the original HOCl solution had a pH of 4.46, what is the value of K_a for HOCl?

   (c) What percent of the HOCl molecules were ionized in the original solution?

   (d) What is the concentration of OCl⁻ ions in the solution at the equivalence point reached in (a)?

   (e) What is the pH of the solution at the equivalence point?

2. A vessel contains 500 milliliters of a 0.100-molar H_2S solution. For H_2S, $K_1 = 1.0 \times 10^{-7}$ and $K_2 = 1.3 \times 10^{-13}$.

   (a) What is the pH of the solution?

   (b) How many milliliters of 0.100-molar NaOH solution must be added to the solution to create a solution with a pH of 7?

   (c) What will be the pH when 800 milliliters of 0.100-molar NaOH has been added?

   (d) What is the value of K_{eq} for the following reaction?

   $$H_2S(aq) \leftrightarrow 2H^+(aq) + S^{2-}(aq)$$

3. A 100 milliliter sample of 0.100-molar NH_4Cl solution was added to 80 milliliters of a 0.200-molar solution of NH_3. The value of K_b for ammonia is 1.79×10^{-5}.

   (a) What is the value of pK_b for ammonia?

   (b) What is the pH of the solution described in the question?

   (c) If 0.200 grams of NaOH were added to the solution, what would be the new pH of the solution? (Assume that the volume of the solution does not change.)

   (d) If equal molar quantities of NH_3 and NH_4^+ were mixed in solution, what would be the pH of the solution?

ESSAYS

4.
$$H_3PO_4 \rightleftharpoons H^+ + H_2PO_4^- \qquad K_1 = 7.5 \times 10^{-3}$$
$$H_2PO_4^- \rightleftharpoons H^+ + HPO_4^{2\pm} \qquad K_2 = 6.2 \times 10^{-8}$$
$$HPO_4^{2-} \rightleftharpoons H^+ + PO_4^{3-} \qquad K_3 = 2.2 \times 10^{-13}$$

(a) Choose an amphoteric species from the reactions listed above, and give its conjugate acid and its conjugate base.

(b) Explain why the dissociation constant decreases with each hydrogen ion lost.

(c) Of the acids listed above, which would be most useful in creating a buffer solution with a pH of 7.5?

(d) Sketch the titration curve that results when H_3PO_4 is titrated with excess NaOH and label the two axes.

5. Use the principles of acid-base theory to answer the following questions:

(a) Predict whether a 0.1-molar solution of sodium acetate, $NaC_2H_3O_2$, will be acidic or basic, and give a reaction occurring with water that supports your conclusion.

(b) Predict whether a 0.1-molar solution of ammonium chloride, NH_4Cl, will be acidic or basic, and give a reaction occurring with water that supports your conclusion.

(c) Explain why buffer solutions are made with weak acids instead of strong acids.

(d) Explain why, although oxygen occupies the same position in both KOH and HBrO, KOH is a base and HBrO is an acid.

6. Identify which one of the acids listed in pairs below is the stronger of the two, and explain why.

(a) HF and HCl

(b) H_2SO_3 and HSO_3^-

(c) HNO_3 and HNO_2

(d) $HClO_4$ and $HBrO_4$

CHAPTER 11 ANSWERS AND EXPLANATIONS

MULTIPLE-CHOICE QUESTIONS

1. **B** At this point, sometimes called the half-equivalence point, enough base has been added to neutralize half of the weak acid. That means that at this point, the concentration of the acid—let's call it HA—will be equal to the concentration of the conjugate base, A$^-$.

   Now let's look at the equilibrium expression.

   $$K_a = \frac{[H^+][A^-]}{[HA]}$$

   If [HA] = [A$^-$], then they cancel and K_a = [H$^+$].

   So pH = pK_a.

2. **D** The point in the middle of the steep rise in a titration curve is the equivalence point. That's the point when exactly enough base has been added to neutralize all of the acid that was originally in the solution.

3. **E** This is the point with the highest pH, so at this point, the solution is most basic.

4. **B** A buffered solution resists changes to its pH, so the flat part of the titration curve is the buffer region because the pH of the solution is changing very little even when base is being added.

   At (B), the solution contains a large quantity of both the acid HA, which absorbs added base, and its conjugate base, A$^-$, which absorbs added acid.

5. **B** In any aqueous solution at 25°C, pH + pOH = 14.

   So pH + 3 = 14 and pH = 3.

6. **A** HCl is a strong acid and dissociates completely, so the hydrogen ion concentration is equal to the molarity.

   So [H$^+$] = 0.1 M

   pH = $-$log[H$^+$] = $-$log(0.1) = 1

7. **B** HNO$_3$ is a strong acid and dissociates completely, so the hydrogen ion concentration is equal to the molarity.

   So [H$^+$] = 0.001 M

   pH = $-$log[H$^+$] = $-$log(0.001) = 3

8. **E** NaOH is a strong acid and dissociates completely, so the hydroxide ion concentration [OH$^-$] is equal to the molarity.

   So [OH$^-$] = 0.1 M

   pOH = $-$log[OH$^-$] = $-$log(0.1) = 1

   Now, pH + pOH = 14

   So pH + 1 = 14 and pH = 13

9. **A** A strong electrolyte is a substance that dissociates to form a lot of ions. HNO$_3$ is the only substance listed that ionizes completely in solution.

10. **E** Water can act as a Lewis base by donating an electron pair to a hydrogen ion to form H_3O^+.

11. **A** HNO_3 is the only strong acid listed, meaning that it is the only acid listed that ionizes completely, so it will have the lowest pH. By the way, in a 0.1 M solution of a strong acid, the hydrogen ion concentration will be 0.1 M, so the pH will be equal to 1.

12. **C** H_2CO_3, carbonic acid, is formed in the reaction between CO_2 and H_2O.

13. **E** NaOH is a strong base, so it can be assumed to dissociate completely. That means that the OH^- concentration will also be 0.01 M.

   $pOH = -\log[OH^-]$, so pOH = 2, but we're looking for the pH.

   In an aqueous solution, pH + pOH = 14, so pH = 12.

14. **B** Every mole of $Mg(OH)_2$ molecules dissociates to produce 2 moles of OH^- ions, so a 0.10 M $Mg(OH)_2$ solution will be a 0.20 M OH^- solution.

   The solution will be neutralized when the number of moles of H^+ ions added is equal to the number of OH^- ions originally in the solution.

   Moles = (molarity)(volume)

   Moles of OH^- = (0.20 M)(50 ml) = 10 millimoles = moles of H^+ added

   $$Volume = \frac{moles}{molarity}$$

   $$Volume \ of \ HCl = \frac{(10\,millimoles)}{(0.05\,M)} = 200\,ml$$

15. **B** You can eliminate (C), (D), and (E) by using common sense. HBrO is a weak acid, so an HBrO solution will be acidic, with a pH of less than 7.

   To choose between (A) and (B) you have to remember that HBrO is a weak acid. If HBrO were a strong acid, it would dissociate completely and $[H^+]$ would be equal to 0.01-molar, for a pH of exactly 2. Because HBrO is a weak acid, it will not dissociate completely and $[H^+]$ will be less than 0.01-molar, which means that the pH will be greater than 2. So by POE, (B) is the answer.

16. **E** To find the lowest pH, we should look for the salt that produces an acidic solution.

   The salt composed of the conjugate of a weak base (NH_4^+ is the conjugate of NH_3) and the conjugate of a strong acid (Cl^- is the conjugate of HCl) will produce an acidic solution.

   As for the other choices, (A), (C), and (D) are composed of conjugates of strong acids and bases and will produce neutral solutions.

   (B) is composed of the conjugate of a strong base (Na^+) and the conjugate of a weak acid ($C_2H_3O_2^-$) and will produce a basic solution.

17. **A** A pH of greater than 7 means that the solution is basic.

   The salt composed of the conjugate of a strong base (Na^+ is the conjugate of NaOH) and the conjugate of a weak acid (CN^- is the conjugate of HCN) will produce a basic solution.

   As for the other choices, (B), (C), and (E) are composed of conjugates of strong acids and bases and will produce neutral solutions.

   (D) is composed of the conjugate of a weak base (NH_4^+) and the conjugate of a strong acid (NO_3^-) and will produce an acidic solution.

18. **C** The best buffered solution occurs when pH = pK_a. That happens when the solution contains equal amounts of acid and conjugate base. If you want to create a buffer with a pH of 5, the best choice would be an acid with a pK_a that is as close to 5 as possible. You shouldn't have to do a calculation to see that the pK_a for choice (C) is much closer to 5 than that of any of the others.

19. **C** An amphoteric species can act either as an acid or a base, gaining or losing a proton.

    HCO_3^- can act as an acid, losing a proton to become CO_3^{2-}, or it can act as a base, gaining a proton to become H_2CO_3.

20. **D** We want to change the hydrogen ion concentration from 0.1 M (pH of 1) to 0.01 M (pH of 2).

    The HCl is completely dissociated, so the number of moles of H^+ will remain constant as we dilute the solution.

    Moles = (molarity)(volume) = Constant

    $(M_1)(V_1) = (M_2)(V_2)$

    $(0.1\ M)(1\ L) = (0.01\ M)(V_2)$

    So, $V_2 = 10$ L, which means that 9 L must be added.

21. **A** A pH of 5 means that $[H^+] = 1 \times 10^{-5}$

    $$K_a = \frac{[H^+][A^-]}{[HA]}$$

    For every HA that dissociates, we get one H^+ and one A^-, so $[H^+] = [A^-] = 1 \times 10^{-5}$.

    The acid is weak, so we can assume that very little HA dissociates and that the concentration of HA remains 1-molar.

    $$K_a = \frac{[H^+][A^-]}{[HA]} = \frac{(1 \times 10^{-5})(1 \times 10^{-5})}{(1)} = 1 \times 10^{-10}.$$

22. **A** For conjugates, $(K_a)(K_b) = K_w = 1 \times 10^{-14}$

    $$K_b = \frac{K_w}{K_a} = \frac{(1 \times 10^{-14})}{(1 \times 10^{-2})} = 1 \times 10^{-12}$$

23. **D** First let's find out how many moles of H^+ ions we need to neutralize.

    Every H_2SO_3 will produce 2 H^+ ions, so for our purposes, we can think of the solution as a 0.4-molar H^+ solution.

    Moles = (molarity)(volume)

    Moles of H^+ = (molarity)(volume) = (0.4 M)(100 ml) = 40 millimoles = moles of OH^- required

    $$\text{Volume of NaOH} = \frac{\text{moles}}{\text{molarity}} = \frac{(40\ \text{millimoles})}{(0.1\ M)} = 400\ \text{ml}$$

24. **C** Let's look at the equilibrium expression for the dissociation of $HC_2H_3O_2$.

    $$K_a = \frac{[H^+][C_2H_3O_2^-]}{[HC_2H_3O_2^-]}$$

    When HCl is added to the solution, H^+ ions are added and $[H^+]$ will increase, so (I) is correct.

    Because of Le Châtelier's law, the equilibrium will shift to consume some of the added H^+ ions, so some of the H^+ ions will combine with $C_2H_3O_2^-$ ions to form more $HC_2H_3O_2$.

    So $[C_2H_3O_2^-]$ will decrease, making (II) wrong, and $[HC_2H_3O_2]$ will increase, making (III) correct.

25. **C** An amphoteric species can act either as a base and gain an H^+ or as an acid and lose an H^+.

HSO_4^- can lose an H^+ to become SO_4^{2-} or it can gain an H^+ to become H_2SO_4.

HNO_3 (A), $HC_2H_3O_2$ (B), and H_3PO_4 (D) can lose only an H^+.

ClO_4^- (E) can gain only an H^+.

26. **D** Every unit of HNO_3 added to the solution will place 1 unit of H^+ ions in the solution. So first find the moles of HNO_3 added.

$$\text{Moles} = \frac{\text{grams}}{\text{MW}}$$

$$\text{Moles of } H^+ = \frac{0.630 \text{ grams}}{63.0 \text{ g / mole}} = 0.01 \text{ moles}$$

Now it's easy to find the H^+ concentration.

$$\text{Molarity} = \frac{\text{moles}}{\text{liters}}$$

$$[H^+] = \frac{0.01 \text{ moles}}{1 \text{ L}} = 0.01 \, M$$

$$pH = -\log [H^+] = -\log(0.01) = 2$$

27. **D** A buffered solution can be prepared by mixing a weak acid with an equal amount of its conjugate or by adding enough strong base to neutralize half of the weak acid present in a solution.

In (I), equal amounts of a strong acid and base are mixed. They'll neutralize each other completely to produce salt water, which is not a buffer.

In (II), enough strong base is added to neutralize half of the weak acid. This will leave equal amounts of weak acid and its conjugate base, producing a buffered solution.

In (III), equal amounts of a weak acid and its conjugate base are mixed. This will produce a buffered solution.

28. **A** A Lewis acid must be able to accept a pair of electrons. By drawing the Lewis structure of $AlBr_3$, shown below, one can see that Al lacks a full octet and therefore is a very strong Lewis acid.

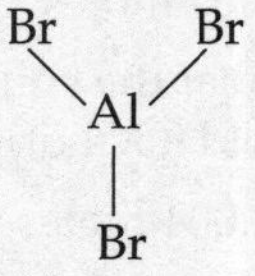

NH_3 is a good Lewis base, since it has a lone pair of electrons to donate. Both CH_4 and Xe have full octets, and will not accept more electrons. The compound PCl_3 looks a lot like NH_3, and has a lone pair of electrons and acts as a Lewis base.

29. **E** The half equivalence point of this titration, shown in the plot below, is around pH = 9.8.

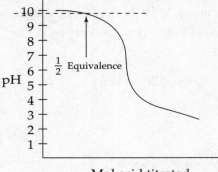

Mol acid titrated

However, this is not the pK_b of the weak base. Remember that the Henderson-Hasselbalch equation relates pH with pK_a, not pK_b. At the half equivalence point shown, the amount of base (which we can call [A⁻] as we do when talking about acids and pH values) is equal to the neutralized base [HA]. When these two quantities are equal, the H-H equation tells us that the pH = the pK_a of the acid form, which when subtracted from 14 gives the pK_b of the base. So, the pK_a of the acid form is 9.8, so the pK_b of the base is 4.2.

30. **E** Use the equilibrium expression

$$K_a = \frac{[H^+][A^\pm]}{[HA]} = \frac{y^2}{(1)}$$

For every HA that dissociates, we get one H⁺ and one A⁻, so [H⁺] = [A⁻] = y.

The acid is weak, so we can assume that very little HA dissociates and that the concentration of HA remains 1-molar.

So [H⁺] = $y = \sqrt{K_a}$

PROBLEMS

1. (a) First let's find out how many moles of NaOH were added.

Moles = (molarity)(volume)

Moles of NaOH = (0.100 M)(.040 L) = 0.004 moles

Every OH⁻ ion neutralizes one H⁺ ion, so at the equivalence point, the number of moles of NaOH added is equal to the number of moles of HOCl originally present.

(moles of NaOH) = (moles of HOCl) = 0.004 moles

Now we can find the original concentration of the HOCl solution.

$$Molarity = \frac{moles}{volume}$$

$$[HOCl] = \frac{(0.004 \text{ mol})}{(0.100 \text{ L})} = 0.040 \ M$$

(b) Use the equilibrium expression.

$$K_a = \frac{\left[H^+\right]\left[OCl^-\right]}{\left[HOCl^-\right]}$$

$[H^+] = 10^{-pH} = 10^{-4.46} = 3.47 \times 10^{-5} \, M$

$[H^+] = [OCl^-] = 3.47 \times 10^{-5} \, M$

$[HOCl] = 0.040 \, M$

$$K_a = \frac{\left(3.47 \times 10^{-5} M\right)^2}{\left(0.040 \, M - 3.47 \times 10^{-5} \, M\right)}$$

0.040 is much larger than 3.47×10^{-5}, so we can simplify the expression.

$$K_a = \frac{\left(3.47 \times 10^{-5} M\right)^2}{\left(0.040 \, M\right)} = \frac{\left(1.20 \times 10^{-9}\right)}{\left(0.040 \, M\right)} M = 3.00 \times 10^{-8} \, M$$

(c) The percent of molecules ionized is given by the following expression:

$$\% \text{ ionized} = \frac{\left[H^+\right]}{\left[HOCl\right]} \times 100\% = \frac{\left(3.47 \times 10^{-5} M\right)}{\left(0.040 M\right)} \times 100\% = 0.087\%$$

(d) At the equivalence point, all of the HOCl initially present has been converted into OCl^- ions.

From (a), we know that there were initially 0.004 moles of HOCl, so at equivalence, there are 0.004 moles of OCl^-.

At equivalence, we have added 40 ml to the 100 ml of solution originally present, so we must take this into account in our concentration calculation.

$$\text{Molarity} = \frac{\text{moles}}{\text{volume}}$$

$$[OCl^-] = \frac{(0.004 \text{ mol})}{(0.100 \text{ L} + 0.040 \text{ L})} = \frac{(0.004 \text{ mol})}{(0.140 \text{ L})} = 0.029 \, M$$

(e) We know the K_a for HOCl, so we can use it to find the K_b for OCl^-.

$$K_b = \frac{1 \times 10^{-14}}{K_a} = \frac{1 \times 10^{-14}}{3 \times 10^{-8}} = 3.33 \quad 10^{-7}$$

Now use the K_b expression.

$$K_b = \frac{\left[HOCl\right]\left[OH^i\right]}{\left[OCl^i\right]}$$

$[HOCl] = [OH^-] = x$

$[OCl^-] = 0.029 \, M$

$$3.33 \times 10^{-7} = \frac{x^2}{\left(0.029 \, M - x\right)}$$

Let's assume that 0.029 M is much larger than x. This simplifies the expression.

$$3.33 \times 10^{-7} = \frac{x^2}{0.029\,M}$$

$$x^2 = (3.33 \times 10^{-7})(0.029\,M) = 9.67 \times 10^{-9}$$

$$x = 9.83 \times 10^{-5} = [OH^-]$$

$$pOH = -\log[OH^-] = 4$$

$$pH = 14 - pOH = 14 - 4 = 10$$

2. (a) Use the equilibrium expression for H_2S to find $[H^+]$.

$$K_1 = \frac{\left[H^+\right]\left[HS^-\right]}{[H_2S]}$$

$[H^+] = [HS^-] = x$

$[H_2S] = 0.100\,M$

$K_1 = 1.0 \times 10^{-7}$

$$1.0 \times 10^{-7} = \frac{x^2}{\left(0.100\,M - x\right)}$$

Let's assume that 0.100 is much larger than x. That simplifies the expression.

$$1.0 \times 10^{-7} = \frac{x^2}{\left(0.100\,M\right)}$$

$x^2 = (1.0 \times 10^{-7})(0.100)\,M^2 = 1.0 \times 10^{-8}\,M^2$

$x = 1.0 \times 10^{-4}\,M = [H^+]$

$pH = -\log[H^+] = -\log(1.0 \times 10^{-4}) = 4.0$

(b) At the midpoint, $pH = pK_1$

Let's look at the Henderson-Hasselbalch expression.

$$pH = pK + \log \frac{\left[HS^-\right]}{[H_2S]}$$

When $pH = pK$, $[HS^-]$ must be equal to $[H_2S]$, making $\frac{\left[HS^-\right]}{[H_2S]}$ equal to one and $\log \frac{\left[HS^-\right]}{[H_2S]}$ equal to zero.

For every molecule of H_2S neutralized, one unit of HS^- is generated, so we must add enough NaOH to neutralize half of the HS^- initially present. (We are assuming that the further dissociation of HS^- into H^+ and S_2^- is negligible and can be ignored.)

Let's find out how many moles of H_2S were initially present.

Moles = (molarity)(volume)

Moles of H_2S = (0.100 M)(0.500 L) = 0.050 moles

We need to neutralize half of that, or 0.025 moles, so we need 0.025 moles of NaOH.

$$\text{Volume} = \frac{\text{moles}}{\text{molarity}}$$

$$\text{Volume of NaOH} = \frac{(0.025 \text{ mol})}{(0.100 \text{ } M)} = 0.250 \text{ L} = 250 \text{ ml}$$

(c) The concentrations of the H_2S solution and NaOH solution are both 0.100 M, so the first 500 ml of NaOH solution will completely neutralize the H_2S. The final 300 ml of NaOH solution will neutralize HS^- molecules.

Initially, there were 0.050 moles of H_2S, so after 500 ml of NaOH solution was added, there were 0.050 moles of HS^-.

By adding 300 ml more of NaOH solution, we added $(0.100 \text{ } M)(0.300 \text{ L}) = 0.030$ moles of NaOH. [Moles = (molarity)(volume)]

Every unit of NaOH added neutralizes one unit of HS^-, so when all the NaOH has been added, we have 0.030 moles of S^{2-} and (0.050 moles) – (0.030 moles) = 0.020 moles of HS^-.

Now we can use the Henderson-Hasselbalch expression to find the pH. We can use the number of moles we just calculated (0.030 for S^{2-} and 0.020 for HS^-) instead of concentrations because in a solution the concentrations are proportional to the number of moles.

$$pH = pK + \log \frac{\left[S^{2-}\right]}{\left[HS^-\right]}$$

$$pK = -\log(1.3 \times 10^{-13}) = 12.9$$

$$\log \frac{\left[S^{2-}\right]}{\left[HS^-\right]} = \log \frac{(0.030)}{(0.020)} = \log(1.5) = 0.18$$

$$pH = 12.9 + 0.18 = 13.1.$$

(d) The reaction ($H_2S \rightleftharpoons 2 \text{ H}^+ + S^{2-}$) is the sum of the two acid dissociation reactions below.

$$H_2S \rightleftharpoons H^+ + HS^-$$

$$HS^- \rightleftharpoons H^+ + S^{2-}$$

If one reaction is the sum of two other reactions, then its equilibrium constant will be the product of the equilibrium constants for the other two reactions.

So $K_{eq} = K_1 K_2 = (1.0 \times 10^{-7})(1.3 \times 10^{-13}) = 1.3 \times 10^{-20}$

3. (a) $pK_b = -\log K_b = -\log(1.79 \times 10^{-5}) = 4.75$

(b) This is a buffered solution, so we'll use the Henderson-Hasselbalch expression.

First let's find $[NH_4^+]$ and $[NH_3]$.

Moles = (molarity)(volume)

Moles of $NH_4^+ = (0.100 \text{ } M)(0.100 \text{ L}) = 0.010$ moles

Moles of $NH_3 = (0.200)(0.080 \text{ L}) = 0.016$ moles

When we mix the solutions, the volume becomes (0.100 L) + (0.080 L) = 0.180 L.

$$\text{Molarity} = \frac{\text{moles}}{\text{volume}}$$

$$[NH_4^+] = \frac{(0.010 \text{ mol})}{(0.180 \text{ L})} = 0.056 \text{ } M$$

$$[NH_3] = \frac{(0.016 \text{ mol})}{(0.180 \text{ mol})} = 0.089 \text{ } M$$

Now we can use the Henderson-Hasselbalch expression for bases.

$$pOH = pK + \log \frac{[NH_4^+]}{[NH_3]}$$

$pK = 4.75$

$$\log \frac{[NH_4^+]}{[NH_3]} = \log \frac{(0.056\ M)}{(0.089\ M)} = -0.20$$

$pOH = 4.75 + (-0.20) = 4.55$

$pH = 14 - pOH = 14 - 4.55 = 9.45$

(c) First let's find out how many moles of NaOH were added.

$$\text{Moles} = \frac{\text{moles}}{\text{MW}}$$

$$\text{Moles of NaOH} = \frac{(0.200\ g)}{(40.0\ g/m)} = 0.005\ mol$$

When NaOH is added to the solution, the following reaction occurs:

$$NH_4^+ + OH^- \rightarrow NH_3 + H_2O$$

So for every unit of NaOH added, one ion of NH_4^+ disappears and one molecule of NH_3 appears. We can use the results of the molar calculations we did in part (b).

Moles of NH_4^+ = (0.010) − (0.005) = 0.005 moles

Moles of NH_3 = (0.016) + (0.005) = 0.021 moles

Now we can use the Henderson-Hasselbalch expression. We can use the number of moles we just calculated (0.005 for NH_4^+ and 0.021 for NH_3) instead of concentrations because in a solution the concentrations will be proportional to the number of moles.

$$pOH = pK + \log \frac{[NH_4^+]}{[NH_3]}$$

$pK = 4.75$

$$\log \frac{[NH_4^+]}{[NH_3]} = \log \frac{(0.005\ mol)}{(0.021\ mol)} = -0.62$$

$pOH = 4.75 + (-0.62) = 4.13$

$pH = 14 - pOH = 14 - 4.13 = 9.87$

(d) When equal quantities of a base (NH_3) and its conjugate acid (NH_4^+) are mixed in a solution, the pOH will be equal to the pK_b.

From (a), $pOH = pK_b = 4.75$

$pH = 14 - pOH = 14 - 4.75 = 9.25$

Essays

4. (a) An amphoteric species can act either as an acid or base.

   $H_2PO_4^{\pm}$ is amphoteric.

   Conjugate base: $HPO_4^{2\pm}$

   Conjugate acid: H_3PO_4

   or

   $HPO_4^{2\pm}$ is amphoteric.

   Conjugate base: PO_4^{3-}

   Conjugate acid: $H_2PO_4^{\pm}$

   (b) The smaller the dissociation constant, the more tightly the acid holds its hydrogen ions and the weaker the acid becomes.

   After a hydrogen ion has been removed, the remaining species has a negative charge that attracts the remaining hydrogen ions more strongly.

   (c) $H_2PO_4^- \rightleftharpoons H^+ + HPO_4^{2\pm}$ $K_2 = 6.2 \times 10^{-8}$

   For $H_2PO_4^-$, $pK_a = -\log(6.2 \times 10^{-8}) = 7.2$

   The best buffer solution is made with an acid whose pK_a is approximately equal to the desired pH.

   (d)

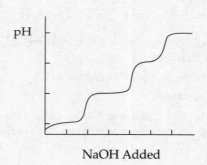

   NaOH Added

   The axes should have the correct labeling (pH and NaOH added).

   The curve should show rising pH, and it should have three bumps.

5. (a) The solution will be basic. A salt composed of the conjugate of a strong base (Na^+) and the conjugate of a weak acid ($C_2H_3O_2^-$) will create a basic solution in water.

   $NaC_2H_3O_2$ dissociates into Na^+ and $C_2H_3O_2^-$.

   Na^+ does not react with water but $C_2H_3O_2^-$ does.

   $C_2H_3O_2^- + H_2O \rightleftharpoons HC_2H_3O_2 + OH^-$

   (b) The solution will be acidic. A salt composed of the conjugate of a strong acid (Cl^-) and the conjugate of a weak base (NH_4^+) will create an acidic solution in water.

   NH_4Cl dissociates into NH_4^+ and Cl^-.

   Cl^- does not react with water but NH_4^+ does.

   $NH_4^+ + H_2O \rightleftharpoons NH_3 + H_3O^+$

(c) For a buffer solution, it is necessary for both an acid (HA) and its conjugate base (A⁻) to be present in the solution.

By definition, strong acids dissociate completely, so the conjugate base will be too weak to accept any hydrogen atoms that are added to the solution.

Weak acids do not dissociate completely, so it is possible to have a solution that contains both undissociated acid (HA) and conjugate base (A⁻). The undissociated acid can react with added hydroxide, and the conjugate base can react with added hydrogen ions.

(d) KOH separates into K^+ and OH^- in water because the ionic bond between O and K in the unit can be broken by water, which is a polar solvent.

HBrO separates into H^+ and BrO^- because the covalent O–Br bond pulls electrons away from the O–H bond. This makes the O–H bond more polar and more easily broken by water, a polar solvent.

6. (a) HCl is the stronger acid. The bond between H and Cl is weaker than the bond between H and F, so HCl dissociates more easily.

(b) H_2SO_3 is stronger. As a polyprotic acid gives up hydrogen ions, it becomes negatively charged, causing it to hang onto its remaining hydrogen ions more tightly, which weakens it as an acid.

(c)

HNO_2	HNO_3

$$H-O-N{=}O \qquad H-O-\overset{\displaystyle \overset{O}{\|}}{N}-O$$

HNO_3 is stronger. As more oxygens are added to the central atom of an oxyacid, the oxidation number of the central atom increases.

In HNO_2, the central atom, N, has an oxidation state of +3.

In HNO_3, N has an oxidation state of +5.

The increased oxidation state of N makes the N–O bond stronger. This draws electrons away from the O–H bond, making it more polar or weaker. The weakness of the O–H bond makes the acid stronger.

(d)

$HClO_4$	$HBrO_4$

$$H-O-\overset{\displaystyle \overset{O}{|}}{\underset{\displaystyle \underset{O}{|}}{Cl}}-O \qquad H-O-\overset{\displaystyle \overset{O}{|}}{\underset{\displaystyle \underset{O}{|}}{Br}}-O$$

$HClO_4$ is stronger. Cl has greater electronegativity than Br, so Cl draws electrons away from O in the Cl–O bond; this makes the O–H bond more polar or weaker. A weak O–H bond makes for a strong acid.

Kinetics

HOW OFTEN DOES KINETICS APPEAR ON THE EXAM?

In the multiple-choice section, this topic appears in about 3 out of 75 questions. In the free-response section, this topic appears almost every year.

In thermodynamics, you determine whether a reaction will occur spontaneously, based on the relative states of the reactants and products. Kinetics deals with the rate at which a reaction occurs between those states. The rate of a chemical reaction is determined experimentally by measuring the rate at which a reactant disappears or a product appears. So reaction rates are generally measured in moles/time or M/time.

THE RATE LAW USING INITIAL CONCENTRATIONS

The rate law for a reaction describes the dependence of the initial rate of a reaction on the concentrations of its reactants. It includes the Arrhenius constant, k, which takes into account the activation energy for the reaction and the temperature at which the reaction occurs. The rate of a reaction is described in terms of the rate of appearance of a product or the rate of disappearance of a reactant. The rate law for a reaction cannot be determined from a balanced equation; it must be determined from experimental data, which is presented on the test in table form.

Here's how it's done
The data below were collected for the following hypothetical reaction:

$$A + 2B + C \rightarrow D$$

| Experiment | Initial Concentration of Reactants (M) | | | Initial Rate of Formation of D (M/sec) |
	[A]	[B]	[C]	
1	0.10	0.10	0.10	0.01
2	0.10	0.10	0.20	0.01
3	0.10	0.20	0.10	0.02
4	0.20	0.20	0.10	0.08

The rate law always takes the following form, using the concentrations of the reactants:

$$\text{Rate} = k[A]^x[B]^y[C]^z$$

The greater the value of a reactant's exponent, the more a change in the concentration of that reactant will affect the rate of the reaction. To find the values for the exponents x, y, and z, we need to examine how changes in the individual reactants affect the rate. The easiest way to find the exponents is to see what happens to the rate when the concentration of an individual reactant is doubled.

Let's look at [A]
From experiment 3 to experiment 4, [A] doubles while the other reactant concentrations remain constant. For this reason, it is useful to use the rate values from these two experiments to calculate x (the order of the reaction with respect to reactant A).

As you can see from the table, the rate quadruples from experiment 3 to experiment 4, going from 0.02 M/sec to 0.08 M/sec.

We need to find a value for the exponent x that relates the doubling of the concentration to the quadrupling of the rate. The value of x can be calculated as follows:

$$(2)^x = 4, \text{ so } x = 2$$

Because the value of x is 2, the reaction is said to be second order with respect to A.

$$\text{Rate} = k[A]^2[B]^y[C]^z$$

Let's look at [B]
From experiment 1 to experiment 3, [B] doubles while the other reactant concentrations remain constant. For this reason it is useful to use the rate values from these two experiments to calculate y (the order of the reaction with respect to reactant B).

As you can see from the table, the rate doubles from experiment 1 to experiment 3, going from 0.01 M/sec to 0.02 M/sec.

We need to find a value for the exponent y that relates the doubling of the concentration to the doubling of the rate. The value of y can be calculated as follows:

$$(2)^y = 2, \text{ so } y = 1$$

Because the value of y is 1, the reaction is said to be first order with respect to B.

$$\text{Rate} = k[A]^2[B][C]^z$$

Let's look at [C]

From experiment 1 to experiment 2, [C] doubles while the other reactant concentrations remain constant.

The rate remains the same at 0.01 M.

The rate change is $(2)^z = 1$, so $z = 0$.

Because the value of z is 0, the reaction is said to be zero order with respect to C.

$$\text{Rate} = k[A]^2[B]$$

Because the sum of the exponents is 3, the reaction is said to be third order overall.

Once the rate law has been determined, the value of the rate constant can be calculated using any of the lines of data on the table. The units of the rate constant are dependent on the order of the reaction, so it's important to carry along units throughout all rate constant calculations.

Let's use experiment 3.

$$k = \frac{\text{Rate}}{[A]^2[B]} = \frac{(0.02\,M/\sec)}{(0.10\,M)^2(0.20\,M)} = 10\left(\frac{(M)}{(M)^3(\sec)}\right) = 10\,M^{-2}\text{-sec}^{-1}$$

You should note that we can tell from the coefficients in the original balanced equation that the rate of appearance of D is equal to the rate of disappearance of A and C because the coefficients of all three are the same. The coefficient of D is half as large as the coefficient of B, however, so the rate at which D appears is half the rate at which B disappears.

THE RATE LAW USING CONCENTRATION AND TIME

It's also useful to have rate laws that relate the rate constant k to the way that concentrations change over time. The rate laws will be different depending on whether the reaction is first, second, or zero order, but each rate law can be expressed as a graph that relates the rate constant, the concentration of a reactant, to the elapsed time.

FIRST-ORDER RATE LAWS

The rate of a first-order reaction depends on the concentration of a single reactant raised to the first power.

$$\text{Rate} = k[A]$$

As the concentration of reactant A is depleted over time, the rate of reaction will decrease with a characteristic half-life. This is the same curve you've seen used for nuclear decay, which is also a first-order process.

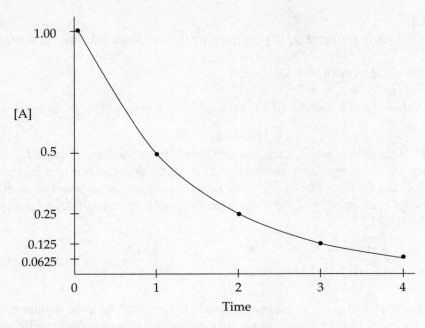

The rate law for a first-order reaction uses natural logarithms.

First-Order Rate Law
$\ln[A]_t - \ln[A]_o = -kt$
$[A]_t$ = concentration of reactant A at time t $[A]_o$ = initial concentration of reactant A k = rate constant t = time elapsed

The use of natural logarithms in the rate law creates a linear graph comparing concentration and time. The slope of the line is given by $-k$ and the y-intercept is given by $\ln[A]_o$.

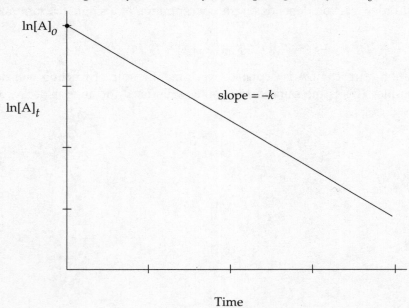

If you are asked to give the half-life of a first-order reaction, you can usually figure it out by looking at the graph or table given in the problem, but it's worth your while to know the equation that gives the half-life in terms of the rate constant.

$$\text{Half-life} = \frac{\ln 2}{k} = \frac{0.693}{k}$$

Let's try an example based on the data below.

[A] (*M*)	Time (min)
2.0	0
1.6	5
1.2	10

(a) Let's find the value of k. We'll use the first two lines of the table.

$\ln[A]_t - \ln[A]_o = -kt$

$\ln(1.6) - \ln(2.0) = -k(5 \text{ min})$

$-0.22 = -(5 \text{ min})k$

$k = 0.045 \text{ min}^{-1}$

(b) Now let's use k to find [A] when 20 minutes have elapsed.

$\ln[A]_t - \ln[A]_o = -kt$

$\ln[A]_t - \ln(2.0) = -(0.045 \text{ min}^{-1})(20 \text{ min})$

$\ln[A]_t = -0.21$

$[A]_t = e^{-0.21} = 0.81 \text{ } M$

(c) Now let's find the half-life of the reaction.

We can look at the answer (b) to see that the concentration dropped by half (1.6 M to 0.8 M) from 5 minutes to 20 minutes. That makes the half-life about 15 minutes. We can confirm this using the half-life equation.

$$\text{Half-life} = \frac{0.693}{k} = \frac{0.693}{0.045 \text{ min}^{-1}} = 15.4 \text{ minutes}$$

SECOND-ORDER RATE LAWS

The rate of a second-order reaction depends on the concentration of a single reactant raised to the second power.

$$\text{Rate} = k[\text{A}]^2$$

The rate law for a second-order reaction uses the inverses of the concentrations.

Second-Order Rate Law
$$\frac{1}{[\text{A}]_t} - \frac{1}{[\text{A}]_o} = kt$$
$[\text{A}]_t$ = concentration of reactant A at time t $[\text{A}]_o$ = initial concentration of reactant A k = rate constant t = time elapsed

The use of inverses in the rate law creates a linear graph comparing concentration and time.

Notice that the line moves upward as the concentration decreases. The slope of the line is given by k and the y-intercept is given by $\frac{1}{[\text{A}]_o}$.

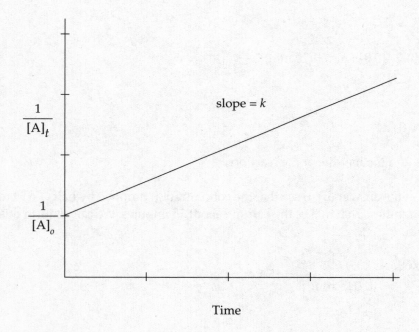

ZERO-ORDER RATE LAWS

The rate of a zero-order reaction does not depend on the concentration of reactants at all, so the rate of a zero-order reaction will always be the same at a given temperature.

$$\text{Rate} = k$$

The graph of the change in concentration of a reactant of a zero-order reaction versus time will be a straight line with a slope equal to $-k$.

COLLISION THEORY

According to collision theory, chemical reactions occur because reactants are constantly moving around and colliding with one another.

When reactants collide with sufficient energy (**activation energy, E_a**), a reaction occurs. These collisions are referred to as effective collisions, because they lead to a chemical reaction. Ineffective collisions do not produce a chemical reaction. At any given time during a reaction, a certain fraction of the reactant molecules will collide with sufficient energy to cause a reaction between them.

Reaction rate increases with increasing concentration of reactants because if there are more reactant molecules moving around in a given volume, then more collisions will occur.

Reaction rate increases with increasing temperature because increasing temperature means that the molecules are moving faster, which means that the molecules have greater average kinetic energy. The higher the temperature, the greater the number of reactant molecules colliding with each other with enough energy (E_a) to cause a reaction.

The Boltzmann distribution diagram below is often used to show that increasing temperature increases the fraction of reactant molecules above the activation energy.

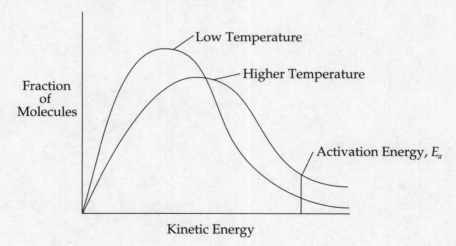

The rate constant will increase when temperature is increased. The relationship between temperature and the rate constant is given by the Arrhenius equation, $k = Ae^{\frac{-E_a}{RT}}$. This equation is generally rewritten using natural logarithms.

The Arrhenius Equation

$$\ln k = -\frac{E_a}{R}\left(\frac{1}{T}\right) + \ln A$$

k = rate constant
E_a = activation energy
R = gas constant, 8.31 J/k-mol
T = absolute temperature (K)
A = a constant that takes into account collision frequency and orientation

---

When $\ln k$ is graphed versus $\frac{1}{T}$, it makes a straight line with a slope of $-\frac{E_a}{R}$. It's a little difficult to see, but the graph below shows that k is increasing as the temperature increases.

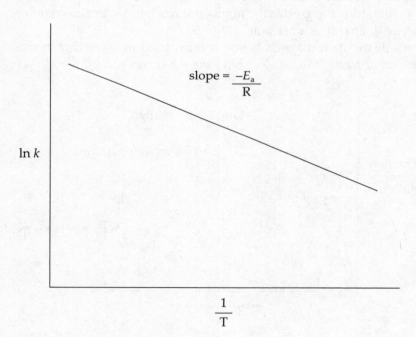

REACTION MECHANISMS

Many chemical reactions are not one-step processes. Rather, the balanced equation is the sum of a series of simple steps. Three molecules will not collide simultaneously very often, so steps of a reaction mechanism involve only one or two reactants at a time.

For instance, the hypothetical reaction

$$2 A + 2 B \rightarrow C + D \qquad\qquad\qquad\qquad \text{Rate} = k[A]^2[B]$$

could take place by the following three-step mechanism:

  I. $A + A \rightleftharpoons X$ (fast)

  II. $X + B \rightarrow C + Y$ (slow)

  III. $Y + B \rightarrow D$ (fast)

Species X and Y are called **intermediates** because they appear in the mechanism, but they cancel out of the balanced equation. The steps of a reaction mechanism must add up to equal the balanced equation, with all intermediates cancelling out.

Let's show that the mechanism above is consistent with the balanced equation by adding up all the steps.

  I. $A + A \rightleftharpoons X$

  II. $X + B \rightarrow C + Y$

  III. $Y + B \rightarrow D$

$$A + A + X + B + Y + B \rightarrow X + C + Y + D$$

Cancel species that appear on both sides.

$$2 A + 2 B \rightarrow C + D$$

By adding up all the steps, we get the balanced equation for the overall reaction, so this mechanism is consistent with the balanced equation.

As in any process where many steps are involved, the speed of the whole process can't be faster than the speed of the slowest step in the process, so the slowest step of a reaction is called the **rate-determining step.** Because the slowest step is the most important step in determining the rate of a reaction, the slowest step and the steps leading up to it are used to see if the mechanism is consistent with the rate law for the overall reaction.

Let's look at the reaction and the three-step mechanism again.

$$2 A + 2 B \rightarrow C + D \qquad\qquad\qquad\qquad \text{Rate} = k[A]^2[B]$$

The reaction above takes place by the following three-step mechanism.

  I. $A + A \rightleftharpoons X$ (fast)

  II. $X + B \rightarrow C + Y$ (slow)

  III. $Y + B \rightarrow D$ (fast)

Let's show that the reaction mechanism is consistent with the rate law (Rate = $k[A]^2[B]$).

The slowest step is the rate-determining step, so we should start with the rate law for step II:

$$\text{Rate} = k_2[X][B]$$

But X is an intermediate, which means that it can't appear in the overall rate law. To eliminate X from the rate law, we need to look at the equilibrium reaction in step I. We can assume that the reaction in step I comes to equilibrium quickly. At equilibrium the rate of the forward reaction is equal to the rate of the reverse reaction, so we get

$$k_f[A][A] = k_r[X]$$

Now we can solve for [X].

$$[X] = \frac{k_f}{k_r}[A]^2$$

Once we have solved our equilibrium rate expression for [X] in terms of [A], we can substitute for [X] in our step II rate law.

$$\text{Rate} = k_2 \frac{k_f}{k_r}[A]^2[B] = k[A]^2[B]$$

Now we have a rate law that contains only reactants from the overall equation and that's consistent with the experimentally derived rate law that we were given. You can always eliminate intermediates from the rate-determining step by this process.

CATALYSTS

As we mentioned earlier in the book, a catalyst increases the rate of a chemical reaction without being consumed in the process; catalysts do not appear in the balanced equation. In some cases, a catalyst is a necessary part of a reaction because in its absence, the reaction would proceed at too slow a rate to be at all useful.

A catalyst increases the rate of a chemical reaction by providing an alternative reaction pathway with a lower activation energy.

KINETICS AND EQUILIBRIUM

There is a relationship between the rate constants for the forward and reverse directions of a particular reaction and the equilibrium constant for that reaction.

$$K_{eq} = \frac{k_f}{k_r}$$

K_{eq} = the equilibrium constant
k_f = the rate constant for the forward reaction
k_r = the rate constant for the reverse reaction

CHAPTER 12 QUESTIONS

MULTIPLE-CHOICE QUESTIONS

<u>Questions 1–3</u>

$$A + B \rightarrow C$$

The following are possible rate laws for the hypothetical reaction given above.

(A)   Rate = $k[A]$
(B)   Rate = $k[A]^2$
(C)   Rate = $k[A][B]$
(D)   Rate = $k[A]^2[B]$
(E)   Rate = $k[A]^2[B]^2$

1. This is the rate law for a first order reaction.

2. This is the rate law for a reaction that is second order with respect to B.

3. This is the rate law for a third order reaction.

<u>Questions 4–6</u>

$$A + B \rightarrow C$$

The following are possible rate laws for the hypothetical reaction given above.

(A)   Rate = $k[A]$
(B)   Rate = $k[B]^2$
(C)   Rate = $k[A][B]$
(D)   Rate = $k[A]^2[B]$
(E)   Rate = $k[A]^2[B]^2$

4. When [A] and [B] are doubled, the initial rate of reaction will increase by a factor of eight.

5. When [A] and [B] are doubled, the initial rate of reaction will increase by a factor of two.

6. When [A] is doubled and [B] is held constant, the initial rate of reaction will not change.

7. A multistep reaction takes place by the following mechanism:

$$A + B \rightarrow C + D$$
$$A + C \rightarrow D + E$$

Which of the species shown above is an intermediate in the reaction?

(A) A
(B) B
(C) C
(D) D
(E) E

8. $$2 \, NOCl \rightarrow 2 \, NO + Cl_2$$

The reaction above takes place with all of the reactants and products in the gaseous phase. Which of the following is true of the relative rates of disappearance of the reactants and appearance of the products?

(A) NO appears at twice the rate that NOCl disappears.
(B) NO appears at the same rate that NOCl disappears.
(C) NO appears at half the rate that NOCl disappears.
(D) Cl_2 appears at the same rate that NOCl disappears.
(E) Cl_2 appears at twice the rate that NOCl disappears.

9. $$H_2(g) + I_2(g) \rightarrow 2 \, HI(g)$$

When the reaction given above takes place in a sealed isothermal container, the rate law is

$$Rate = k[H_2][I_2]$$

If a mole of H_2 gas is added to the reaction chamber, which of the following will be true?

(A) The rate of reaction and the rate constant will increase.
(B) The rate of reaction and the rate constant will not change.
(C) The rate of reaction will increase and the rate constant will decrease.
(D) The rate of reaction will increase and the rate constant will not change.
(E) The rate of reaction will not change and the rate constant will increase.

10. $$A + B \rightarrow C$$

When the reaction given above takes place, the rate law is

$$Rate = k[A]$$

If the temperature of the reaction chamber were increased, which of the following would be true?

(A) The rate of reaction and the rate constant will increase.
(B) The rate of reaction and the rate constant will not change.
(C) The rate of reaction will increase and the rate constant will decrease.
(D) The rate of reaction will increase and the rate constant will not change.
(E) The rate of reaction will not change and the rate constant will increase.

11. $$A + B \rightarrow C$$

Based on the following experimental data, what is the rate law for the hypothetical reaction given above?

Experiment	[A] (M)	[B] (M)	Initial Rate of Formation of C (mol/L-sec)
1	0.20	0.10	3×10^{-2}
2	0.20	0.20	6×10^{-2}
3	0.40	0.20	6×10^{-2}

(A) Rate = $k[A]$
(B) Rate = $k[A]^2$
(C) Rate = $k[B]$
(D) Rate = $k[B]^2$
(E) Rate = $k[A][B]$

12.
$$A + B \rightarrow C + D$$

The rate law for the hypothetical reaction shown above is as follows:

$$Rate = k[A]$$

Which of the following changes to the system will increase the rate of the reaction?

I. An increase in the concentration of A
II. An increase in the concentration of B
III. An increase in the temperature

(A) I only
(B) I and II only
(C) I and III only
(D) II and III only
(E) I, II, and III

13.
$$A + B \rightarrow C$$

Based on the following experimental data, what is the rate law for the hypothetical reaction given above?

Experiment	[A] (M)	[B] (M)	Initial Rate of Formation of C (M/sec)
1	0.20	0.10	2.0×10^{-6}
2	0.20	0.20	4.0×10^{-6}
3	0.40	0.40	1.6×10^{-5}

(A) Rate = $k[A]$
(B) Rate = $k[A]^2$
(C) Rate = $k[B]$
(D) Rate = $k[B]^2$
(E) Rate = $k[A][B]$

14.
$$A + B \rightarrow C$$

Based on the following experimental data, what is the rate law for the hypothetical reaction given above?

Experiment	[A] (M)	[B] (M)	Initial Rate of Formation of C (M/sec)
1	0.10	0.10	1.5×10^{-3}
2	0.40	0.10	6.0×10^{-3}
3	0.40	0.20	2.4×10^{-2}

(A) Rate = $k[A]$
(B) Rate = $k[A]^2$
(C) Rate = $k[A][B]^2$
(D) Rate = $k[B]^2$
(E) Rate = $k[A]^2[B]^2$

15.

Time (Hours)	[A] M
0	0.40
1	0.20
2	0.10
3	0.05

Reactant A underwent a decomposition reaction. The concentration of A was measured periodically and recorded in the chart above. Based on the data in the chart, which of the following is the rate law for the reaction?

(A) Rate = $k[A]$

(B) Rate = $k[A]^2$

(C) Rate = $2k[A]$

(D) Rate = $\frac{1}{2}k[A]$

(E) Rate = k

PROBLEMS

1.
$$A + 2B \rightarrow 2C$$

The following results were obtained in experiments designed to study the rate of the reaction above:

Experiment	Initial Concentration (mol/L)		Initial Rate of Disappearance of A (M/sec)
	[A]	[B]	
1	0.05	0.05	3.0×10^{-3}
2	0.05	0.10	6.0×10^{-3}
3	0.10	0.10	$1.2 \ 10^2$
4	0.20	0.10	$2.4 \ 10^2$

(a) Determine the order of the reaction with respect to each of the reactants, and write the rate law for the reaction.

(b) Calculate the value of the rate constant, k, for the reaction. Include the units.

(c) If another experiment is attempted with [A] and [B], both 0.02-molar, what would be the initial rate of disappearance of A?

(d) The following reaction mechanism was proposed for the reaction above:
$$A + B \rightarrow C + D$$
$$D + B \rightarrow C$$

   (i) Show that the mechanism is consistent with the balanced reaction.

   (ii) Show which step is the rate-determining step, and explain your choice.

2.
$$2 NO(g) + Br_2(g) \rightarrow 2 NOBr(g)$$

The following results were obtained in experiments designed to study the rate of the reaction above:

Experiment	Initial Concentration (mol/L)		Initial Rate of Appearance of NOBr (M/sec)
	[NO]	[Br$_2$]	
1	0.02	0.02	9.6×10^{-2}
2	0.04	0.02	3.8×10^{-1}
3	0.02	0.04	$1.9 \ 10^1$

(a) Write the rate law for the reaction.

(b) Calculate the value of the rate constant, k, for the reaction. Include the units.

(c) In experiment 2, what was the concentration of NO remaining when half of the original amount of Br$_2$ was consumed?

(d) Which of the following reaction mechanisms is consistent with the rate law established in (a)? Explain your choice.

I. $NO + NO \rightleftharpoons N_2O_2$ (fast)

    $N_2O_2 + Br_2 \rightarrow 2\ NOBr$ (slow)

II. $Br_2 \rightarrow Br + Br$ (slow)

    $2(NO + Br \rightarrow NOBr)$ (fast)

3.
$$N_2O_5(g) \rightarrow 4NO_2(g) + O_2(g)$$

Dinitrogen pentoxide gas decomposes according to the equation above. The first-order reaction was allowed to proceed at 40°C and the data below were collected.

$[N_2O_5]$ (M)	Time (min)
0.400	0.0
0.289	20.0
0.209	40.0
0.151	60.0
0.109	80.0

(a) Calculate the rate constant for the reaction using the values for concentration and time given in the table. Include units with your answer.

(b) After how many minutes will $[N_2O_5]$ be equal to 0.350 M?

(c) What will be the concentration of N_2O_3 after 100 minutes have elapsed?

(d) Calculate the initial rate of the reaction. Include units with your answer.

(e) What is the half-life of the reaction?

4.
$$2\ A + B \rightarrow C + D$$

The following results were obtained in experiments designed to study the rate of the reaction above:

Experiment	Initial Concentration (moles/L)		Initial Rate of Formation of D (M/min)
	[A]	[B]	
1	0.10	0.10	1.5×10^{-3}
2	0.20	0.20	3.0×10^{-3}
3	0.20	0.40	$6.0\ 10^3$

(a) Write the rate law for the reaction.

(b) Calculate the value of the rate constant, k, for the reaction. Include the units.

(c) If experiment 2 goes to completion, what will be the final concentration of D? Assume that the volume is unchanged over the course of the reaction and that no D was present at the start of the experiment.

(d) Which of the following possible reaction mechanisms is consistent with the rate law found in (a)?

    I. $A + B \rightarrow C + E$ (slow)

         $A + E \rightarrow D$ (fast)

    II. $B \rightarrow C + E$ (slow)

         $A + E \rightarrow F$ (fast)

         $A + F \rightarrow D$ (fast)

(e) Calculate the half-life of reactant B.

ESSAYS

5. $$A(g) + B(g) \rightarrow C(g)$$

The reaction above is second order with respect to A and zero order with respect to B. Reactants A and B are present in a closed container. Predict how each of the following changes to the reaction system will affect the rate and rate constant and explain why.

(a) More gas A is added to the container.

(b) More gas B is added to the container.

(c) The temperature is increased.

(d) An inert gas D is added to the container.

(e) The volume of the container is decreased.

6. Use your knowledge of kinetics to answer the following questions. Justify your answers.

(a)

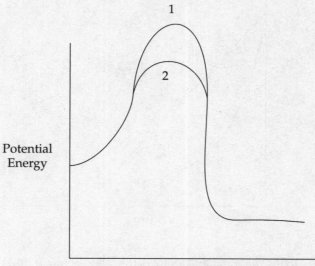

Reaction Coordinate

The two lines in the diagram above show different reaction pathways for the same reaction. Which of the two lines shows the reaction when a catalyst has been added?

(b)

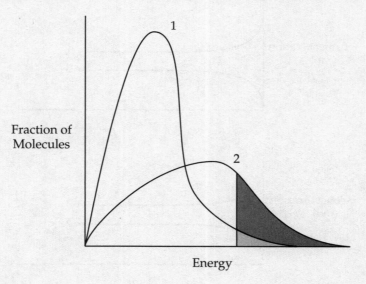

Energy

Which of the two lines in the energy distribution diagram shows the conditions at a higher temperature?

(c)

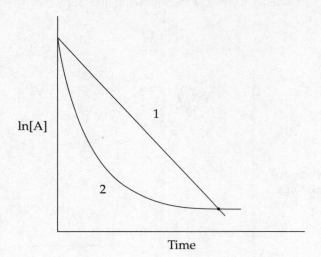

Which of the two lines in the diagram above shows the relationship of ln[A] to time for a first order reaction with the following rate law?

$$\text{Rate} = k[A]$$

(d)

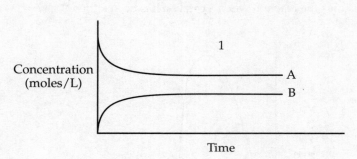

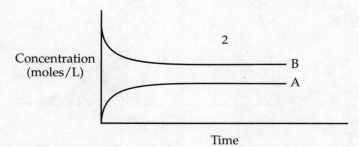

Which of the two graphs above shows the changes in concentration over time for the following reaction?

$$A \rightarrow B$$

(e)

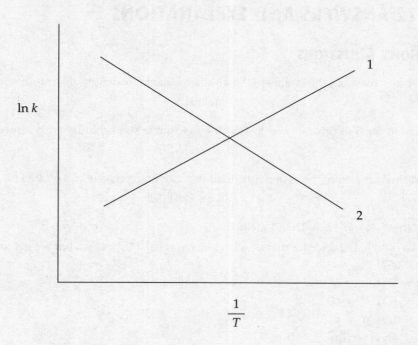

Which of the two lines in the diagram above shows the relationship of ln k to $\dfrac{1}{T}$ for a reaction? How is the slope of the line related to the activation energy for the reaction?

7. Use your knowledge of kinetics to explain each of the following statements:

   (a) An increase in the temperature at which a reaction takes place causes an increase in reaction rate.

   (b) The addition of a catalyst increases the rate at which a reaction will take place.

   (c) A catalyst that has been ground into powder will be more effective than a solid block of the same catalyst.

   (d) Increasing the concentration of reactants increases the rate of a reaction.

CHAPTER 12 ANSWERS AND EXPLANATIONS

MULTIPLE-CHOICE QUESTIONS

1. **A** In a first-order reaction, the exponents of all the reactants present in the rate law add up to 1.

$$\text{Rate} = k[A]^1$$

2. **E** The exponent for B in this rate law is 2, so the reaction is second order with respect to B.

$$\text{Rate} = k[A]^2[B]^2$$

3. **D** In a third-order reaction, the exponents of all the reactants present in the rate law add up to 3.

$$\text{Rate} = k[A]^2[B]^1$$

4. **D** Let's say that $[A] = [B] = 1$. Then Rate $= k$.

    Now if we double $[A]$ and $[B]$, that is, we make $[A] = [B] = 2$, here's what we get for each of the answer choices.

    (A) Rate $= k(2) = 2k$

    (B) Rate $= k(2)^2 = 4k$

    (C) Rate $= k(2)(2) = 4k$

    (D) Rate $= k(2)^2(2) = 8k$

    (E) Rate $= k(2)^2(2)^2 = 16k$

    So in (D), the rate increases by a factor of 8.

5. **A** Let's say that $[A] = [B] = 1$. Then Rate $= k$.

    Now if we double $[A]$ and $[B]$, that is, we make $[A] = [B] = 2$, here's what we get for each of the answer choices.

    (A) Rate $= k(2) = 2k$

    (B) Rate $= k(2)^2 = 4k$

    (C) Rate $= k(2)(2) = 4k$

    (D) Rate $= k(2)^2(2) = 8k$

    (E) Rate $= k(2)^2(2)^2 = 16k$

    So in (A), the rate increases by a factor of 2.

6. **B** We need to find the rate law that is independent of changes in $[A]$.

    The only choice listed that does not include $[A]$ is choice (B), Rate $= k[B]^2$.

7. **C** C is created and used up in the reaction, so it will not be present in the balanced equation.

8. **B** For every two NO molecules that form, two NOCl molecules must disappear, so NO is appearing at the same rate that NOCl is disappearing. Choices (D) and (E) are wrong because for every mole of Cl_2 that forms, two moles of NOCl are disappearing, so Cl_2 is appearing at *half* the rate that NOCl is disappearing.

9. **D** From the rate law given in the question (Rate $= k[H_2][I_2]$), we can see that increasing the concentration of H_2 will increase the rate of reaction. The rate constant, k, is not affected by changes in the concentration of the reactants.

10. **A** When temperature increases, the rate constant increases to reflect the fact that more reactant molecules are likely to have enough energy to react at any given time. So both the rate constant and the rate of reaction will increase.

11. **C** From a comparison of experiments 1 and 2, when [B] is doubled while [A] is held constant, the rate doubles. That means that the reaction is first order with respect to B.

    From a comparison of experiments 2 and 3, when [A] is doubled while [B] is held constant, the rate doesn't change. That means that the reaction is zero order with respect to A and that A will not appear in the rate law.

    So the rate law is Rate = k[B].

12. **C** An increase in the concentration of A will increase the rate, as shown in the rate law for the reaction, so (I) is correct. Reactant B is not included in the rate law, so an increase in the concentration of B will not affect the rate; therefore, (II) is wrong. An increase in temperature causes more collisions with greater energy among reactants and always increases the rate of a reaction, so (III) is correct.

13. **E** From a comparison of experiments 1 and 2, when [B] is doubled while [A] is held constant, the rate doubles. That means that the reaction is first order with respect to B.

    From a comparison of experiments 2 and 3, when both [A] and [B] are doubled, the rate increases by a factor of 4. We would expect the rate to double based on the change in B; because the rate is in fact multiplied by 4, the doubling of A must also change the rate by a factor of 2, so the reaction is also first order with respect to A.

    So the rate law is Rate = k[A][B].

14. **C** From a comparison of experiments 1 and 2, when [A] is quadrupled while [B] is held constant, the rate quadruples. That means that the reaction is first order with respect to A.

    From a comparison of experiments 2 and 3, when [B] is doubled while [A] is held constant, the rate quadruples. That means that the reaction is second order with respect to B.

    So the rate law is Rate = k[A][B]^2.

15. **A** The key to this question is to recognize that reactant A is disappearing with a characteristic half-life. This is a signal that the reaction is first order with respect to A. So the rate law must be Rate = k[A].

PROBLEMS

1. (a) When we compare the results of experiments 3 and 4, we see that when [A] doubles, the rate doubles, so the reaction is first order with respect to A.

   When we compare the results of experiments 1 and 2, we see that when [B] doubles, the rate doubles, so the reaction is first order with respect to B.

   Rate = k[A][B]

   (b) Use the values from experiment 3, just because they look the simplest.

   $$k = \frac{\text{Rate}}{[A][B]} = \frac{\left(1.2 \times 10^{-2}\ M/\text{sec}\right)}{(0.10\ M)(0.10\ M)} = 1.2\ M^{-1}\text{sec}^{-1} = 1.2\ \text{L/mol-sec}$$

   (c) Use the rate law.

   Rate = k[A][B]

   Rate = $(1.2\ M^{-1}\text{sec}^{-1})(0.02\ M)(0.02\ M) = 4.8\ \ 10^{-4}\ M/\text{sec}$

(d) (i) $A + B \rightarrow C + D$

        $D + B \rightarrow C$

The two reactions add up to

        $A + 2B + D \rightarrow 2C + D$

D's cancel, and we're left with the balanced equation.

        $A + 2B \rightarrow 2C$

(ii) $A + B \rightarrow C + D$ (slow)

        $D + B \rightarrow C$ (fast)

The first part of the mechanism is the slow, rate-determining step because its rate law is the same as the experimentally determined rate law.

2. (a) When we compare the results of experiments 1 and 2, we see that when [NO] doubles, the rate quadruples, so the reaction is second order with respect to NO.

When we compare the results of experiments 1 and 3, we see that when $[Br_2]$ doubles, the rate doubles, so the reaction is first order with respect to Br_2.

Rate = $k[NO]^2[Br_2]$

(b) Use the values from experiment 1, just because they look the simplest.

$$k = \frac{\text{Rate}}{[NO]^2[Br_2]} = \frac{\left(9.6 \times 10^{-2} \, M/\text{sec}\right)}{\left(0.02 \, M\right)^2 \left(0.02 \, M\right)} = 1.2 \times 10^4 \, M^{-2}\text{sec}^{-1} = 1.2 \times 10^4 \, L^2/\text{mol}^2\text{-sec}$$

(c) In experiment 2, we started with $[Br_2] = 0.02 \, M$, so $0.01 \, M$ was consumed.

From the balanced equation, 2 moles of NO are consumed for every mole of Br_2 consumed. So $0.02 \, M$ of NO are consumed.

[NO] remaining = $0.04 \, M - 0.02 \, M = 0.02 \, M$

(d) Choice (I) agrees with the rate law.

        $NO + NO \leftrightarrow N_2O_2$ (fast)

        $N_2O_2 + Br_2 \rightarrow 2 \, NOBr$ (slow)

The slow step is the rate-determining step, with the following rate law:

Rate = $k[N_2O_2][Br_2]$

We can replace the intermediate (N_2O_2) by assuming that the first step reaches equilibrium instantaneously and remembering that at equilibrium, the rates of the forward and reverse reactions are equal.

$k_f[NO]^2 = k_r[N_2O_2]$

Now solve for $[N_2O_2]$.

$[N_2O_2] = \dfrac{k_f}{k_r}[NO]^2$

Now we can substitute the rate law for the rate determining step.

Rate = $k\dfrac{k_f}{k_r}[NO]^2[Br_2]$

This matches the experimentally determined rate law. By the way, the mechanism in choice (II) would have a rate law of Rate = $k[Br_2]$.

3. (a) Use the first-order rate law and insert the first two lines from the table.

$$\ln[N_2O_5]_t - \ln[N_2O_5]_o = -kt$$

$$\ln(0.289) - \ln(0.400) = -k(20.0 \text{ min})$$

$$-.325 = -k(20.0 \text{ min})$$

$$k = 0.0163 \text{ min}^{-1}$$

(b) Use the first-order rate law.

$$\ln[N_2O_5]_t - \ln[N_2O_5]_o = -kt$$

$$\ln(0.350) - \ln(0.400) = -(0.0163 \text{ min}^{-1})t$$

$$-0.134 = -(0.0163 \text{ min}^{-1})t$$

$$t = 8.19 \text{ min}$$

(c) Use the first-order rate law.

$$\ln[N_2O_5] - \ln[N_2O_5]_o = -kt$$

$$\ln[N_2O_5] - \ln(0.400) = -(0.0163 \text{ min}^{-1})(100 \text{ min})$$

$$\ln[N_2O_5] + 0.916 = 1.63$$

$$\ln[N_2O_5] = -2.55$$

$$[N_2O_5] = e^{-2.55} M = 0.078 M$$

(d) For a first-order reaction, Rate = $k[N_2O_5]$ = (0.0163 min^{-1})((0.400 M) = 0.0652 M/min)

(e) You can see from the numbers on the table that the half-life is slightly over 40 min. To calculate it exactly, use the formula

$$\text{Half-life} = \frac{0.693}{k} = \frac{0.693}{0.0163 \text{ min}^{-1}} = 42.5 \text{min}$$

4. (a) When we compare the results of experiments 2 and 3, we see that when [B] doubles, the rate doubles, so the reaction is first order with respect to B.

Experiments 2 and 3 prove that the rate must double when [B] doubles. Knowing this, we can see that when the rate doubles from experiment 1 to 2, it must be because of B, and the change in A has no effect. So the reaction is zero order with respect to A.

Rate = $k[B]$

(b) Take the values from experiment 1.

$$k = \frac{\text{Rate}}{[B]} = \frac{\left(1.5 \times 10^{-3} \ M/\text{min}^{-1}\right)}{(0.10 \ M)} = 1.5 \times 10^{-2} \text{min}^{-1}$$

(c) In experiment 2, there are equal concentrations of A and B, but A is consumed twice as fast, so A is the limiting reagent.

0.2 M of A is consumed.

According to the balanced equation, for every 2 moles of A consumed, 1 mole of D is produced.

So if 0.2 M of A is consumed, 0.1 M of D is produced.

(d) The reaction mechanism in choice (II) is consistent with the rate law.

$$B \rightarrow C + E \qquad \text{(slow)}$$
$$A + E \rightarrow F \qquad \text{(fast)}$$
$$A + F \rightarrow D \qquad \text{(fast)}$$

The slow, rate-determining step gives the proper rate law:

Rate = $k[B]$

By the way, choice (I) gives a rate law of Rate = $k[A][B]$

(e) The reaction is first order with respect to B, so use the half-life formula.

$$\text{Half-Life} = \frac{0.693}{k} = \frac{0.693}{1.5 \times 10^2 \, \text{min}^{-1}} = 46 \, \text{min}$$

ESSAYS

5. (a) The rate of the reaction will increase because the rate depends on the concentration of A as given in the rate law: Rate = $k[A]^2$.

    The rate constant is independent of the concentration of the reactants and will not change.

    (b) The rate of the reaction will not change. If the reaction is zero order with respect to B, then the rate is independent of the concentration of B.

    The rate constant is independent of the concentration of the reactants and will not change.

    (c) The rate of the reaction will increase with increasing temperature because the rate constant increases with increasing temperature.

    The rate constant increases with increasing temperature because at a higher temperature more gas molecules will collide with enough energy to overcome the activation energy for the reaction.

    (d) Neither the rate nor the rate constant will be affected by the addition of an inert gas.

    (e) The rate of the reaction will increase because decreasing the volume of the container will increase the concentration of A. Rate = $k[A]^2$

    The rate constant is independent of the concentration of the reactants and will not change.

6. (a) Line 2 is the catalyzed reaction. Adding a catalyst lowers the activation energy of the reaction, making it easier for the reaction to occur.

    (b) Line 2 shows the higher temperature distribution. At a higher temperature, more of the molecules will be at higher energies, causing the distribution to flatten out and shift to the right.

(c) Line 1 is correct. ln[reactant] for a first order reaction changes in a linear fashion over time, as shown in the following equation.

$$\ln[A]_t = -kt + \ln[A]_o$$

$$y = mx + b$$

Notice the similarity to the slope-intercept form for a linear equation.

(d) Graph 1 is correct, showing a decrease in the concentration of A as it is consumed in the reaction, and a corresponding increase in the concentration of B as it is produced.

(e) Line 2 is correct. The rate constant k increases with increasing temperature, so the line will slope downward when graphed against $\frac{1}{T}$. The slope of the line is proportional to the activation energy (slope $= -\frac{E_a}{R}$).

7. (a) An increase in temperature means an increase in the energy of the molecules present. If the molecules have more energy, then more of them will collide more often with enough energy to overcome the activation energy required for a reaction to take place, causing the reaction to proceed more quickly.

(b) A catalyst offers a reaction an alternative pathway with a lower activation energy. If the activation energy is lowered, then more molecular collisions will occur with enough energy to overcome the activation energy, causing the reaction to proceed more quickly.

(c) The effectiveness of a solid catalyst depends on the surface area of the catalyst that is exposed to the reactants. Grinding a solid into powder greatly increases its surface area.

(d) Increasing the concentration of reactants crowds the reactants more closely together, making it more likely that they will collide with one another. The more collisions that occur, the more likely that collisions that will result in reactions will occur.

13

Oxidation-Reduction and Electrochemistry

HOW OFTEN DO OXIDATION-REDUCTION AND ELECTROCHEMISTRY APPEAR ON THE EXAM?

In the multiple-choice section, these topics appears in about 5 out of 75 questions. In the free-response section, these topics appears almost every year.

OXIDATION STATES

The **oxidation state** (or oxidation number) of an atom indicates the number of electrons that it gains or loses when it forms a bond. For instance, upon forming a bond with another atom, oxygen generally gains two electrons, which are negatively charged, so the oxidation state of oxygen in a bond is –2. Similarly, sodium generally loses one electron when it bonds to another atom, so its oxidation state in a bond is +1.

Here are three important things you have to keep in mind when dealing with oxidation numbers.

- The oxidation state of an atom that is not bonded to an atom of another element is zero. That means either an atom that is not bonded to any other atom or an atom that is bonded to another atom of the same element (like the oxygen atoms in O_2).

- The oxidation numbers for all the atoms in a molecule must add up to zero.

- The oxidation numbers for all the atoms in a polyatomic ion must add up to the charge on the ion.

Most elements have different oxidation numbers that can vary depending on the molecule that they are a part of. The following chart shows some elements that consistently take the same oxidation numbers as their corresponding molecules.

Element	Oxidation Number
Alkali metals (Li, Na, ...)	+1
Alkaline earths (Be, Mg, ...)	+2
Group 3A (B, Al, ...)	+3
Oxygen	–2
Halogens (F, Cl, ...)	–1

Transition metals can have several oxidation states, which are differentiated from one another by a Roman numeral in the name of the compound. For example, in copper (II) sulfate ($CuSO_4$), the oxidation state for copper is +2, and in lead (II) oxide (PbO), the oxidation state for lead is +2. However, copper can also have an oxidation number of +1, and Pb can range from –4 to +4. In general, transition metals or d-block metals are characterized by variable or changing oxidation states.

You should be familiar with the following polyatomic ions and their charges.

Hydroxide	OH^-
Nitrate	NO_3^-
Perchlorate	ClO_4^-
Acetate	$C_2H_3O_2^-$
Carbonate	CO_3^{2-}
Sulfate	SO_4^{2-}
Phosphate	PO_4^{3-}

OXIDATION-REDUCTION REACTIONS

In an oxidation-reduction (or redox, for short) reaction, electrons are exchanged by the reactants, and the oxidation states of some of the reactants are changed over the course of the reaction. Look at the following reaction:

$$Fe + 2\ HCl \rightarrow FeCl_2 + H_2$$

The oxidation state of Fe changes from 0 to +2.
The oxidation state of H changes from +1 to 0.

- *When an atom gains electrons, its oxidation number decreases, and it is said to have been reduced.*

In the reaction above, H was reduced.

- *When an atom loses electrons, its oxidation number increases, and it is said to have been oxidized.*

In the reaction above, Fe was oxidized.

Here's a mnemonic device that might be useful.

> LEO the lion says GER
> **LEO:** you **L**ose **E**lectrons in **O**xidation
> **GER:** you **G**ain **E**lectrons in **R**eduction

Oxidation and reduction go hand in hand. If one atom is losing electrons, another atom must be gaining them.

- If an atom is losing electrons and being oxidized, it must be giving the electrons to another atom, which is being reduced. So if a reactant contains an atom that is being oxidized, it is a **reducing agent** or **reductant**.

- If an atom is taking electrons and being reduced, it must be taking electrons away from another atom, which is being oxidized. So if a reactant contains an atom that is being reduced, it is an **oxidizing agent** or **oxidant**.

An oxidation-reduction reaction can be written as two **half-reactions**: one for the reduction and one for the oxidation. For example, the reaction

$$Fe + 2\ HCl \rightarrow FeCl_2 + H_2$$

can be written as

$Fe \rightarrow Fe^{2+} + 2\ e^-$	Oxidation
$2\ H^+ + 2\ e^- \rightarrow H_2$	Reduction

REDUCTION POTENTIALS

Every half-reaction has a potential, or voltage, associated with it. For Section II of the test, you'll be given a table of standard reduction potentials like the one below. The potentials are given as reduction half-reactions, but you can read them in reverse and change the sign on the voltage to get oxidation potentials.

Standard Reduction Potentials, E, in Water Solution at 25 C (in V)			
$F_2(g) + 2\,e^-$	$\rightarrow$	$2\,F^-$	2.87
$Co^{3+} + e^-$	$\rightarrow$	Co^{2+}	1.82
$Au^{3+} + 3e^-$	$\rightarrow$	$Au(s)$	1.50
$Cl_2(g) + 2\,e^-$	$\rightarrow$	$2\,Cl^-$	1.36
$O_2(g) + 4H^+ + 4\,e^-$	$\rightarrow$	$2\,H_2O$	1.23
$Br_2(l) + 2\,e^-$	$\rightarrow$	$2\,Br^-$	1.07
$2\,Hg^{2+} + 2\,e^-$	$\rightarrow$	Hg_2^{2+}	0.92
$Hg^{2+} + 2\,e^-$	$\rightarrow$	$Hg(l)$	0.85
$Ag^+ + e^-$	$\rightarrow$	$Ag(s)$	0.80
$Hg_2^{2+} + 2\,e^-$	$\rightarrow$	$2\,Hg(l)$	0.79
$Fe^{3+} + e^-$	$\rightarrow$	Fe^{2+}	0.77
$I_2(s) + 2\,e^-$	$\rightarrow$	$2\,I^-$	0.53
$Cu^+ + e^-$	$\rightarrow$	$Cu(s)$	0.52
$Cu^{2+} + 2\,e^-$	$\rightarrow$	$Cu(s)$	0.34
$Cu^{2+} + e^-$	$\rightarrow$	Cu^+	0.15
$Sn^{4+} + 2\,e^-$	$\rightarrow$	Sn^{2+}	0.15
$S(s) + 2\,H^+ + 2\,e^-$	$\rightarrow$	H_2S	0.14
$2\,H^+ + 2\,e^-$	$\rightarrow$	$H_2(g)$	0.00
$Pb^{2+} + 2\,e^-$	$\rightarrow$	$Pb(s)$	−0.13
$Sn^{2+} + 2\,e^-$	$\rightarrow$	$Sn(s)$	−0.14
$Ni^{2+} + 2\,e^-$	$\rightarrow$	$Ni(s)$	−0.25
$Co^{2+} + 2\,e^-$	$\rightarrow$	$Co(s)$	−0.28
$Tl^+ + e^-$	$\rightarrow$	$Tl(s)$	−0.34
$Cd^{2+} + 2\,e^-$	$\rightarrow$	$Cd(s)$	−0.40
$Cr^{3+} + e^-$	$\rightarrow$	Cr^{2+}	−0.41
$Fe^2 + 2\,e^-$	$\rightarrow$	$Fe(s)$	−0.44
$Cr^{3+} + 3\,e^-$	$\rightarrow$	$Cr(s)$	−0.74
$Zn^{2+} + 2\,e^-$	$\rightarrow$	$Zn(s)$	−0.76
$Mn^{2+} + 2\,e^-$	$\rightarrow$	$Mn(s)$	−1.18
$Al^{3+} + 3\,e^-$	$\rightarrow$	$Al(s)$	−1.66
$Be^{2+} + 2\,e^-$	$\rightarrow$	$Be(s)$	−1.70
$Mg^{2+} + 2\,e^-$	$\rightarrow$	$Mg(s)$	−2.37
$Na^+ + e^-$	$\rightarrow$	$Na(s)$	−2.71
$Ca^{2+} + 2\,e^-$	$\rightarrow$	$Ca(s)$	−2.87
$Sr^{2+} + 2\,e^-$	$\rightarrow$	$Sr(s)$	−2.89
$Ba^{2+} + 2\,e^-$	$\rightarrow$	$Ba(s)$	−2.90
$Rb^+ + e^-$	$\rightarrow$	$Rb(s)$	−2.92
$K^+ + e^-$	$\rightarrow$	$K(s)$	−2.92
$Cs^+ + e^-$	$\rightarrow$	$Cs(s)$	−2.92
$Li^+ + e^-$	$\rightarrow$	$Li(s)$	−3.05

Look at the reduction potential for Zn^{2+}.

$$Zn^{2+} + 2\ e^- \rightarrow Zn \qquad E = -0.76\ V$$

Read the reduction half-reaction in reverse and change the sign on the voltage to get the oxidation potential for Zn.

$$Zn \rightarrow Zn^{2+} + 2\ e^- \qquad E = +0.76\ V$$

The larger the potential for a half-reaction, the more likely it is to occur; for instance, let's look at the top of the table of half-reactions.

$$F_2(g) + 2e^- \rightarrow 2\ F^- \qquad E = +2.87\ V$$

$F_2(g)$ has a very large reduction potential, so it is likely to gain electrons and be reduced, making it a strong oxidizing agent.

Now let's look at the bottom of the table. We need to look at the reverse (oxidation) reaction to get a positive potential.

$$Li(s) \rightarrow Li^+ + e^- \qquad E = +3.05\ V$$

$Li(s)$ has a very large oxidation potential, making it very likely to lose electrons and be oxidized, so $Li(s)$ is a very strong reducing agent.

You can calculate the potential of a redox reaction if you know the potentials for the two half-reactions that constitute it. There are two important things to remember when calculating the potential of a redox reaction.

- Add the potential for the oxidation half-reaction to the potential for the reduction half-reaction.

- Never multiply the potential for a half-reaction by a coefficient.

Let's look at the following reaction:

$$Zn + 2\ Ag^+ \rightarrow Zn^{2+} + 2\ Ag$$

The two half-reactions are

| Oxidation: | $Zn \rightarrow Zn^{2+} + 2\ e^-$ | $E = +0.76\ V$ |
| Reduction: | $Ag^+ + e^- \rightarrow Ag$ | $E = +0.80\ V$ |

$E = E_{oxidation} + E_{reduction}$

$E = 0.76\ V + 0.80\ V = 1.56\ V$

Notice that we ignored that silver has a coefficient of 2 in the balanced equation.

VOLTAGE AND SPONTANEITY

A redox reaction will occur spontaneously if its potential has a positive value. We also know from thermodynamics that a reaction that occurs spontaneously has a negative value for free-energy change. The relationship between reaction potential and free energy for a redox reaction is given by the equation below, which serves as a bridge between thermodynamics and electrochemistry.

$$\Delta G° = -nFE°$$

$\Delta G°$ = Standard Gibbs free energy change (kJ/mol)

n = the number of moles of electrons exchanged in the reaction (mol)

F = Faraday's constant, 96,500 coulombs/mole (that is, 1 mole of electrons has a charge of 96,500 coulombs).

E = Standard reaction potential (V)

From this equation we can see a few important things. *If $E°$ is positive, $\Delta G°$ is negative and the reaction is spontaneous, and if $E°$ is negative, $\Delta G°$ is positive and the reaction is nonspontaneous.*

Look at the reaction we saw on the previous page.

$$Zn + 2\,Ag^+ \rightarrow Zn^{2+} + 2\,Ag \qquad\qquad E = +1.56\ V$$

The reaction potential is positive, so the free energy change is negative and the reaction is spontaneous.

VOLTAGE AND EQUILIBRIUM

The standard reaction potential is related to the equilibrium constant for a reaction by the equation below, which is another bridge between two important concepts.

$$E = \frac{RT}{nF}\ln K$$

E = standard reduction potential

R = the gas constant, 8.31 (volt-coulomb)/(mol-K)

T = absolute temperature (K)

n = the number of moles of electrons exchanged in the reaction (mol)

F = Faraday's constant, 96,500 coulombs/mole

K = equilibrium constant

At 25°C (298 K), this simplifies to

$$\log K = \frac{nE°}{0.0592}$$

If $E°$ is positive, then K is greater than 1 and the forward reaction is favored, and if $E°$ is negative, then K is less than 1 and the reverse reaction is favored.

GALVANIC CELLS

In a **galvanic cell** (also called a voltaic cell), a spontaneous redox reaction is used to generate a flow of current.

Look at the following spontaneous redox reaction:

$$Zn(s) + Cu^{2+}(aq) \rightarrow Zn^{2+}(aq) + Cu(s) \qquad E = 1.10 \text{ V}$$

Oxidation:	$Zn(s) \rightarrow Zn^{2+}(aq) + 2\text{ e}^-$	$E = 0.76$ V
Reduction:	$Cu^{2+}(aq) + 2\text{ e}^- \rightarrow Cu(s)$	$E = 0.34$ V

A galvanic cell using this reaction is shown below.

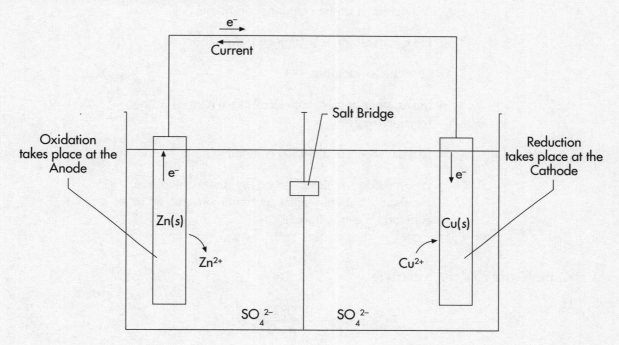

In a galvanic cell, the two half-reactions take place in separate chambers, and the electrons that are released by the oxidation reaction pass through a wire to the chamber where they are consumed in the reduction reaction. That's how the current is created. **Current** is defined as the flow of positive charge, so current is always in the opposite direction from the flow of electrons.

In any electric cell (either a galvanic cell or an electrolytic cell, which we'll discuss a little later) oxidation takes place at the electrode called the **anode**. Reduction takes place at the electrode called the **cathode**.

There's a mnemonic device to remember that.

AN OX
RED CAT

The salt bridge maintains electrical neutrality in the system by providing enough negative ions to equal the positive ions being created at the anode (during oxidation) and providing positive ions to replace the Cu^{2+} ions being used up at the cathode (during reduction). The salt bridge can be an actual salt, or it can be a slim passage that allows ions to move between the two chambers.

Under standard conditions (when all concentrations are 1 M) the voltage of the cell is the same as the total voltage of the redox reaction. Under nonstandard conditions, the cell voltage can be computed by using the Nernst equation.

The Nernst Equation
$$E_{cell} = E_{cell} - \frac{RT}{nF} \ln Q$$
E_{cell} = cell potential under nonstandard conditions (V)
E_{cell} = cell potential under standard conditions (V)
R = the gas constant, 8.31 (volt-coulomb)/(mol-K)
T = absolute temperature (K)
n = the number of moles of electrons exchanged in the reaction (mol)
F = Faraday's constant, 96,500 coulombs/mole
Q = the reaction quotient (same as the equilibrium expression, but with initial concentrations instead of equilibrium concentrations)

At 25°C, the Nernst equation reduces to

$$E_{cell} = E_{cell} - \frac{0.0592}{n} \log Q$$

Here's the most important thing you should understand from the Nernst equation: *As the concentration of the products of a redox reaction increases, the voltage decreases; and as the concentration of the reactants in a redox reaction increases, the voltage increases.* This is a direct result of Le Châtelier's law.

ELECTROLYTIC CELLS

In an electrolytic cell, an outside source of voltage is used to force a nonspontaneous redox reaction to take place. Let's look at the electrolysis of molten NaCl.

$$2\ Na^+(molten) + 2\ Cl^-(molten) \rightarrow 2\ Na(l) + Cl_2(g) \qquad\qquad E = -4.07\ V$$

Oxidation: $\quad 2\ Cl^-(molten) \rightarrow Cl_2(g) + 2\ e^- \qquad\qquad E = -1.36\ V$

Reduction: $\quad 2\ Na^+(molten) + 2\ e^- \rightarrow 2\ Na(l) \qquad\qquad E = -2.71\ V$

An electrolytic cell that forces this reaction to take place is shown below.

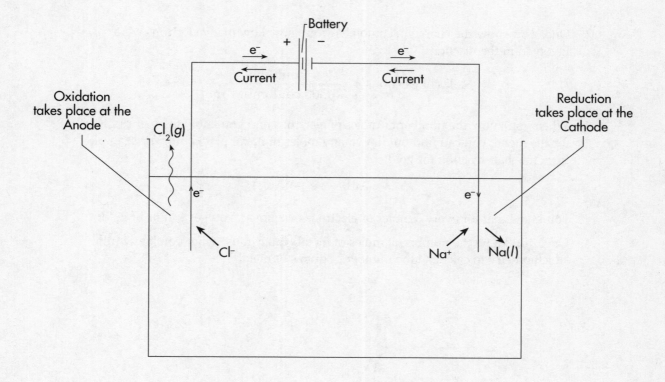

In this process, pure liquid sodium and pure chlorine gas are generated from molten sodium chloride. Notice that the process does not take place in an aqueous solution. That's because water is more easily reduced than Na^+, so water would be reduced instead of sodium ions in an aqueous solution.

The AN OX/RED CAT rule applies to the electrolytic cell in the same way that it applies to the galvanic cell.

Electrolytic cells are used for electroplating. You may see a question on the test that gives you an electrical current and asks you how much metal "plates out."

There are roughly four steps for figuring out electrolysis problems.

1. If you know the current and the time, you can calculate the charge in coulombs.

Current
$$I = \frac{q}{t}$$ I = current (amperes, A) q = charge (coulombs, C) t = time (sec)

2. Once you know the charge in coulombs, you know how many electrons were involved in the reaction.

$$\text{Moles of electrons} = \frac{\text{coulombs}}{96,500 \text{ coulombs / mol}}$$

3. When you know the number of moles of electrons and you know the half-reaction for the metal, you can find out how many moles of metal plated out. For example, from this half-reaction for gold

$$Au^{3+} + 3 \text{ e}^- \rightarrow Au(s)$$

you know that for every 3 moles of electrons consumed, you get 1 mole of gold.

4. Once you know the number of moles of metal, you can use what you know from stoichiometry to calculate the number of grams of metal.

CHAPTER 13 QUESTIONS

MULTIPLE-CHOICE QUESTIONS

<u>Questions 1–3</u>

    (A) MnO_4^-
    (B) H^+
    (C) H_2
    (D) Na
    (E) Na^+

1. This is a very strong reducing agent.

2. By definition, the reduction potential for this species is equal to zero.

3. This is a very strong oxidizing agent.

<u>Questions 4–7</u>

The choices listed below refer to n, the number of moles of electrons transferred in a reaction.

    (A) $n = 4$
    (B) $n = 3$
    (C) $n = 2$
    (D) $n = 1$
    (E) $n = 0$

4. $2\ Fe^{3+} + Mg \rightarrow 2\ Fe^{2+} + Mg^{2+}$

5. $F_2 + 2\ Br^- \rightarrow 2\ F^- + Br_2$

6. $NH_3 + H_2O \rightarrow NH_4^+ + OH^-$

7. $MnO_4^- + Cr + 2\ H_2O \rightarrow$
$$MnO_2 + Cr^{3+} + 4OH^-$$

8. $Cr_2O_7^{2-} + 6I^- + 14 H^+ \rightarrow 2Cr^{3+} + 3I_2 + 7H_2O$

Which of the following statements about the reaction given above is NOT true?

(A) The oxidation number of chromium changes from +6 to +3.
(B) The oxidation number of iodine changes from –1 to 0.
(C) The oxidation number of hydrogen changes from +1 to 0.
(D) The oxidation number of oxygen remains the same.
(E) The reaction takes place in acidic solution.

9. $Al^{3+} + 3e^- \rightarrow Al(s)$ $E = -1.66$ V

   $Cr^{3+} + 3e^- \rightarrow Cr(s)$ $E = -0.74$ V

The standard reduction potentials for two half-reactions are shown above. Which of the statements listed below will be true for the following reaction taking place under standard conditions?

   $Al(s) + Cr^{3+} \rightarrow Al^{3+} + Cr(s)$

(A) $E° = 2.40$ V, and the reaction is not spontaneous.
(B) $E° = 0.92$ V, and the reaction is spontaneous.
(C) $E° = -0.92$ V, and the reaction is not spontaneous.
(D) $E° = -0.92$ V, and the reaction is spontaneous.
(E) $E° = -2.40$ V, and the reaction is not spontaneous.

10. In which of the following molecules does hydrogen have an oxidation state of –1?

(A) H_2O
(B) NH_3
(C) CaH_2
(D) CH_4
(E) H_2

11. Oxygen takes the oxidation state –1 in hydrogen peroxide, H_2O_2. The equation for the decomposition of H_2O_2 is shown below.

   $2 H_2O_2 \rightarrow 2 H_2O + O_2$

Which of the following statements about the reaction shown above is true?

(A) Oxygen is reduced, and hydrogen is oxidized.
(B) Oxygen is oxidized, and hydrogen is reduced.
(C) Oxygen is both oxidized and reduced.
(D) Hydrogen is both oxidized and reduced.
(E) Neither oxygen nor hydrogen changes oxidation state.

12. When solid iron is brought into contact with water and oxygen, it undergoes the following half-reaction:

   $Fe(s) \rightarrow Fe^{2+}(aq) + 2e^-$

This half-reaction is instrumental in the corrosion of iron. When iron is coated with solid zinc, the half-reaction above is impeded, even if the zinc coating is incomplete. This is most likely because

(A) $Zn(s)$ is more easily reduced than $Fe(s)$.
(B) $Zn(s)$ is more easily oxidized than $Fe(s)$.
(C) $Zn^{2+}(aq)$ is more easily reduced than $Fe(s)$.
(D) $Zn^{2+}(aq)$ is more easily oxidized than $Fe(s)$.
(E) $Zn(s)$ is more easily reduced than $Fe^{2+}(aq)$.

13. $2 H_2O(l) + 2e^- \rightarrow$

$$H_2(g) + 2 OH^-(aq) \qquad E = -0.8 \text{ V}$$

$$Na^+(aq) + e^- \rightarrow Na(s) \qquad E = -2.7 \text{ V}$$

$$Cl_2(g) + 2e^- \rightarrow Cl^-(aq) \qquad E = +1.4 \text{ V}$$

Based on the reduction potentials given above, which of the following would be expected to occur when electrodes connected to the terminals of a 2.0 V battery are immersed in a solution of sodium chloride in water?

(A) Solid sodium will appear at the anode, and chlorine gas will appear at the cathode.

(B) Chlorine gas will appear at the anode, and solid sodium will appear at the cathode.

(C) Hydrogen gas will appear at the cathode, and solid sodium will appear at the anode.

(D) Hydrogen gas will appear at the anode, and chlorine gas will appear at the cathode.

(E) Chlorine gas will appear at the anode, and hydrogen gas will appear at the cathode.

14. $Cu^{2+} + 2e^- \rightarrow Cu \qquad E = +0.3 \text{ V}$

$Fe^{2+} + 2e^- \rightarrow Fe \qquad E = -0.4 \text{ V}$

Based on the reduction potentials given above, what is the reaction potential for the following reaction?

$$Fe^{2+} + Cu \rightarrow Fe + Cu^{2+}$$

(A) −0.7 V
(B) −0.1 V
(C) +0.1 V
(D) +0.7 V
(E) +1.4 V

15. $Cu^{2+} + 2e^- \rightarrow Cu \qquad E = +0.3 \text{ V}$

$Zn^{2+} + 2e^- \rightarrow Zn \qquad E = -0.8 \text{ V}$

$Mn^{2+} + 2e^- \rightarrow Mn \qquad E = -1.2 \text{ V}$

Based on the reduction potentials given above, which of the following reactions will occur spontaneously?

(A) $Mn^{2+} + Cu \rightarrow Mn + Cu^{2+}$
(B) $Mn^{2+} + Zn \rightarrow Mn + Zn^{2+}$
(C) $Zn^{2+} + Cu \rightarrow Zn + Cu^{2+}$
(D) $Zn^{2+} + Mn \rightarrow Zn + Mn^{2+}$
(E) $Cu^{2+} + Zn^{2+} \rightarrow Cu + Zn$

16. Which of the following is true about the oxidation-reduction reaction that takes place in a galvanic cell under standard conditions?

(A) $G°$ and $E°$ are positive, and K_{eq} is greater than 1.

(B) $G°$ is negative, $E°$ is positive, and K_{eq} is greater than 1.

(C) $G°$ is positive, $E°$ is negative, and K_{eq} is less than 1.

(D) $G°$ and $E°$ are negative, and K_{eq} is greater than 1.

(E) $G°$ and $E°$ are negative, and K_{eq} is less than 1.

17. $F_2 + 2e^- \rightarrow 2F^-$ $E° = 2.87$ V

     $Ag^{2+} + e^- \rightarrow Ag^+$ $E° = 1.99$ V

     $Au^{3+} + 3e^- \rightarrow 3Au°$ $E° = 1.50$ V

     $Cl_2 + 2e^- \rightarrow 2Cl^-$ $E° = 1.36$ V

     $Ag^+ + 3e^- \rightarrow 3Ag°$ $E° = 0.80$ V

In two separate experiments, pieces of metallic gold and metallic silver are reacted. In the first experiment both are reacted an inert solvent with Cl_2, while in the second they are reacted with F_2. Which of the following represents the metallic species present after the second reaction, which were not present after the first?

(A) Ag^+ only
(B) Ag^{2+} and Au^+
(C) Ag^+ and Ag^{2+}
(D) Ag^{2+} and Au^{3+}
(E) Ag^{2+} only

18. Many metallic ions in the +1 oxidation state (generically stated, M^{+1}) are known to undergo *disproportionation* reactions in which two M^{+1} ions electrochemically react to form one M^{2+} and one $M°$. If this reaction for the generic metal M^+ is spontaneous at room temperature, which of the following 1 electron standard potentials *must* sum to a positive number?

(A) the reduction potential of M^{2+} and the reduction potential of M^+
(B the reduction potential of M^+ and the oxidation potential of M^+
(C) the oxidation potential of M^{2+} and the oxidation potential of M^+
(D) the reduction potential of $M°$ and the reduction potential of M^+
(E) the oxidation potential of $M°$ and the reduction potential of M^{2+}

19. Molten $AlCl_3$ is electrolyzed with a constant current of 5.00 amperes over a period of 600.0 seconds. Which of the following expressions is equal to the maximum mass of Al(s) that plates out? (1 faraday = 96,500 coulombs)

(A) $\dfrac{(600)(5.00)}{(96,500)(3)(27.0)}$ grams

(B) $\dfrac{(600)(5.00)(3)(27.0)}{(96,500)}$ grams

(C) $\dfrac{(600)(5.00)(27.0)}{(96,500)(3)}$ grams

(D) $\dfrac{(96,500)(3)(27.0)}{(600)(5.00)}$ grams

(E) $\dfrac{(96,500)(3)}{(600)(5.00)(27.0)}$ grams

20. The half-reaction at the anode of a galvanic cell is as follows:

$$Zn(s) \rightarrow Zn^{2+} + 2e^-$$

What is the maximum charge, in coulombs, that can be delivered by a cell with an anode composed of 6.54 grams of zinc? (1 faraday = 96,500 coulombs)

(A) 4,820 coulombs
(B) 9,650 coulombs
(C) 19,300 coulombs
(D) 38,600 coulombs
(E) 48,200 coulombs

PROBLEMS

1. An electrochemical cell was created by placing a zinc electrode in a 1.00-molar solution of $ZnSO_4$ and placing a copper electrode in a 1.00-molar solution of $CuSO_4$. The two compartments were connected by a salt bridge, and the following reaction occurred at 25°C:

$$Zn(s) + Cu^{2+} \rightarrow Zn^{2+} + Cu(s)$$

   (a) What is the standard potential for the cell?

   (b) What is the value of $G°$ for the cell?

   (c) What is the value of K_{eq} for the reaction?

   (d) At a certain point in the progress of the reaction, $[Cu^{2+}]$ drops to 0.10 molar and $[Zn^{2+}]$ increases to 1.90 molar. What is the cell potential at this point?

2. An 800-milliliter sample of 0.080-molar Ag^+ solution was electrolyzed, resulting in the formation of solid silver and oxygen gas. The solution was subjected to a current of 2.00 amperes for 10.0 minutes. The solution became progressively more acidic as the reaction progressed.

   (a) Write the two half-reactions that occur, stating which takes place at the anode and which at the cathode.

   (b) If the oxygen gas produced in the electrolysis was collected at standard temperature and pressure, what was its volume?

   (c) What was the mass of solid silver produced in the electrolysis?

   (d) If the solution was neutral at the start of the electrolysis, what was the pH of the solution when the process was completed?

ESSAYS

3. Use the principles of electrochemistry to answer each of the following:

   (a) Explain why a salt bridge connecting the two compartments of a galvanic cell is necessary for the operation of the cell.

   (b) Explain why, when an iron nail is placed in hydrochloric acid, a reaction occurs, but when a copper penny is placed in hydrochloric acid, no reaction occurs.

   (c) $Ag^+ + Sn(s) \rightarrow Ag(s) + Sn^{2+}$

       (i) Give the standard cell potential for the reaction above.

       (ii) What happens to the cell potential in (i) when $[Ag^+]$ is increased?

       (iii) What happens to the cell potential when the amount of $Sn(s)$ in the cell is increased?

4. $M(s) + Cd^{2+} \rightarrow M^+ + Cd(s)$

   The reaction above proceeds spontaneously at 25°C in an electrochemical cell.

   (a) The reduction potential for M(s) must be less than a certain value. What is that value?

   (b) Write the two half-reactions, and identify which takes place at the anode and which takes place at the cathode.

   (c) Describe the changes that occur in the quantities below as the reaction proceeds in the forward direction.

       (i) $[M^+]$

       (ii) E_{cell}

       (iii) G

5. Molten sodium chloride can be electrolyzed to form pure sodium and chlorine.

   (a) Write the balanced equations for the half-reactions that take place at the electrodes during the electrolysis. Indicate which half-reaction takes place at which electrode.

   (b) Describe how the pure sodium and chlorine are separated.

   (c) For the electrolyzation of molten sodium chloride, will the value of each of the following be positive or negative?

       (i) $E_{reaction}$

       (ii) ΔG

       (iii) ΔS

   (d) Explain why, when the molten sodium chloride is replaced with an aqueous solution of sodium chloride, pure sodium is not produced.

CHAPTER 13 ANSWERS AND EXPLANATIONS

MULTIPLE-CHOICE QUESTIONS

1. **D** Na likes to give up an electron (oxidation) in the reaction below.

   $$Na \rightarrow Na^+ + e^-$$

   When a substance likes to be oxidized, it is a strong reducing agent.

2. **B** H^+ is reduced in the reaction below.

   $$2\,H^+ + 2e^- \rightarrow H_2$$

   By definition, the potential for this reaction is zero, so accordingly, the oxidation potential for H_2 is also zero.

3. **A** MnO_4^- likes to gain electrons (reduction) in either of the following reactions:

   $$MnO_4^- + 8\,H^+ + 5e^- \rightarrow Mn_2^+ + 2\,H_2O$$

   $$MnO_4^- + 2\,H_2O + 3e^- \rightarrow MnO_2 + 4\,OH^-$$

   When a substance likes to be reduced, it is a strong oxidizing agent.

4. **C** The two half-reactions are

   $$2\,Fe^{3+} + 2e^- \rightarrow 2\,Fe^{2+}$$

   $$Mg \rightarrow Mg^{2+} + 2e^-$$

5. **C** The two half-reactions are

   $$F_2 + 2e^- \rightarrow 2\,F^-$$

   $$2\,Br^- \rightarrow 2e^- + Br_2$$

6. **E** None of the elements involved in this acid-base reaction changes its oxidation state, so no electrons are transferred.

7. **B** The two half-reactions are

   $$Mn^{7+} + 3e^- \rightarrow Mn^{4+}$$

   $$Cr \rightarrow Cr^{3+} + 3e^-$$

8. **C** The oxidation numbers of the reactants are Cr^{6+}, O^{2-}, I^-, and H^+, and the oxidation numbers of the products are Cr^{3+}, O^{2-}, I^0, and H^+.

   Chromium gains electrons and is reduced, and iodine loses electrons and is oxidized; the oxidation states of oxygen and hydrogen are not changed.

9. **B** $E°$ for a redox reaction is given by the expression below.

   $$E° = E°_{ox} + E°_{red}$$

   $Al(s)$ loses 3 electrons in the reaction, so it is oxidized (LEO), and we use the voltage given for the reduction half-reaction, but we change the sign. So $E°_{ox} = 1.66$ V.

   Cr^{3+} gains 3 electrons in the reaction, so it is reduced (GER), and we use the voltage given for the reduction half-reaction. So $E°_{red} = -0.74$ V.

   So, $E° = 1.66$ V $+ (-0.74$ V$) = 0.92$ V.

   From the relationship, $\Delta G° = -nFE°$, we know that if $E°$ is positive, then $\Delta G°$ is negative, and if $\Delta G°$ is negative, then the reaction is spontaneous under standard conditions.

10. **C** In a molecule, the more electronegative element takes the negative oxidation state. Hydrogen is more electronegative than calcium, so the oxidation state for Ca is +2 and the oxidation state for H is –1.

In choices (A), (B), and (D), hydrogen is the less electronegative element, and it takes the oxidation state of +1; in choice (E), the oxidation state for hydrogen is 0.

11. **C** The oxidation state of hydrogen remains +1 throughout the reaction.

We have O^- at the start.

When water forms, oxygen has gained an electron and been reduced to O^{2-} (GER). When O_2 forms, oxygen has been oxidized to O^0 through the loss of an electron (LEO). As you can see, in this process, oxygen is both oxidized and reduced.

12. **B** The half-reaction given above is for the oxidation of Fe (LEO). If the presence of Zn impedes this process it is because Zn is oxidized instead of Fe. Iron nails are often coated with zinc to help keep them from rusting.

13. **E** A voltage of 2.0 V is not enough to cause the reduction of Na^+ (more than 2.7 V is needed), so no solid sodium will form. A voltage of 2.0 V will cause the reduction of H_2O, which forms hydrogen gas, and the oxidation of Cl^-, which forms chlorine gas.

From AN OX and RED CAT, we know that hydrogen gas is formed at the cathode and that chlorine gas is formed at the anode.

14. **A** We add the reduction potential for Fe^{2+} (–0.4 V) to the oxidation potential for Cu (–0.3 V, the reverse of the reduction potential) to get –0.7 V.

15. **D** To get the reaction potential, we add the reduction potential for the reduction half-reaction to the oxidation potential for the oxidation half–reaction.

Here are the reaction potentials for choices (A)–(D).

(A) (–1.2 V) + (–0.3 V) = –1.5 V

(B) (–1.2 V) + (+0.8 V) = –0.4 V

(C) (–0.8 V) + (–0.3 V) = –1.1 V

(D) (–0.8 V) + (+1.2 V) = +0.4 V

(E) $Cu^{2+} + Zn^{2+} \rightarrow Cu + Zn$

Choice (E) is a combination of two reduction half-reactions, which can't happen.

Choice (D) is the only reaction with a positive voltage, which means that it is the only reaction listed that occurs spontaneously.

16. **B** The oxidation-reduction reaction in a galvanic cell occurs spontaneously and provides a voltage that causes a flow of electrons, so $\Delta G°$ is negative and $E°$ is positive. Also, from the equation $\Delta G° = -nFE°$, we know that $\Delta G°$ and $E°$ always have opposite signs.

We know that K_{eq} must be greater than 1 from either of the following equations:

$\Delta G° = -2.303RT\log K$, which tells us that if $\Delta G°$ is to be negative, then log K must be positive, which means that K must be greater than 1.

Or

$\log K = \dfrac{nE°}{0.0592}$, which tells us that if $E°$ is to be positive, then $\log K$ must be positive, which means that K must be greater than 1.

17. **D** In order for the metallic species to be oxidized the reduction potential of the oxidant (F_2 or Cl_2) must be greater than the reduction potential of the metallic species. Cl_2 is a strong enough oxidant to take Ag° to Ag^+, but not strong enough to take Ag^+ to Ag^{2+} or Au° to Au^{3+}. On the other hand, F_2 *is* a strong enough oxidant, as it has the highest reduction potential listed, to do all of the above electrochemical reactions. Therefore Au° will be oxidized to Au^{3+} and Ag° will be oxidized all the way to Ag^{2+}.

18. **B** The spontaneity of the disproportionation reaction means that the standard potentials of the following two half reactions must sum to a positive number:

$$M^+ + 1e^- \rightarrow M^\circ$$

$$M^+ \rightarrow 1e^- + M^{2+}$$

The two potentials here are the reduction potential and oxidation potential of M^+. Without knowing the relative values and signs of the two individual reactions we cannot say for sure any of the rest of the listed pairs of potentials would sum to a positive.

19. **C** First let's find out how many electrons are provided by the current.

Coulombs = (seconds)(amperes) = (600)(5.00)

$$\text{Moles of electrons} = \frac{(\text{coulombs})}{(96,500)} = \frac{(600)(5.00)}{(96,500)} \text{ moles}$$

In $AlCl_3$, we have Al^{3+}, so the half-reaction for the plating of aluminum is

$$Al^{3+} + 3e^- \rightarrow Al(s)$$

So we get $\frac{1}{3}$ as many moles of $Al(s)$ as we have moles of electrons.

$$\text{Moles of Al}(s) = (\text{moles of electrons})(\frac{1}{3}) = \frac{(600)(5.00)}{(96,500)(3)} \text{ moles}$$

Now, grams = (moles)(MW)

$$\text{Grams of Al}(s) = \frac{(600)(5.00)}{(96,500)(3)}(27.0) = \frac{(600)(5.00)(27.0)}{(96,500)(3)} \text{ grams}$$

20. **C** First let's find out how many moles of zinc we have.

$$\text{Moles} = \frac{\text{grams}}{\text{MW}}$$

$$\text{Moles of Zn} = \frac{(6.54 \text{ g})}{(65.4 \text{ g / mol})} = 0.100 \text{ mole}$$

Based on the half-reaction, we can see that for every mole of Zn consumed, we get 2 moles of electrons. So we have (2)(0.100) = 0.200 moles of electrons.

$$\text{Moles of electrons} = \frac{\text{coulombs}}{96,500}$$

So, coulombs = (moles of electrons)(96,500)

= (0.200)(96,500) = a little less than 20,000 = exactly 19,300

PROBLEMS

1. (a) Use the reduction potentials from the reduction potential chart.

$$Zn(s) \rightarrow Zn^{2+} + 2e^- \qquad\qquad E = +0.76\ V$$
$$Cu^{2+} + 2e^- \rightarrow Cu(s) \qquad\qquad E = \underline{+0.34\ V}$$
$$\qquad\qquad\qquad\qquad\qquad\qquad\qquad +1.10\ V$$

(b) Use the following expression:

$\Delta G = -nFE$

$n = 2$ because two moles of electrons are exchanged in the redox reaction.

$E = 1.10$ V, from part (a)

$F = 96,500$ C/mol

$\Delta G = -(2)(96,500\ C/mol)(1.10\ V) = -212,300\ C\text{-}V/mol = -212,300\ J/mol = -212\ kJ/mol$

(c) You have two options.

$$\log K = \frac{nE^\circ}{0.0592} = \frac{(2)(1.10)}{(0.0592)} = 37.1$$

$K = 1.45 \times 10^{37}$

Or

$$\log K = \frac{\Delta G^\circ}{-2.303\ RT} = \frac{(-212,000\ J/mol)}{(-2.303)(8.31\ J/mol\text{-}K)(298\ K)} = 37.2$$

$K = 1.49 \times 10^{37}$

(d) Use the Nernst equation for 25°C.

$$E_{cell} = E - \frac{0.0592}{n} \log Q$$

$$E_{cell} = E - \frac{0.0592}{2} \log \frac{\left[Zn^{2+} \right]}{\left[Cu^{2+} \right]}$$

$$E_{cell} = 1.10\ V - \frac{0.0592}{2} \log \frac{(1.90)}{(0.10)}\ V$$

$$E_{cell} = 1.10\ V - 0.04\ V = 1.06\ V$$

2. (a) Silver half-reaction: $Ag^+ + e^- \rightarrow Ag(s)$

Silver gains an electron, so it is reduced (GER). Reduction takes place at the cathode (RED CAT).

Water half-reaction: $2\ H_2O \rightarrow O_2 + 4\ H^+ + 4e^-$

Water gives up electrons, so it is oxidized (LEO). Oxidation takes place at the anode (AN OX).

(b) First let's find out how many electrons were supplied by the current.

Coulombs = (amperes)(seconds)

Coulombs = (2.00 A)(10.0 min)(60 sec/min) = 1,200 C

$$\text{Moles of electrons} = \frac{\text{coulombs}}{96,500} = \frac{1,200}{96,500} = 0.0124\ \text{moles}$$

From the water half-reaction we know that one mole of oxygen gas is produced for every 4 electrons given up, so there will be $\frac{1}{4}$ as many moles of O_2 as there are electrons.

$$\text{Moles of } O_2 = (\tfrac{1}{4})(\text{moles of electrons}) = (\tfrac{1}{4})(0.0124\ \text{mol}) = 3.10 \times 10^{-3}\ \text{moles}$$

At STP, volume of gas = (moles)(22.4 L/mol)

Volume of O_2 = $(3.10 \times 10^{-3}$ mol$)(22.4$ L/mol$)$ = 0.0694 L

(c) From (b), we know that the current provided 0.0124 moles of electrons.

From the silver half-reaction, we know that for every mole of electrons consumed, one mole of Ag(s) is produced. So we have 0.0124 moles of Ag(s).

Grams = (moles)(MW)

Grams of Ag(s) = (0.0124 mol)(107.87 g/mol) = 1.34 grams

(d) From (b), we know that the current provided 0.0124 moles of electrons. From the water half-reaction we know that 4 moles of H^+ were produced for every 4 electrons given up, so the number of moles of H^+ will be the same as the number of electrons.

Moles of H^+ = 0.0124

We can assume that the volume of solution doesn't change during the electrolysis.

$$Molarity = \frac{moles}{volume}$$

$$[H^+] = \frac{(0.0124 \text{ mol})}{(0.800 \text{ L})} = 0.0155 \text{ } M$$

$$pH = -\log[H^+] = -\log(0.0155) = 1.81$$

ESSAYS

3. (a) The salt bridge is necessary to maintain electrical neutrality in the two compartments where the half-reactions are taking place. It does this by allowing anions (–) to move into the anode compartment and cations (+) to move into the cathode compartment.

(b) From the table of standard reduction potentials, we can see that Fe wants to be oxidized (reading the table from right to left gives a positive oxidation potential), so it can act as the reducing agent for H^+ ions.

Cu does not want to be oxidized (reading the table [from right to left] gives a negative oxidation potential), so it will not reduce the H^+ ions.

(c) (i) Use the table of standard reduction potentials.

$Ag^+ + e^- \rightarrow Ag(s)$ E = +0.80 V

$Sn(s) \rightarrow Sn^{2+} + e^-$ E = +0.14 V

                        +0.94 V

(ii) Use the Nernst equation.

$$E_{cell} = E - \frac{0.0592}{n} \log Q$$

$$E_{cell} = E - \frac{0.0592}{n} \log \frac{\left[Sn^{2+}\right]}{\left[Ag^+\right]}$$

When $[Ag^+]$ is increased, $\log \dfrac{\left[Sn^{2+}\right]}{\left[Ag^+\right]}$ decreases, which means that a smaller number is subtracted from E, which means that E_{cell} increases.

You can also use Le Châtelier's law. Increasing $[Ag^+]$ increases the concentration of the reactants, which favors the forward reaction and increases the voltage of the cell.

(iii) The cell potential changes with changing concentrations. The concentration of a solid is constant, so changing the amount of solid Sn present in the cell will have no effect on the cell potential.

4. (a) The reduction potential for Cd^{2+} is equal to –0.40 V. For the reaction to proceed spontaneously, the cell potential must be greater than zero, so the oxidation potential for M(s) must be greater than +0.40 V. That means that the reduction potential must be less than –0.40V.

(b) $M(s) \rightarrow M^+ + e^-$

M(s) loses an electron and is oxidized (LEO), so this half-reaction takes place at the anode (AN OX).

$Cd^{2+} + 2e^- \rightarrow Cd(s)$

Cd^{2+} gains electrons and is reduced (GER), so this half-reaction takes place at the cathode (RED CAT).

(c) (i) As the reaction proceeds in the forward direction, more M^+ will be generated, so $[M^+]$ will increase.

(ii) Use the Nernst equation.

$$E_{cell} = E - \frac{0.0592}{n} \log Q$$

$$E_{cell} = E - \frac{0.0592}{n} \log \frac{[M^+]}{[Cd^{2+}]}$$

As the reaction proceeds, $[M^+]$ increases and $[Cd^{2+}]$ decreases, thereby increasing the log value, which means that progressively larger numbers are subtracted from E, which means that E_{cell} decreases.

(iii) Use the equation

$\Delta G = -nFE$

From (ii), we know that E is decreasing. As E decreases, G becomes less negative, so it increases. That means that as the reaction proceeds, it becomes less spontaneous.

5. (a) $Na^+ + e^- \rightarrow Na$

Na^+ gains an electron, so it is reduced (GER).

Reduction takes place at the cathode (RED CAT).

$2\ Cl^- \rightarrow Cl_2 + 2e^-$

Cl^- loses an electron, so it is oxidized (LEO).

Oxidation takes place at the anode (AN OX).

(b) Chlorine gas (Cl_2) bubbles out of the molten mixture, leaving pure molten sodium.

(c) (i) $E_{reaction}$ will be negative because E for both half-reactions is negative. The voltage source that drives the reaction must be strong enough to overcome the negative voltage for the reaction.

(ii) ΔG will be positive because the reaction is not spontaneous. Also, from the relationship $\Delta G = -nFE$ we know that ΔG and E always have opposite signs.

(iii) ΔS will be positive because the reactant is a liquid and the products are a liquid and a gas.

(d) Water molecules are more easily reduced than sodium ions, so the hydrogen in H_2O will be reduced at the cathode instead of Na^+.

14
Nuclear Decay

HOW OFTEN DOES NUCLEAR DECAY APPEAR ON THE EXAM?

In the multiple-choice section, this topic appears in about 3 out of 75 questions. In the free-response section, this topic appears occasionally.

A nucleus is held together by a nonelectrical, nongravitational force called the **nuclear force.**

Some nuclei are more stable than others. When a nucleus is unstable, it can attempt to increase its stability by altering its number of neutrons and protons. This is the process of radioactive decay. Elements are naturally radioactive after atomic number 83, Bismuth.

TYPES OF NUCLEAR DECAY

ALPHA EMISSION ($_2^4\alpha$)

In **alpha decay**, the nucleus emits a particle that has the same constitution as a helium molecule, with 2 protons and 2 neutrons. Alpha particles are the least penetrating of the products of nuclear decay.

$$_2^4\alpha = _2^4\text{He}$$

When a nucleus undergoes alpha decay

- subtract 4 from the **mass number**
- subtract 2 from the atomic number

$$\,^{238}_{92}U \rightarrow \,^{4}_{2}\alpha + \,^{234}_{90}Th$$

BETA EMISSION ($\,^{0}_{-1}\beta$)

A beta particle is identical to an electron. In beta decay, the nucleus changes a neutron into a proton and an electron and emits the electron.

$$\,^{1}_{0}n \rightarrow \,^{0}_{-1}\beta + \,^{1}_{1}p$$

When a nucleus undergoes beta decay

- the mass number remains the same
- add 1 to the atomic number

$$\,^{14}_{6}C \rightarrow \,^{0}_{-1}\beta + \,^{14}_{7}N$$

POSITRON EMISSION ($\,^{0}_{+1}\beta$)

A positron is like an electron with a positive charge. In positron emission, the nucleus changes a proton into a neutron and a positron and emits the positron.

$$\,^{1}_{1}p \rightarrow \,^{0}_{+1}\beta + \,^{1}_{0}n$$

When a nucleus undergoes positron emission

- the mass number remains the same
- subtract 1 from the atomic number

$$\,^{8}_{5}B \rightarrow \,^{0}_{+1}\beta + \,^{8}_{4}Be$$

ELECTRON CAPTURE ($\,^{0}_{-1}e$)

In electron capture, the nucleus captures a low energy electron and combines it with a proton to form a neutron.

$$\,^{0}_{-1}e + \,^{1}_{1}p \rightarrow \,^{1}_{0}n$$

When a nucleus undergoes electron capture

- the mass number remains the same
- subtract 1 from the atomic number

$$\,^{18}_{9}F + \,^{0}_{-1}e \rightarrow \,^{18}_{8}O$$

GAMMA RAYS ($\,^{0}_{0}\gamma$)

Gamma rays are electromagnetic radiation and have no mass and no charge. Gamma rays usually accompany other forms of nuclear decay, and are the most penetrating of the nuclear decay products.

NUCLEAR STABILITY

Nuclei undergo decay to achieve greater stability. You can use the periodic table to predict the kind of decay that an isotope will undergo.

- If an isotope's *mass number is greater than its atomic weight* ($^{16}_{6}C$, for instance), the nucleus will try to gain protons and lose neutrons; therefore, if its mass number is greater than its atomic weight, you can *expect beta decay*.

- If an isotope's *mass number is less than its atomic weight* ($^{11}_{6}C$, for instance), the nucleus will try to lose protons and gain neutrons; therefore, if its mass number is less than its atomic weight, you can *expect positron emission* or *electron capture*.

- Alpha emission is seen mainly in very large nuclei, usually with atomic numbers of 60 or greater.

HALF-LIFE

The half-life of a radioactive substance is the time it takes for half of the substance to decay. Most half-life problems can be solved by using a simple chart.

Time	Sample
0	100%
1 half-life	50%
2 half-lives	25%
3 half-lives	12.5%

So a sample with a mass of 120 grams and a half-life of 3 years will decay as follows: (Don't forget that the chart should start with the time at zero.)

Time	Sample
0	120 g
3	60 g
6	30 g
9	15 g

The fact that you're not allowed to use a calculator for Section I and that you're not given any half-life formulas for Section II means that you should be able to solve any half-life problem that comes up by using the chart and POE.

MASS DEFECT AND BINDING ENERGY

When protons and neutrons come together to form a nucleus, the mass of the nucleus is less than the sum of the masses of its constituent protons and neutrons. This difference in mass is called the **mass defect**. The mass lost in this process is released in the form of energy. If we reverse the process this is the same amount of energy, called the **binding energy**, required to decompose the nucleus back into protons and neutrons. The relationship between mass and energy is given by Einstein's famous equation.

$$E = mc^2$$

E = energy (J)
m = mass (kg)
c = the speed of light, 3×10^8 m/sec

You can see that because c^2 is such a large number, a very small change in mass results in a very large change in energy.

CHAPTER 14 QUESTIONS

MULTIPLE-CHOICE QUESTIONS

Questions 1–4

   (A) Alpha decay
   (B) Beta (β^-) decay
   (C) Electron capture
   (D) Gamma radiation
   (E) Mass defect

1. In this process the number of protons in a nuclide is increased while the mass number remains constant.

2. In this process a nuclide releases a particle that is the equivalent of a helium nucleus.

3. This is the difference between the mass of a nucleus and the sum of its component nucleons.

4. In this process the number of protons in a nuclide is decreased while the mass number remains constant.

Questions 5–8

   (A) $^{26}_{13}$Al, $^{26}_{12}$Mg

   (B) $^{58}_{28}$Ni, $^{64}_{28}$Ni

   (C) $^{12}_{5}$B, $^{8}_{3}$Li

   (D) $^{19}_{9}$F, $^{20}_{10}$Ne

   (E) $^{39}_{18}$Ar, $^{39}_{20}$Ca

5. This pair could be the only two elements present in a sample undergoing alpha decay.

6. This pair could be the only two elements present in a sample undergoing β^+ decay.

7. This pair are isotopes of each other.

8. This pair could be the only two elements present in a sample undergoing electron capture.

9. The nuclide $^{128}_{50}Sn$ is unstable. Which of the following decay types would $^{128}_{50}Sn$ be expected to undergo?

   I. Beta decay (β^-) to increase the ratio of protons to neutrons.
   II. Positron emission (β^+) to decrease the ratio of protons to neutrons.
   III. Electron capture to decrease the ratio of protons to neutrons.

   (A) I only
   (B) II only
   (C) I and III only
   (D) II and III only
   (E) I, II, and III

10. Which of the following statements is true regarding the mass and magnitude of charge of alpha particles and beta particles?

   (A) Alpha particles are more highly charged and have greater mass than beta particles.
   (B) Beta particles are more highly charged and have greater mass than alpha particles.
   (C) Alpha particles are more highly charged than beta particles, but beta particles have greater mass.
   (D) Beta particles are more highly charged than alpha particles, but alpha particles have greater mass.
   (E) Alpha particles are more highly charged than beta particles, but the masses of the two types of particles are the same.

11. Strontium-90 decays through the emission of beta particles. It has a half-life of 29 years. How long does it take for 80 percent of a sample of strontium-90 to decay?

   (A) 9.3 years
   (B) 21 years
   (C) 38 years
   (D) 67 years
   (E) 96 years

12. A sample of radioactive material undergoing nuclear decay is found to contain only potassium and calcium. The sample could be undergoing which of the following decay processes?

   I. Beta (β^-) decay
   II. Alpha decay
   III. Electron capture

   (A) I only
   (B) II only
   (C) I and III only
   (D) II and III only
   (E) I, II, and III

13. A sample of radioactive material undergoing nuclear decay is found to contain only $_{84}$Po and $_{82}$Pb. The sample could be undergoing which of the following decay processes?

    I. Beta (β^-) decay
   II. Alpha decay
  III. Electron capture

(A) I only
(B) II only
(C) I and III only
(D) II and III only
(E) I, II, and III

14. A $_{84}^{214}$Po nuclide emits two alpha particles and two beta (β^-) particles. The resulting nuclide is

(A) $_{82}^{210}$Pb

(B) $_{83}^{210}$Bi

(C) $_{84}^{210}$Po

(D) $_{82}^{206}$Pb

(E) $_{84}^{206}$Po

15. After 44 minutes, a sample of $_{19}^{44}$K is found to have decayed to 25 percent of the original amount present. What is the half-life of $_{19}^{44}$K?

(A) 11 minutes
(B) 22 minutes
(C) 44 minutes
(D) 66 minutes
(E) 88 minutes

ESSAYS

1. Use the principles of nuclear chemistry to explain each of the following:

   (a) Gamma radiation penetrates farther into the human body than either alpha or beta radiation.

   (b) A nucleus weighs less than the sum of the weights of its neutrons and protons.

   (c) Arsenic-81 is unstable. What form of radioactive decay would it be expected to undergo?

   (d) Potassium-38 decays by electron capture. Write the balanced nuclear reaction for this process.

   (e) In a nuclear explosion, a small mass has enormous destructive power.

2.

Isotope	Half-Life	Mode of Decay
$^{21}_{11}$Na	23 sec	β^+
$^{23}_{11}$Na	—	—
$^{25}_{11}$Na	60 sec	β^-
$^{26}_{11}$Na	1.0 sec	β^-

   (a) Write the balanced nuclear reaction for sodium-21.

   (b) Write the balanced nuclear reaction for sodium-26.

   (c) A sample of sodium-25 is observed to decay for 5 minutes. After this time, no change in the mass of the sample is detected. Explain.

   (d) Describe alpha, beta (β^-), and gamma radiation in terms of mass and charge.

3. Neon-23 is an unstable nuclide with a half-life of 38 seconds.

   (a) What form of decay would neon-23 be expected to undergo? Write the balanced nuclear reaction for that process.

   (b) Explain why the nucleus of a neon-23 atom weighs less than the sum of its constituent protons and neutrons.

   (c) Describe the change that a sample of neon-23 undergoes in 38 seconds.

   (d) Describe alpha, beta (β^-), and gamma radiation in terms of mass and charge.

   (e) The decay of neon-23 is shown on the graph below.

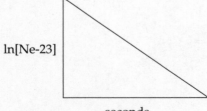

   ln[Ne-23]

   seconds

   (i)   What is the order of the nuclear reaction?
   (ii)   What is represented by the slope of the graph?
   (iii) What are the units of the rate constant for this reaction?

CHAPTER 14 ANSWERS AND EXPLANATIONS

MULTIPLE-CHOICE QUESTIONS

1. **B** In beta decay, a nuclide releases a β^- particle (which is the same as an electron) and converts a neutron to a proton.

   $$_0^1\text{n} \rightarrow {}_1^1\text{p} + {}_{-1}^{0}\beta$$

2. **A** An alpha particle ($_2^4\alpha$) has two protons and two neutrons, the same as a helium nucleus.

3. **E** Mass defect is the mass that seems to disappear from neutrons and protons when they are brought together to form a nucleus.

   Mass defect is the m in Einstein's famous equation for the equivalence of mass and energy, $E = mc^2$.

4. **C** In electron capture, a nuclide absorbs one of its low energy electrons and combines it with a proton to form a neutron.

   $$_1^1\text{p} + {}_{-1}^{0}\text{e} \rightarrow {}_0^1\text{n}$$

5. **C** In alpha decay, a particle with the mass (4) and charge (+2) of a helium nucleus is given off, so the two nuclides must differ by a mass of 4 and a charge of 2.

   $$_5^{12}\text{B} \rightarrow {}_3^8\text{Li} + {}_2^4\alpha$$

6. **A** In β^+ decay, the mass number remains constant and the proton number decreases by one, so the two nuclides must have the same mass number and differ by one in their proton numbers.

   $$_{13}^{26}\text{Al} \rightarrow {}_{12}^{26}\text{Mg} + {}_{+1}^{0}\beta$$

7. **B** Isotopes are atoms of the same element (same number of protons) with differing mass numbers.

8. **A** In electron capture, the mass number remains constant and the proton number decreases by one, so the two nuclides must have the same mass number and differ by one in their proton numbers.

   $$_{13}^{26}\text{Al} + \text{e}^- \rightarrow {}_{12}^{26}\text{Mg}$$

9. **A** From the periodic table, we can see that Sn has an atomic mass of 118.71, so 128 is a very large mass number for Sn. The nuclide will decay to increase its proton to neutron ratio, thereby making itself more stable.

10. **A** An alpha particle is a helium nucleus, so it has a mass of 4 g/mol and a charge of +2.

    A beta particle is an electron, so its mass is very small compared with the masses of the protons and neutrons in the alpha particle. The charge on a beta particle is the same as the charge on an electron, –1.

11. **D** Make a chart. Always start at time = 0. We're looking for the time it takes for 80 percent of the stuff to decay. That's the time when 20 percent of the stuff remains.

Half-Lives	Time	Stuff
0	0	100%
1	29 years	50%
2	58 years	25%
3	87 years	12.5%

It takes between 2 and 3 half-lives to get to the point when 20 percent remains. The correct answer is the only choice between 58 years and 87 years.

12. **C** Since decay is occurring and only potassium and calcium are present, one must be decaying to produce the other. Potassium and calcium differ by one proton, so the decay process must either add or subtract a proton.

Beta decay (I) is possible.

$$_{19}K \rightarrow {_{20}}Ca + {_{-1}^{0}}\beta$$

Electron capture (III) is possible.

$$_{20}Ca + {_{-1}^{0}}e^- \rightarrow {_{19}}K$$

Alpha decay (II) causes the loss of two protons, so it can't be occurring if only K and Ca are present.

13. **B** Since decay is occurring and only Po and Pb are present, one must be decaying to produce the other. Po has 84 protons and Pb has 82. The only way that a decay process can involve only these two elements is if Po undergoes alpha decay to produce Pb.

Alpha decay (II)

$$_{84}^{214}Po \rightarrow {_{82}^{210}}Pb + {_{2}^{4}}\alpha$$

If beta decay (I) or electron capture (III) were occurring, the proton count would be changing one at a time and $_{83}Bi$ would also be present in the sample.

14. **D** Let's do the math; you set it up like this:

$$_{84}^{214}Po - {_{2}^{4}}\alpha - {_{2}^{4}}\alpha - {_{-1}^{0}}\beta - {_{-1}^{0}}\beta =$$

For the mass number we have

$$214 - 4 - 4 = 206$$

For the proton number we have

$$84 - 2 - 2 - (-1) - (-1) = 84 - 4 + 2 = 82$$

So the answer is $_{82}^{206}Pb$.

15. **B** Make a chart. Always start at time = 0.

Half-Lives	Time	Stuff
0	0	100%
1	X	50%
2	44 min.	25%

It takes two half-lives for the amount of $_{19}^{44}K$ to decrease to 25 percent. If two half-lives takes 44 minutes, one half-life must be 22 minutes.

ESSAYS

1. (a) Alpha and beta particles have mass and volume, so the surface of the body can often stop or slow them down. Gamma rays are electromagnetic radiation, which doesn't undergo collisions and can travel farther into the body.

   (b) Some of the mass of the neutrons and protons is lost to nuclear binding energy, which is released by the stable nucleus.

   (c) Stable arsenic has an atomic weight of 75, so arsenic-81 has too many neutrons. It will undergo β^- decay, which converts a neutron into a proton.

   $$^{75}_{33}\text{As} \rightarrow {}^{75}_{34}\text{Se} + {}^{0}_{-1}\beta$$

   (d) In electron capture, the nucleus captures an electron from a lower energy level and combines it with a proton to form a neutron.

   $$^{38}_{19}\text{K} + {}^{0}_{-1}e \rightarrow {}^{38}_{18}\text{Ar}$$

   (e) Remember Einstein's equation $E = mc^2$.

   This means that the energy released by a nuclear process is 9×10^{16} times as large as the mass consumed.

2. (a) In β^+ decay, a proton is converted into a neutron and a positron is emitted.

   $$^{22}_{11}\text{Na} \rightarrow {}^{21}_{10}\text{Ne} + {}^{0}_{+1}\beta$$

   (b) In β^- decay, a neutron is converted into a proton, and a β^- particle (electron) is emitted.

   $$^{26}_{11}\text{Na} \rightarrow {}^{26}_{12}\text{Mg} + {}^{0}_{-1}\beta$$

   (c) Over the course of the decay process, only β^- particles are emitted. Neutrons are converted into protons, but there is no change in the mass numbers of the nuclides undergoing decay. Beta particles are virtually massless, so the loss of beta particles may not be detected.

   (d) Alpha particles have the mass (4) and charge (+2) of a helium nucleus.

   Beta (β^-) particles have the mass (insignificant) and charge (−1) of an electron.

   Gamma radiation has neither mass nor charge. Gamma radiation is electromagnetic radiation.

3. (a) Stable neon has an atomic weight of 20, so neon-23 has too many neutrons. It will undergo β^- decay, which converts a neutron to a proton.

   $$^{23}_{10}\text{Ne} \rightarrow {}^{23}_{11}\text{Na} + {}^{0}_{+1}\beta$$

(b) Some of the mass of the protons and neutrons is converted to nuclear binding energy ($E = mc^2$), which is given off when the nucleus forms.

(c) The half-life of neon-23 is 38 seconds. That means that half of any sample of neon-23 will be converted into sodium-23 after 38 seconds.

(d) Alpha particles have the mass (4) and the charge (+2) of a helium nucleus.

Beta (β^-) particles have insignificant mass and the charge (−1) of an electron.

Gamma radiation has neither mass nor charge. Gamma radiation is electromagnetic radiation.

(e) (i) This nuclear reaction is first order. The graph of ln[Ne-23] versus time is linear.

   (ii) The slope is the negative of the rate constant, k.

   (iii) Units are sec^{-1}.

15

Laboratory

HOW OFTEN DOES LABORATORY APPEAR ON THE EXAM?

In the multiple-choice section, this topic appears in about 7 out of 75 questions. In the free-response section, this topic appears every year.

There will be some questions on the test that are specifically about lab technique, but most "lab" questions will be about specific chemistry topics, placed in a lab setting to make them seem more intimidating. You just need to remember some basic rules and combine them with science and common sense. The free-response questions may require you to remember some specifics about some basic experiments and laboratory techniques. The best way to prepare for this is to review your lab notebooks.

RECOMMENDED EXPERIMENTS

These are the 22 experiments that the College Board recommends that AP Chemistry classes perform during the school year.

1. Determination of the formula of a compound

2. Determination of the percentage of water in a hydrate

3. Determination of molar mass by vapor density

4. Determination of molar mass by freezing-point depression

5. Determination of the molar volume of a gas

6. Standardization of a solution using a primary standard

7. Determination of concentration by acid-base titration, including a weak acid or weak base

8. Determination of concentration by oxidation-reduction titration

9. Determination of mass and mole relationship in a chemical reaction

10. Determination of the equilibrium constant for a chemical reaction

11. Determination of appropriate indicators for various acid-base titrations and determining pH

12. Determination of the rate of a reaction and its order

13. Determination of enthalpy change associated with a reaction

14. Separation and qualitative analysis of anions and cations

15. Synthesis of a coordination compound and its chemical analysis

16. Analytical gravimetric determination

17. Colorimetric or spectrophotometric analysis

18. Separation by chromatography

19. Preparation and properties of buffer solutions

20. Determination of electrochemical series

21. Measurements using electrochemical cells and electroplating

22. Synthesis, purification, and analysis of an organic compound

SAFETY

Here are some basic safety rules that might turn up in test questions.

- Don't put chemicals in your mouth. You were told this when you were four years old, and it still holds true for the AP Chemistry Exam.

- When diluting an acid, always add the acid to the water. This is to avoid the spattering of hot solution.

- Always work with good ventilation; many common chemicals are toxic.

- When heating substances, do it slowly. When you heat things too quickly, they can spatter, burn, or explode.

ACCURACY

Here are some rules for ensuring the accuracy of experimental results.

- When titrating, rinse the buret with the solution to be used in the titration instead of with water. If you rinse the buret with water, you might dilute the solution, which will cause the volume added from the buret to be too large.

- Allow hot objects to return to room temperature before weighing. Hot objects on a scale create convection currents that may make the object seem lighter than it is.

- Don't weigh reagents directly on a scale. Use a glass or porcelain container to prevent corrosion of the balance pan.

- When collecting a gas over water, remember to take into account the pressure and volume of the water vapor.

- Don't contaminate your chemicals. Never insert another piece of equipment into a bottle containing a chemical. Instead you should always pour the chemical into another clean container. Also, don't let the inside of the stopper for a bottle containing a chemical touch another surface.

- When mixing chemicals, stir slowly to ensure even distribution.

- Be conscious of significant figures when you record your results. The number of significant figures you use should indicate the accuracy of your results.

- Be aware of the difference between accuracy and precision. A measurement is accurate if it is close to the accepted value. A series of measurements is precise if the values of all of the measurements are close together.

LAB PROCEDURES

METHODS OF SEPARATION

Filtration—In filtration, solids are separated from liquids when the mixture is passed through a filter. Typically, porous paper is used as the filter. To find the amount of solid that is filtered out of a mixture, the filter paper containing the solid is allowed to dry and then weighed. The initial weight of the clean, dry filter is then subtracted from the weight of the dried filter paper and solid.

Distillation—In distillation, the differences in the boiling points of liquids are used to separate them. The temperature of the mixture is raised to a temperature that is greater than the boiling point of the more volatile substance and lower than the boiling point of the less volatile substance. The more volatile substance will vaporize, leaving the less volatile substance.

Chromatography—In chromatography, substances are separated by the differences in the degree to which they are adsorbed onto a surface. The substances are passed over the adsorbing surface, and the ones that stick to the surface with greater attraction will move slower than the substances that are less attracted to the surface. This difference in speeds separates the substances. The name chromatography came about because the process is used to separate pigments.

MEASURING CONCENTRATION

Titration is one of the most important laboratory procedures. In titration, an acid-base neutralization reaction is used to find the concentration of an unknown acid or base. It takes exactly one mole of hydroxide ions (base) to neutralize one mole of hydrogen ions (acid), so the concentration of an unknown acid solution can be found by calculating how much of a known basic solution is required to neutralize a sample of given volume. The most important formula in titration experiments is derived from the definition of molarity.

Molarity = Moles/Liters

Moles = (Molarity)(Liters)

The moment when exactly enough base has been added to the sample to neutralize the acid present is called the equivalence point. In the lab, an indicator is used to tell when the equivalence point has been reached. An indicator is a substance that is one color in acid solution and a different color in basic solution. Two popular indicators are phenolphthalein, which is clear in acidic solution and pink in basic solution, and litmus, which is pink in acidic solution and blue in basic solution.

Spectrophotometer—A spectrophotometer measures slight variations in color. It can be used to measure the concentration of ions that produce colored solutions.

The concentration of the ions will be directly proportional to the absorbance of the solution measured by the spectrophotometer. The relationship between absorbance and concentration is given by **Beer's law**.

Beer's Law
$A = abc$
A = absorbance measured by spectrophotometer
a = molar absorptivity, a constant whose value depends on the solution and the wavelength of light used
b = path length, the distance the light travels through the solution
c = concentration of the solution (mol/L)

IDENTIFYING CHEMICALS IN SOLUTION

PRECIPITATION

One of the most useful ways of identifying unknown ions in solution is precipitation. You can use the solubility rules given in the solubility chapter to see how the addition of certain ions to solution will cause the specific precipitation of other ions. For instance, the fact that $BaSO_4$ is insoluble can be used to identify either Ba^{2+} or SO_4^{2-} in solution; if the solution contains Ba^{2+} ions, then the addition of SO_4^{2-} will cause a precipitation reaction. The inverse is true for a solution that contains SO_4^{2-} ions.

In the same way, the insolubility of AgCl can be used to identify either Ag^+ or Cl^- in solution.

CONDUCTION

You can tell whether a solution contains ions by checking to see if the solution conducts electricity. Ionic solutes conduct electricity in solution; nonionic solutes do not.

FLAME TESTS

Some ions burn with distinctly colored flames. Flame tests can be used to identify Li^+ (red), Na^+ (yellow), and K^+ (purple), as well as the other alkali metals. The alkaline earths, including Ba^{2+} (green), Sr^{2+} (red), and Ca^{2+} (red) also burn with colored flames.

ACID-BASE REACTION

- When a base is added to an NH_4^+ solution, the distinctive odor of ammonia can be detected.

- When an acid is added to a solution containing S^{2-}, the rotten-egg odor of H_2S can be detected.

- When acid is added to a solution containing CO_3^{2-}, CO_2 gas is produced.

COLORED SOLUTIONS

- Many of the transition metals form ions with distinctive colors, for example, Cu^{2+} (blue) and Ni^{2+} (green).

- Bromine and iodine will show a dark brown color when placed in a nonpolar solvent.

- Permanganate ion (MnO_4^-), a strong oxidizing agent, turns a solution purple.

- Dichromate ion ($Cr_2O_7^{2-}$), also a strong oxidizing agent, turns a solution orange. Chromate (CrO_4^{2-}) turns a solution yellow.

LABORATORY EQUIPMENT

The pictures below show some standard chemistry lab equipment.

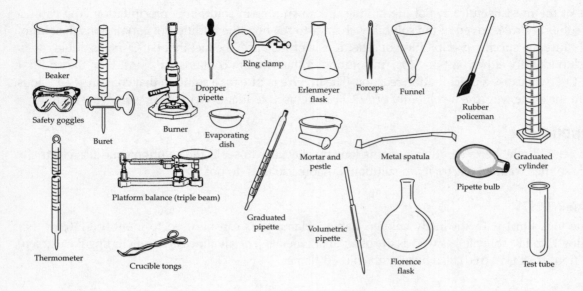

Beaker

Safety goggles

Buret

Burner

Dropper pipette

Ring clamp

Erlenmeyer flask

Forceps

Funnel

Rubber policeman

Evaporating dish

Graduated cylinder

Thermometer

Platform balance (triple beam)

Crucible tongs

Graduated pipette

Mortar and pestle

Metal spatula

Pipette bulb

Volumetric pipette

Florence flask

Test tube

CHAPTER 15 QUESTIONS

MULTIPLE-CHOICE QUESTIONS

Questions 1–4

    (A) Oxidation-reduction
    (B) Neutralization
    (C) Fusion
    (D) Combination
    (E) Decomposition

Which of the reaction types listed above best describes each of these processes?

1. $CO_2(g) + CaO(s) \rightarrow CaCO_3(s)$

2. $2\,Fe^{3+}(aq) + 2\,I^-(aq) \rightarrow 2\,Fe^{2+}(aq) + I_2(aq)$

3. $CH_3COOH(aq) + NaOH(aq) \rightarrow$
                $CH_3COONa(aq) + H_2O(l)$

4. $CH_4(g) + 2\,O_2(g) \rightarrow CO_2(g) + 2\,H_2O(g)$

Questions 5–8

    (A) Na^+
    (B) Cu^{2+}
    (C) Ag^+
    (D) Al^{3+}
    (E) NH_4^+

5. This ion turns an aqueous solution deep blue.

6. This ion forms a white precipitate when added to a solution containing chloride ions.

7. This ion produces a yellow flame when burned.

8. This ion produces a strong odor when added to a basic solution.

9. Which of the following indicators would be most useful in identifying the equivalence point of a titration for a solution that has a hydrogen ion concentration of $7 \times 10^{-4} M$ at the equivalence point?

(A) Methyl violet (pH range for color change is 0.1–2.0)
(B) Methyl yellow (pH range for color change is 1.2–2.3)
(C) Methyl orange (pH range for color change is 2.9–4.0)
(D) Methyl red (pH range for color change is 4.3–6.2)
(E) Bromthymol blue (pH range for color change is 6.1–7.6)

10. The volume of a liquid is to be measured. Which of the following cylindrical flasks would take the most accurate measurement?

(A) A flask with 1 ml gradations and a diameter of 1 cm
(B) A flask with 1 ml gradations and a diameter of 3 cm
(C) A flask with 5 ml gradations and a diameter of 1 cm
(D) A flask with 5 ml gradations and a diameter of 3 cm
(E) A flask with 10 ml gradations and a diameter of 1 cm

11. Which of the following is (are) considered to be proper laboratory procedure?

I. Reading the height of a fluid in a buret from a point level with the fluid's meniscus.
II. Placing a sample to be weighed directly on the pan of the balance.
III. Stirring a solution constantly during a titration.

(A) I only
(B) III only
(C) I and III only
(D) II and III only
(E) I, II, and III

12. A 0.1-molar NaOH solution is to be released from a buret in a titration experiment to measure the hydrogen ion concentration of an unknown acid. Which of the following laboratory procedures would cause an error in the measure of the concentration of the acid?

I. The buret was rinsed with the NaOH solution immediately before the titration.
II. The buret was rinsed with distilled water immediately before the titration.
III. The buret was rinsed with the unknown acid immediately before the titration.

(A) I only
(B) III only
(C) I and II only
(D) I and III only
(E) II and III only

13. Which of the following is NOT proper procedure for transferring a solution with a pipette?

(A) Rinsing the pipette with the solution to be transferred.
(B) Using your mouth to draw the solution into the pipette.
(C) Covering the top of the pipette with your index finger to keep the solution from escaping.
(D) Draining the solution into a waste beaker until the meniscus drops to the calibration mark.
(E) Touching the pipette to the side of the destination beaker at the end of the transfer.

14. An object that was weighed on a balance was later found to be slightly heavier than the weight that was recorded by the balance. Which of the following could have caused the discrepancy?

    I. There was some foreign matter on the weighing paper along with the object.

    II. The object was hot when it was weighed.

    III. The experimenter neglected to account for the weight of the weighing paper.

(A) I only
(B) II only
(C) I and II only
(D) I and III only
(E) I, II, and III

15. An experimenter wishes to use test paper to find the pH of a solution. Which of the following is part of the proper procedure for this process?

(A) Dipping the test paper in the solution while stirring.
(B) Dipping the test paper in the solution without stirring.
(C) Pouring some of the solution onto the dry test paper.
(D) Dipping the test paper in distilled water and slowly adding the solution to the water while stirring.
(E) Dipping the test paper in distilled water and slowly adding the solution to the water without stirring.

ESSAYS

1. A titration experiment was conducted to determine the pH of a known volume of a strong monoprotic acid solution. A hydroxide solution of known concentration was poured into a buret and then titrated into the acid solution. The volume of hydroxide solution titrated into the acid solution was measured at the equivalence point and used to calculate the concentration of the acid solution. Which of the following situations would cause an error in the calculated value of the pH of the acid solution? Explain each answer.

    (a) The buret was rinsed with the hydroxide solution before the solution was poured into the buret.

    (b) The experimenter did not notice that a few drops of hydroxide solution spattered outside the acid solution container during the titration.

    (c) The buret was rinsed with distilled water before the hydroxide solution was added.

    (d) The experimenter read the hydroxide solution level from the top of the fluid instead of the bottom of the meniscus both before the titration and at the equivalence point.

    (e) Some hydroxide solution was spilled while the experimenter was pouring it into the buret.

2. An experiment was conducted to determine the molecular weight of a pure salt sample. The mass of the salt sample was known. The salt was dissolved in a container of water of known mass and the freezing point of the solution was measured. The molecular weight was calculated by the freezing point depression method. How would the calculated value of the molecular weight be affected by each of the following?

    (a) The experimenter failed to take the dissociation of the salt into account.

    (b) The experimenter mistook molarity for molality, and used liters of solution instead of kilograms of solvent in the calculation to find the number of moles of solute.

    (c) The container used for the experiment was not rinsed and contained dust particles.

    (d) The experimenter misread the thermometer and recorded a freezing point that was higher than the true value.

    (e) The experimenter did not notice that some solid salt did not completely dissolve.

3. Use your knowledge of chemical principles to answer or explain each of the following:

    (a) When helium gas is to be collected in a jar by the displacement of air, the opening of the jar must be directed downward. When carbon dioxide gas is to be collected in a jar by the displacement of air, the opening of the jar must be directed upward.

    (b) Will the molar quantity calculated for a gas collected over water be too large or too small if the experimenter fails to take into account the vapor pressure of water?

    (c) Why is it easier to separate oxygen gas from hydrogen gas by the method of successive effusion than it is to separate oxygen gas from nitrogen gas by the same method?

    (d) Give an explanation for why an attempt to separate two liquids by distillation may fail.

CHAPTER 15 ANSWERS AND EXPLANATIONS

MULTIPLE-CHOICE QUESTIONS

1. **D** In a combination (or composition, or synthesis) reaction, two substances combine to form a more complex substance.

   A combination reaction is the opposite of a decomposition reaction, so if this reaction occurred in reverse, it would be a decomposition reaction.

2. **A** In an oxidation-reduction reaction, electrons are transferred between the reactants, causing the oxidation state of the element that is oxidized to increase and the oxidation state of the element that is reduced to decrease. In this reaction, Fe^{3+} is reduced to Fe^{2+} and I^- is oxidized to I^0.

3. **B** In a neutralization reaction, an acid (in this case, CH_3COOH) and a base (NaOH) react to form water and a salt (CH_3COONa).

4. **A** This reaction, the combustion of an organic compound, is also an oxidation-reduction reaction. As we said above, in a redox reaction, electrons are transferred between the reactants, causing the oxidation state of the element that is oxidized to increase and the oxidation state of the element that is reduced to decrease. In this reaction, H^- is oxidized to H^+ and O^0 is reduced to O^{2-}.

5. **B** Copper, along with most of the transition metals, forms colored solutions with water. This is true because the d electrons of the transition metals are constantly changing energy levels and emitting radiation in the visible spectrum.

6. **C** Silver is the only ion listed that forms an insoluble chloride.

7. **A** Sodium, along with the other Group IA elements, produces a colored flame in the flame test.

8. **E** Ammonium ion reacts with hydroxide ion to form ammonia, which has a strong, distinct odor. This reaction is shown below.

$$NH_4^+ + OH^- \rightarrow NH_3 + H_2O$$

9. **C** If $[H^+]$ is 7×10^{-4} M, then the pH must be between 3 and 4. Only methyl orange changes color between 3 and 4.

10. **A** Volume = (height)(cross-sectional area)

    The smaller the gradations, the more accurately the height can be measured. The smaller the area, the farther apart the 1 ml gradations will be and the more accurately the height of the fluid can be measured.

11. **C** Choices (I) and (III) are proper experimental procedures.

    (II) is not; a sample should always be weighed in a glass or porcelain container to prevent a reaction with the balance pan.

12. **E** Rinsing the buret with the NaOH solution (I) is proper procedure and will not change the concentration of the NaOH solution and will not cause an error. Rinsing the buret with distilled water (II) will dilute the NaOH solution, lowering the concentration and causing an error, and rinsing the buret with the unknown acid (III) will cause some of the NaOH solution to be neutralized in the buret, lowering its concentration and causing an error.

13. **B** You should never use your mouth to draw solution into a pipette. Instead, you should use a rubber suction bulb.

All of the other choices are part of the proper procedure.

14. **B** When a hot object is weighed (II), convection currents around the object can reduce the apparent mass measured by the balance.

Choices (I) and (III) would both cause the weight measured by the balance to be greater than the actual weight of the object. We're looking for the opposite effect.

15. **C** (A) and (B) are wrong because there is a danger of contaminating the solution by adding the paper.

(D) and (E) are wrong because adding the solution to distilled water completely changes the solution and defeats the purpose of testing it.

So pouring the solution onto the dry test paper (C) is the proper procedure.

ESSAYS

1. A note for the answers:

In this experiment, the volume of OH^- added is directly measured.

Moles = (molarity)(volume) is used to find the moles of OH^- added.

The moles of OH^- added to reach the equivalence point is equal to the number of moles of H^+ originally present in the acid solution.

Molarity = $\dfrac{moles}{volume}$ is used to find $[H^+]$ of the acid solution.

pH = $-\log[H^+]$ is used to calculate the pH.

(a) This is proper experimental procedure and will have no adverse effect on the calculated value of the acid solution.

(b) This will make the measured volume of the hydroxide solution larger than the actual amount added, the calculated value for the moles of OH^- and H^+ will be too large, the calculated value of $[H^+]$ will be too large, and the calculated pH will be too small.

(c) This will dilute the hydroxide solution, which means that too large a volume of hydroxide solution will be added, the calculated value for the moles of OH^- and H^+ will be too large, the calculated $[H^+]$ will be too large, and the calculated pH will be too small.

(d) The levels were read consistently although they were read from the wrong spot. The two errors should cancel, and the calculated pH should be correct.

(e) This will not affect the concentration of the hydroxide solution or the measurement of the volume poured into the acid solution, so the calculated pH should not be affected.

2. A note for the answers:

In this experiment, the freezing point of the solution is measured, and from the freezing point the freezing-point depression, ΔT, is calculated.

The equation $m = \dfrac{\Delta T}{kx}$ is used to calculate the molality of the solution.

Moles = (molality)(kg of solvent) is used to calculate the number of moles of salt.

MW = $\dfrac{moles}{grams}$ is used to calculate the molecular weight of the salt.

(a) The experimenter makes $x = 1$, instead of 2 or 3. So the calculated value of m will be too large, the calculated value of moles of salt will be too large, and the calculated MW will be too small.

(b) Moles = (molality)(kg of solvent)   Moles = (molarity)(liters of solution)

Because the solvent is water, the distinction between kilograms and liters is not important (for water, 1kg = 1 L), but the distinction between solvent and solution may make a difference.

Liters of solution will be a little larger than kilograms of solvent, so the calculated value for moles of salt will be too large, and the calculated MW will be too small.

(c) Extra particles in the solution will cause the measured freezing point to be too low. ΔT will be too large, m will be too large, and the calculated MW will be too small.

(d) If the freezing point is too high, the calculated ΔT will be too small, m will be too small, and the calculated MW will be too large.

(e) If some salt does not dissolve, then the grams of salt used in the calculation will be larger than the amount actually in the solution, and the calculated MW will be too large.

3. (a) Helium (MW = 4 g/mol) is less dense than air, so it will rise to the top of the jar, displacing air downward.

Carbon dioxide (MW = 44 g/mol) is denser than air, so it will sink to the bottom of the jar, displacing air upward.

(b) If the vapor pressure from water is ignored, the pressure of the gas used in the calculation will be too large.

$$n = \frac{PV}{RT}$$

If P is too large, then n, the calculated molar quantity of gas, will be too large.

(c) Separation of gases by successive effusion depends on Graham's law.

$$\frac{v_1}{v_2} = \sqrt{\frac{MW_2}{MW_1}}$$

The greater the difference in molecular weights, the greater the difference in average molecular speeds, and the greater the difference in rates of effusion.

Oxygen gas (32 g/mol) and nitrogen gas (28 g/mol) have similar molecular weights.

Oxygen gas (32 g/mol) and hydrogen gas (2 g/mol) have very different molecular weights.

(d) The most likely reason for the failure of separation by distillation would be that the boiling points of the two liquids are too close together.

16

Organic Chemistry

HOW OFTEN DOES ORGANIC CHEMISTRY APPEAR ON THE EXAM?

In the multiple-choice section, this topic appears in about 1 out of 75 questions. In the free-response section, this topic appears occasionally.

Organic chemistry is the study of **carbon compounds**, which are important because all living things on Earth are based on carbon compounds. Each carbon atom can form up to four bonds in a compound; the type of bonding that exists in organic compounds is almost always covalent with little polarity.

This means that organic compounds are much more soluble in nonpolar solvents than in polar solvents. Remember: Like dissolves like; because carbon compounds are generally nonpolar, they will be soluble in nonpolar solvents. This means that organic substances are not very soluble in water, which is highly polar.

Organic compounds don't dissociate in solution, because there are no ionic bonds in organic compounds. That means that organic solutions are poor conductors of electricity, and organic compounds do not behave as electrolytes in solution. Generally, the larger and more complicated the molecules of an organic substance, the higher its boiling and melting points will be.

Here's a review of some of the basic organic compounds.

HYDROCARBONS

Hydrocarbons are compounds that contain only carbon and hydrogen.

ALKANES

Alkanes are hydrocarbons that contain only single bonds. They are also known as saturated hydrocarbons.

Alkane (C_nH_{2n+2})	Formula
Methane	CH_4
Ethane	C_2H_6
Propane	C_3H_8
Butane	C_4H_{10}
Pentane	C_5H_{12}

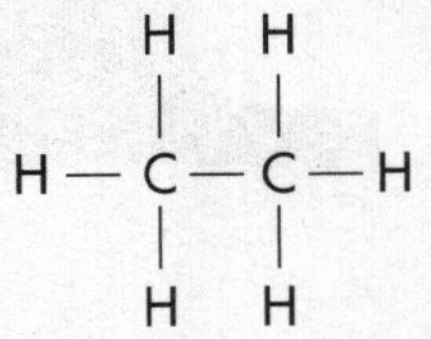

ethane

ALKENES

Alkenes are hydrocarbons that contain double bonds. They are examples of unsaturated hydrocarbons.

Alkene (C_nH_{2n})	Formula
Ethene (ethylene)	C_2H_4
Propene	C_3H_6
Butene	C_4H_8
Pentene	C_5H_{10}

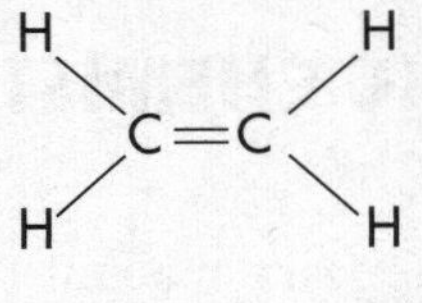

ethylene

ALKYNES

Alkynes are hydrocarbons that contain triple bonds. They are also examples of unsaturated hydrocarbons.

Alkyne (C_nH_{2n-2})	Formula
Ethyne	C_2H_2
Propyne	C_3H_4
Butyne	C_4H_6
Pentyne	C_5H_8

HYDROCARBON RINGS

Many hydrocarbons form rings instead of chains. One of the most important classes of these compounds is the aromatic hydrocarbons, the simplest of which is benzene, C_6H_6.

SOME FUNCTIONAL GROUPS

The presence of certain groups of atoms (called functional groups) in organic compounds can give the compounds specific chemical properties.

ALCOHOLS

Alcohols are organic compounds in which a hydrogen has been replaced with a hydroxyl group (OH). The hydroxyl group makes alcohols polar and able to form hydrogen bonds, which means that alcohols are more soluble in water than most organic compounds.

Alcohol	Formula
Methanol	CH_3OH
Ethanol	C_2H_5OH
Propanol	C_3H_7OH

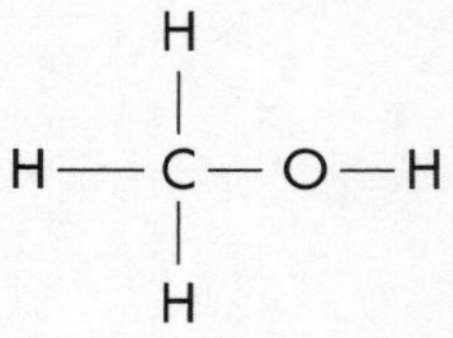

methanol

ORGANIC ACIDS

Organic acids are organic compounds in which a hydrogen has been replaced with a carboxyl group (COOH).

Organic Acid	Formula
Methanoic Acid	HCOOH
Ethanoic (Acetic) Acid	CH_3COOH
Propanoic Acid	C_2H_5COOH

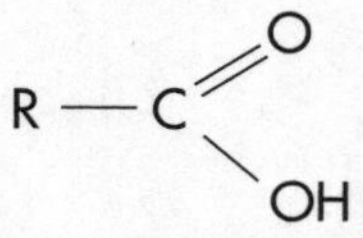

organic acid

The "R" in the diagram stands for the rest of the hydrocarbon chain.

HALIDES

Halides are organic compounds in which one or more hydrogens have been replaced with a halide (F, Cl, Br, I).

Halide	Formula
Chloromethane	CH_3Cl
Chloroethane	C_2H_5Cl
Chloropropane	C_3H_7Cl

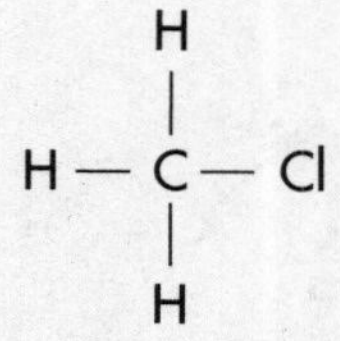

```
        H
        |
  H  —  C  —  Cl
        |
        H
```

chloromethane

AMINES

In an amine, a hydrogen atom has been replaced by an amino group (NH_2).

Amine	Formula
Aminomethane (Methyl amine)	CH_3NH_2
Aminoethane (Ethyl amine)	$C_2H_5NH_2$
Aminopropane (Propyl amine)	$C_3H_7NH_2$

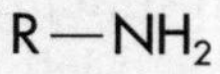

$$R — NH_2$$

amine

ALDEHYDES

An aldehyde contains a carbonyl group (C=O) connected to at least one hydrogen atom.

Aldehyde	Formula
Methanal (Formaldehyde)	HCHO
Ethanal	CH_3CHO
Propanal	C_2H_5CHO

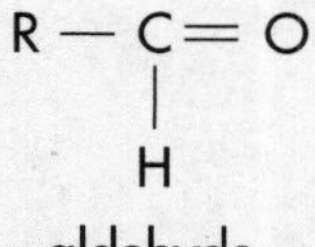

```
  R  —  C  =  O
         |
         H
```

aldehyde

KETONES

A ketone is similar to an aldehyde in that it also contains a carbonyl group (C=O), but in a ketone, the carbon in the carbonyl group is not connected to any hydrogen atoms.

Ketone	Formula
Propanone	C_3H_6O
Butanone	C_4H_8O
Pentanone	$C_5H_{10}O$

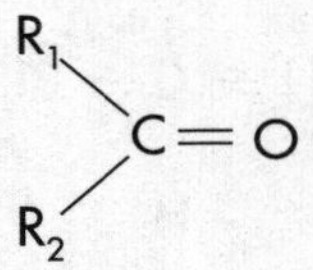

ketone

ETHERS

In an ether, an oxygen atom serves as a link in a hydrocarbon chain. The name of an ether is determined by the lengths of the chains on either side of oxygen.

Ether	Formula
Methoxymethane (Dimethyl ether)	CH_3OCH_3
Methoxyethane (Methyl ethyl ether)	$CH_3OC_2H_5$
Methoxybutane (Methyl butyl ether)	$CH_3OC_4H_9$
Ethoxypropane (Ethyl propyl ether)	$C_2H_5OC_3H_7$

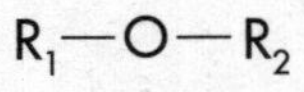

$R_1 - O - R_2$

ether

Esters

In an ester, an ester group (COO) serves as a link in a hydrocarbon chain. An ester is formed in a reaction between an alcohol and an organic acid, and its name is derived from these reactants.

Ester	Formula
Methyl methanoate	$HCOOCH_3$
Methyl ethanoate	CH_3COOCH_3
Methyl propanoate	$C_2H_5COOCH_3$
Ethyl methanoate	$HCOOC_2H_5$
Propyl methanoate	$HCOOC_3H_7$

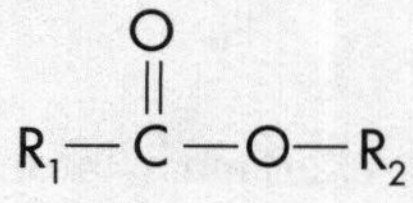

$$R_1 - \overset{\overset{\displaystyle O}{\|}}{C} - O - R_2$$

ester

ISOMERS

Among organic compounds, it is common to find two or more molecules with the same molecular formula, but with different arrangements of atoms and different chemical properties. These molecules are called isomers. For instance, ethanol (C_2H_5OH) and dimethyl ether (CH_3-O-CH_3) are isomers because each compound contains 2 carbon atoms, 6 hydrogen atoms, and 1 oxygen atom.

You may be asked to draw isomers of an organic molecule, so here are some examples to give you an idea of how it works.

Draw three isomers of pentane.

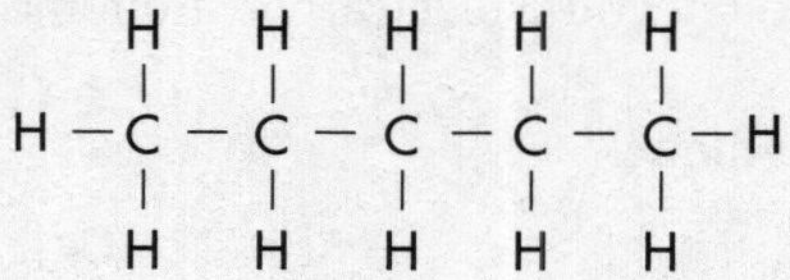

pentane

This is the standard chain version of pentane.

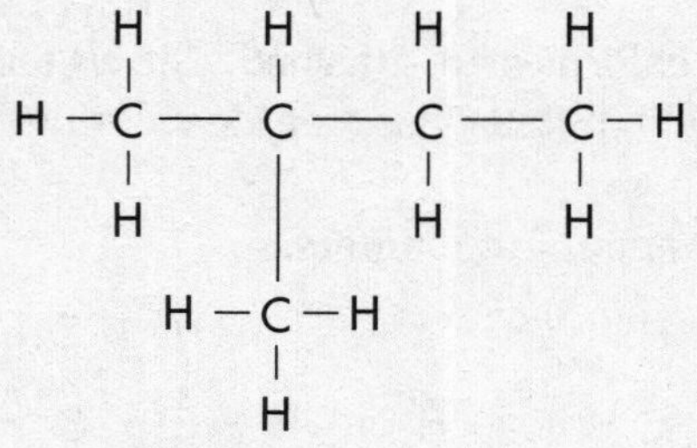

2-methylbutane

This isomer is created by moving one methyl group (CH$_3$) to the middle of the chain. The new name is now butane because the chain is now only four carbons long. The 2-methyl tells you that the methyl group is attached to the second carbon from the end.

$$
\begin{array}{c}
\text{H} \\
| \\
\text{H} - \text{C} - \text{H} \\
\end{array}
$$

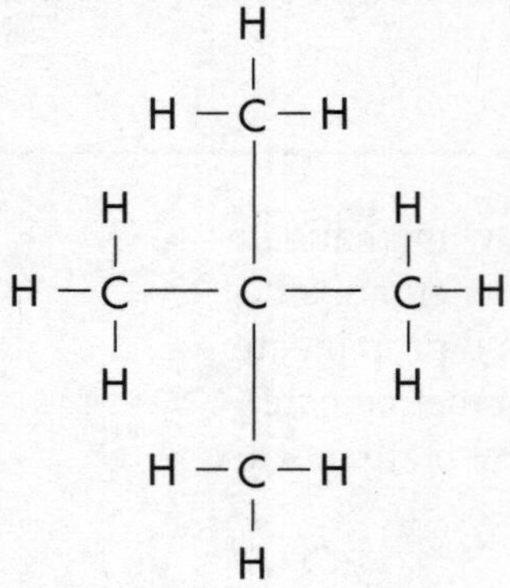

2,2 dimethylpropane

Now we've moved a second methyl group to the inside. The new name is propane because the chain is now only three carbons long. The 2,2 dimethyl tells you that both methyl groups are attached to the second carbon from the end. You should take a moment to convince yourself that any other rearrangements that don't involve rings will be redundant.

Draw two isomers of propanol.

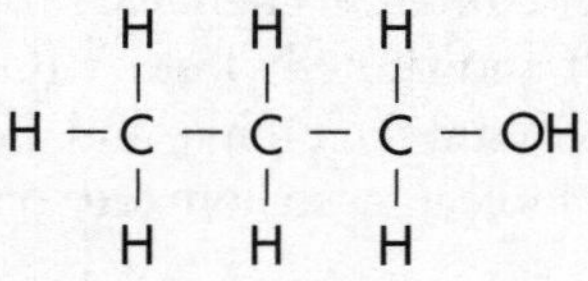

1-propanol 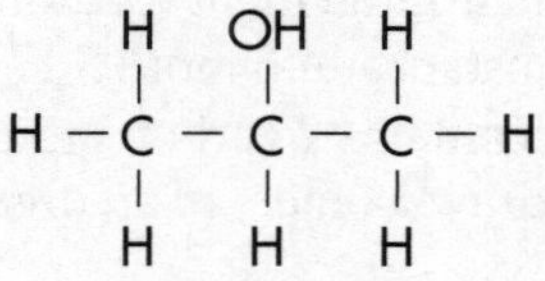   2-propanol

The 1 and 2 in front are there to tell you which carbon atom the hydroxyl group is attached to.

Draw three isomers of dichloroethene.

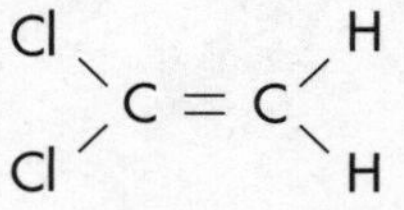

1,1 dicloroethene 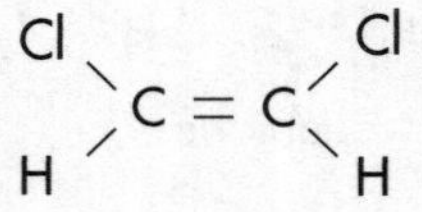   cis-1,2 dicloroethene 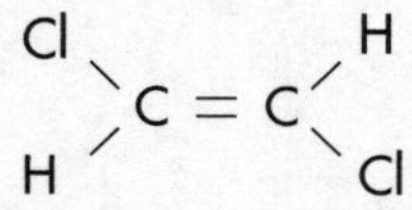   trans-1,2 dicloroethene

The numbers show where the chlorines are attached. The *cis* tells you that both chlorines are on the same side of the molecule. The *trans* tells you that the chlorines are on opposite sides.

Show that propanone and 2-propanal are isomers.

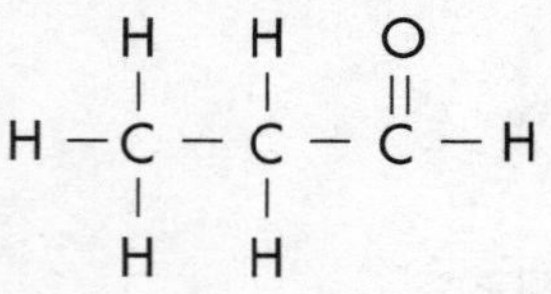

propanal 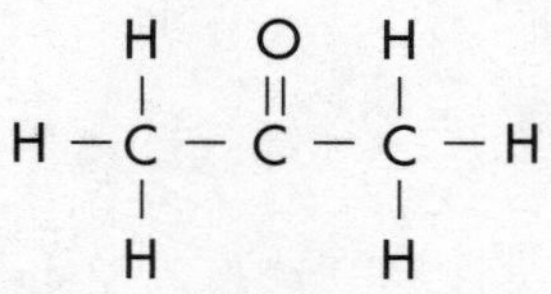   2-propanone

The only difference between the two molecules is the placement of oxygen.

These are NOT isomers.

$$H-\underset{\underset{H}{|}}{\overset{\overset{H}{|}}{C}}-\underset{\underset{H}{|}}{\overset{\overset{H}{|}}{C}}-\underset{\underset{H}{|}}{\overset{\overset{H}{|}}{C}}-OH \qquad H-\underset{\underset{H}{|}}{\overset{\overset{H}{|}}{C}}-\underset{\underset{H}{|}}{\overset{\overset{H}{|}}{C}}-\underset{\underset{H}{|}}{\overset{\overset{OH}{|}}{C}}-H$$

$$HO-\underset{\underset{H}{|}}{\overset{\overset{H}{|}}{C}}-\underset{\underset{H}{|}}{\overset{\overset{H}{|}}{C}}-\underset{\underset{H}{|}}{\overset{\overset{H}{|}}{C}}-H \qquad H-\underset{\underset{OH}{|}}{\overset{\overset{H}{|}}{C}}-\underset{\underset{H}{|}}{\overset{\overset{H}{|}}{C}}-\underset{\underset{H}{|}}{\overset{\overset{H}{|}}{C}}-H$$

If all four of these propanol molecules were drawn in 3-D, they would be identical.

ORGANIC REACTIONS

You should be familiar with a handful of organic reactions in which the organic substances are created or combined.

Addition—In an addition reaction, a carbon–carbon double bond is broken down into a single bond, freeing each of the two carbons to bond with another element. In the same way, a triple bond can be converted into a double bond in an addition reaction.

Substitution—In a substitution, one atom or group in a compound is replaced with another atom or group. Typically, a hydrogen atom will be replaced by one of the functional groups in a substitution reaction.

Polymerization—In polymerization, two smaller compounds, called monomers, are joined to form a much larger third compound. In condensation polymerization, two monomers are joined in a reaction that produces water as a product.

Cracking—In cracking, a larger compound is broken down into smaller compounds.

Oxidation—An organic compound can react with oxygen at high temperatures to form carbon dioxide and water. This reaction should be familiar to you as combustion or burning.

Esterification—In esterification, an organic acid reacts with an alcohol to produce an ester and water.

Fermentation—In fermentation, an organic compound reacts in the absence of oxygen to produce an alcohol and carbon dioxide. Wine is produced through fermentation.

CHAPTER 16 QUESTIONS

MULTIPLE-CHOICE QUESTIONS

Questions 1–4

(A) $CH_3CH_2CH_2CH_2CH_2CH_3$
(B) $CH_3CH_2CH_2CH_2COCH_3$
(C) $CH_3CHCHCH_2CH_2CH_3$
(D) $CH_3CH_2CH_2CH_2CH_2CHO$
(E) $CH_3CH_2CH_2CH_2CH_2CH_2OH$

1. 1-hexanol

2. Hexane

3. 2-hexanone

4. Hexene

Questions 5–7

(A) Alkyne
(B) Alcohol
(C) Aldehyde
(D) Amine
(E) Acid

5. Contains a carboxyl group

6. Contains only hydrogen and carbon atoms

7. Contains nitrogen

8. Which of the following is true of the bonds contained in a molecule of ethene, C_2H_4?

(A) There is one pi (π), and there are two sigma (σ) bonds.

(B) There is one pi (π), and there are four sigma (σ) bonds.

(C) There is one pi (π), and there are five sigma (σ) bonds.

(D) There are two pi (π) and two sigma (σ) bonds.

(E) There are two pi (π) and four sigma (σ) bonds.

9.

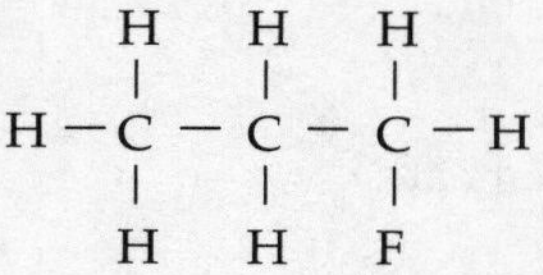

```
     H   H   H
     |   |   |
H —  C — C — C — H
     |   |   |
     H   H   F
```

Which of the structural formulas shown below is an isomer of the compound represented by the structural formula shown above?

(A)
```
     H   H   H
     |   |   |
H —  C — C — C — F
     |   |   |
     H   H   H
```

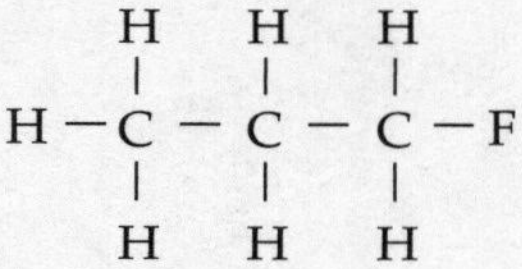

(B)
```
     F   H   H
     |   |   |
H —  C — C — C — H
     |   |   |
     H   H   F
```

(C)
```
     H   H   H
     |   |   |
H —  C — C — C — H
     |   |   |
     F   H   H
```

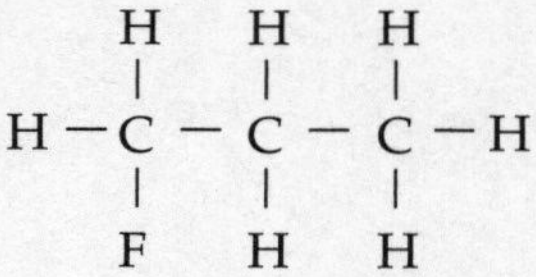

(D)
```
     H   H   H
     |   |   |
H —  C — C — C — H
     |   |   |
     H   F   H
```

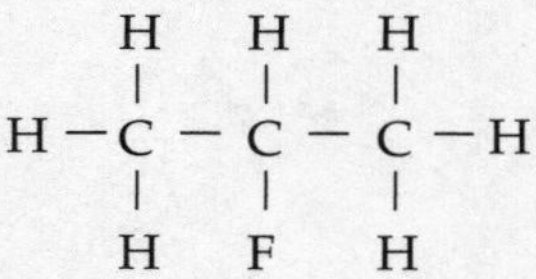

(E)
```
     H   H   H   H
     |   |   |   |
H —  C — C — C — C — H
     |   |   |   |
     H   H   H   F
```

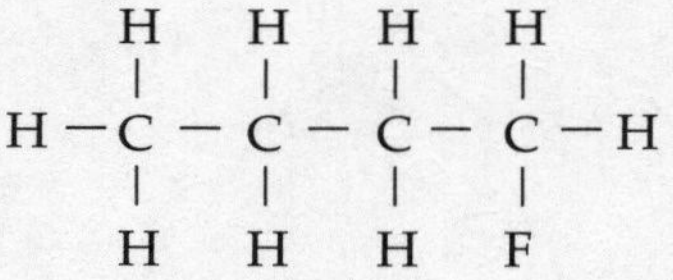

10. Which of the following substances does not contain pi (π) bonding?

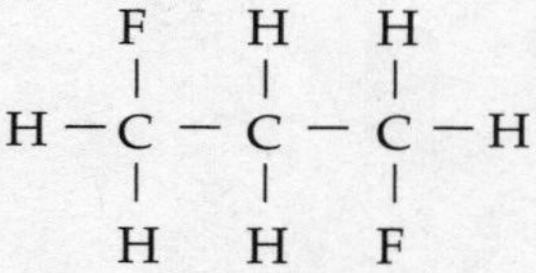

(A) Propanol
(B) Propene
(C) Propanal
(D) Propanone
(E) Propyne

11. Which of the following structural formulas shows a compound that is an isomer of butane?

(A)
```
     H   H   H
     |   |   |
H —  C — C — C — H
     |   |   |
     H   H   H
```

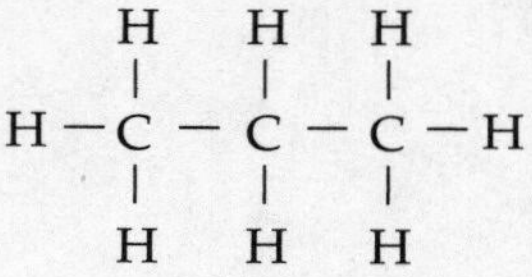

(B)
```
     H   H
     |   |
H —  C — C — H
     |   |
H —  C — C — H
     |   |
     H   H
```

(C)
```
     H   H   H
     |   |   |
H —  C — C — C — H
     |       |
     H       H
         |
     H — C — H
         |
         H
```

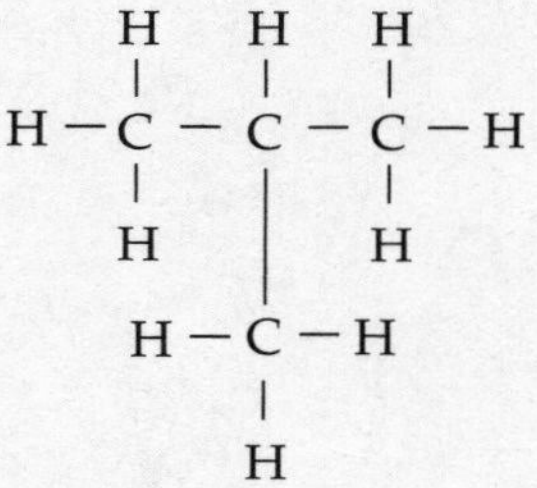

(D)
```
     H   H   H   H
     |   |   |   |
H —  C — C = C — C — H
     |           |
     H           H
```

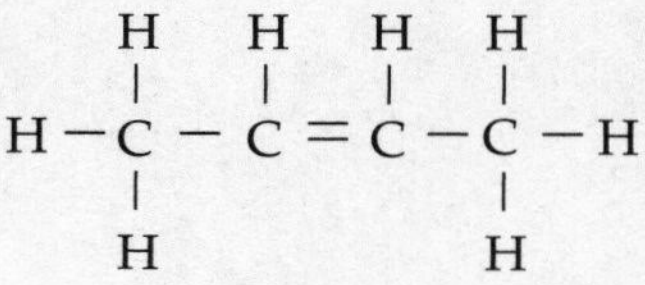

(E)
```
     H   H   H   H
     |   |   |   |
H —  C — C — C — C — OH
     |   |   |   |
     H   H   H   H
```

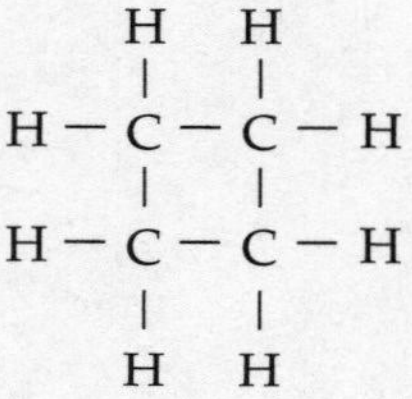

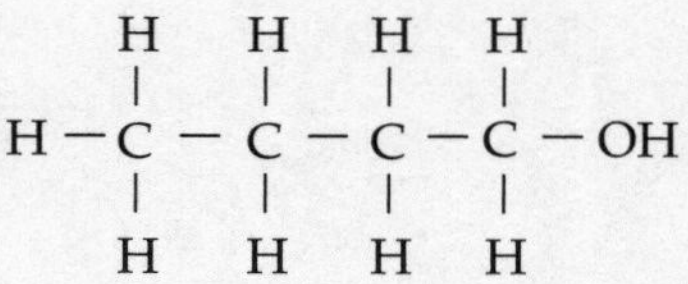

12. Which of the following pairs of compounds are isomers of each other?

   (A) Pentane and pentene
   (B) Pentanol and pentyne
   (C) Chloropentane and fluoropentane
   (D) Pentanoic acid and aminopentane
   (E) Pentanal and pentanone

13. In ethyne, C_2H_2, the two carbons are connected by a triple bond. Which of the following indicates the hybridization of the carbon atoms?

   (A) sp
   (B) sp^2
   (C) sp^3
   (D) dsp^3
   (E) d^2sp^3

14. Which of the following substances will have the greatest value for the heat of vaporization?

   (A) Octane
   (B) Hexane
   (C) Butane
   (D) Propane
   (E) Methane

15. Which of the substances listed below will exhibit hydrogen bonding?

   (A) Butane
   (B) Butanone
   (C) Butanal
   (D) Butanoic acid
   (E) Methyl butanoate

ESSAYS

1. $CH_3CH_2COOH + CH_3CH_2OH \rightarrow CH_3CH_2COOCH_2CH_3 + H_2O$

   An organic acid combines with an alcohol to form an ester and water, as shown above.

   (a) Draw the structural formulas and give the IUPAC names for the following:

       (i) the organic acid

       (ii) the alcohol

       (iii) the ester

   (b) For organic acid that you drew in part (a)(i)

       (i) State the hybridization of the carbon atom in the O–C–O bond.

       (ii) Predict the bond angle in the O–C–O bond.

       (iii) Indicate the number of pi (π) and sigma (σ) bonds in the molecule.

   (c) Explain why methanol and methanoic acid are soluble in water, while methane is not.

   (d) Dibromoethene, $C_2H_2Br_2$, exists in three isomers. Draw two of them.

Compound Name	Compound Formula	Boiling Point (K)
Butane	$CH_3CH_2CH_2CH_3$	273
1-butyne	$CH_3CHCHCH_3$	281
1-chlorobutane	$CH_3CH_2CH_2CH_2Cl$	352
2-butanone	$CH_3COCH_2CH_3$	353
1-butanol	$CH_3CH_2CH_2CH_2OH$	391
Pentane	$CH_3CH_2CH_2CH_2CH_3$	309

2. Use the information in the table above to answer the following questions:

   (a) Draw the structural formulas for the following:

       (i) 1-chlorobutane

       (ii) 2-butanone

   (b) For each pair of compounds below, explain the differences in boiling point. (Include specific information about each compound in your answers.)

       (i) Butane and Pentane

       (ii) Butane and 2-butanone

       (iii) 1-chlorobutane and 1-butanol

   (c) Each of the compounds below exists in two different forms. Draw the structural formula for both isomers of each compound.

       (i) Butane

       (ii) Butyne

   (d) A researcher is trying to separate a solution of 1-chlorobutane and 2-butanone. Why would distillation be a poor choice?

CHAPTER 16 ANSWERS AND EXPLANATIONS

MULTIPLE-CHOICE QUESTIONS

1. **E** This molecule has a hydroxyl (OH) group on one end.

2. **A** This molecule contains only carbon and hydrogen in the ratio C_6H_{14}.

3. **B** The second carbon from the end of this molecule is part of a carbonyl group. Choice (D) is incorrect because the carbonyl group is at the end, making choice (D) an aldehyde, hexanal.

4. **C** This molecule contains only carbon and hydrogen in the ratio C_6H_{12}.

5. **E** An organic acid contains a carboxyl group (COOH).

6. **A** An alkyne contains only carbon and hydrogen atoms. All of the other choices contain functional groups containing other atoms.

7. **D** An amine contains nitrogen in its amino group (NH_2).

8. **C** In ethene, the carbons are joined by a double bond, which contains one pi bond and one sigma bond. Each carbon is joined to two hydrogen atoms, each by a sigma bond. That makes a total of 1 pi bond and 5 sigma bonds.

9. **D** This choice shows a structure with the same molecular formula but with the hydroxyl group attached to a different carbon. The pictured formula shows 1-propanol and the answer choice shows 2-propanol.

10. **A** Of the substances listed, only propanol contains no double bonds.

11. **C** Butane has the molecular formula C_4H_{10}. Only choice (C), 2-methylpropane, has the same number of carbons and hydrogens.

12. **E** Pentanal is and aldehyde and pentanone is a ketone. They both contain a carbonyl group (CO) and they both have the molecular formula $C_5H_{10}O$. The difference is that pentanal has the carbonyl group at one end and pentanone has the carbonyl group in the middle.

13. **A** Each carbon forms 2 sigma bonds and 2 pi bonds. Since the pi bonds are not counted in hybridization, the hybridization is *sp*.

14. **A** Octane is the largest of these non-polar molecules, so it will exhibit the strongest dispersion forces. Stronger intermolecular forces make for a higher heat of vaporization.

15. **D** Butanoic acid contains a carboxyl group (COOH), where hydrogen is bonded to oxygen, so it will form hydrogen bonds. None of the other choices contains a hydrogen atom bonded to a highly electronegative element.

ESSAYS

1. (a) (i)

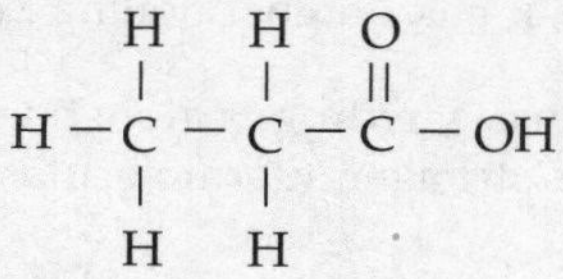

$$H - C(H)(H) - C(H)(H) - C(=O) - OH$$

propanoic acid

(ii)

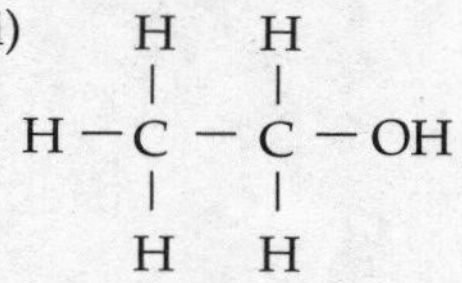

$$H - C(H)(H) - C(H)(H) - OH$$

ethanol

(iii)

$$H - C(H)(H) - C(H)(H) - C(=O) - O - C(H)(H) - C(H)(H) - H$$

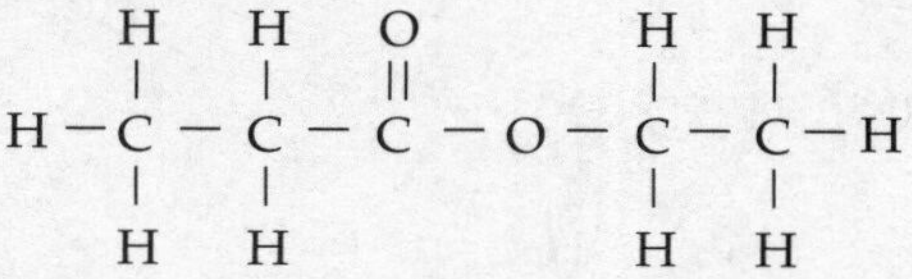

ethyl propanoate

(b) (i) The carbon is bonded to three atoms, so the hybridization is sp^2.

(ii) For sp^2 hybridization, the approximate bond angle is 120 degrees.

(iii) There are 9 bonds, 1 of which is a double bond. So there is 1 pi bond, and there are 9 sigma bonds.

(c) Methanol (CH_3OH) and methanoic acid (HCOOH) are each polar, with a functional group that can form hydrogen bonds in water. Methane (CH_4) is nonpolar, so it is not easily dissolved by water, a polar solvent.

(d)

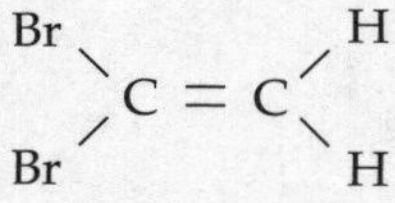

1,1-dibromoethene 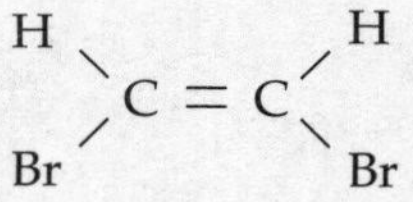   1,2-cis-dibromoethene 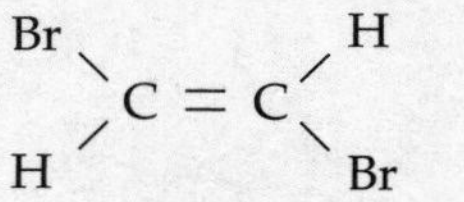   1,2-trans-dibromoethene

2. (a) (i)

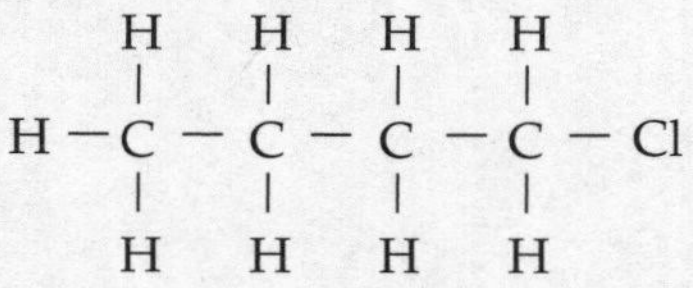

$$H - C(H)(H) - C(H)(H) - C(H)(H) - C(H)(H) - Cl$$

(ii)

$$H - C(H)(H) - C(=O) - C(H)(H) - C(H)(H) - H$$

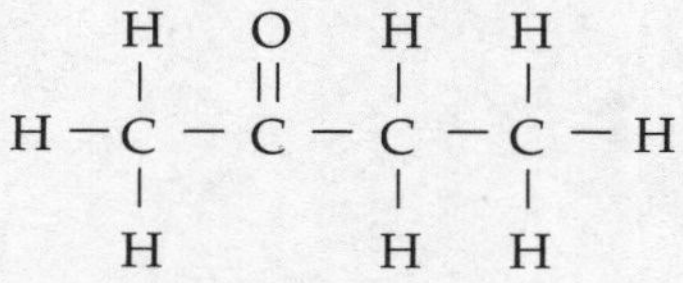

(b) (i) Both butane and pentane are nonpolar molecules that exhibit only dispersion forces. Pentane is larger than butane and contains more electrons, making it more easily polarizable. Because pentane has stronger intermolecular forces, it has a higher boiling point.

(ii) Butane is nonpolar, while the oxygen in 2-butanone makes it slightly polar. So while butane exhibits only dispersion forces, 2-butanone will have both dispersion forces and dipole–dipole attractions. Since 2-butanone has stronger intermolecular forces, it has a higher boiling point.

(iii) Both 1-chlorobutane and 1-butanol are polar, but 1-butanol has a hydrogen atom bonded to oxygen, so it will exhibit hydrogen bonding. Since 1-butanol has stronger intermolecular forces, it has a higher boiling point.

(c) (i)

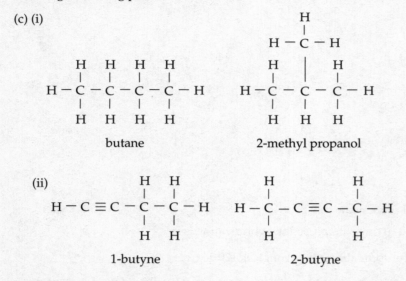

butane

2-methyl propanol

(ii)

1-butyne

2-butyne

(d) In distillation, two substances are separated by increasing the temperature until one substance has reached its boiling point. Since the boiling points are about the same, the two substances cannot be separated this way.

17

Descriptive Chemistry

You'll see these questions on Section II of the test. You should take about 10 minutes to write the equations for three chemical reactions and answer the accompanying questions.

THE BASICS

Rather than wasting your time trying to memorize an infinite number of possible chemical reactions, your best bet is to understand some general rules for approaching this section, which we give you in this chapter.

You have to write balanced equations

Make sure that you balance the equation with lowest whole number coefficients. You don't have to write the phases (*s*, *l*, *g*, *aq*, and so on) of the reactants or products.
Here's an example:

If Pb^{2+} and Cl^- ions are placed in solution, $PbCl_2$ will precipitate. You *could* write

$$Pb^{2+}(aq) + 2\ Cl^-(aq) \rightarrow PbCl_2(s)$$

but instead you *should* write

$$Pb^{2+} + 2\ Cl^- \rightarrow PbCl_2$$

Substances that dissociate extensively should be written as ions

All the reactions are assumed to take place in water unless otherwise stated, so things that ionize in water should be written as ions.

Consider these examples:

$NaNO_3$ ionizes, so it should be written as Na^+ and NO_3^-.
AgCl does not ionize, so it should be written as AgCl.
$Ca(OH)_2$ ionizes to some extent, so either $Ca(OH)_2$, or Ca^{2+} and $2\ OH^-$ are okay.

Leave out any molecules or ions that are not changed in the reaction

Let's say you're given this situation:

Solutions of magnesium chloride and silver nitrate are mixed.

You know from your solubility rules that silver chloride precipitates.
The full reaction is

$$Mg^{2+} + 2\ Cl^- + 2\ Ag^+ + 2\ NO_3^- \rightarrow Mg^{2+} + 2\ NO_3^- + 2\ AgCl$$

Leaving out things that are not changed results in

$$Ag^+ + Cl^- \rightarrow AgCl$$

Notice that the final reaction uses the lowest whole number coefficients.

SCORING

The descriptive chemistry part of the exam is worth 10 percent of Section II, or 5 percent of the overall test score.

Each of the 3 questions that you answer on this section is worth 5 points, for a total of 15. On each question, you get 1 point for correct reactants, 2 points for correct products, 1 point for balancing the equation, and 1 point for answering the accompanying question.

Don't forget partial credit. Even if you're not exactly sure what will happen when two reactants react, you can still get a point for writing them down correctly and you may still be able to answer the accompanying question.

CRACKING THE DESCRIPTIVE CHEMISTRY SECTION

The best way to approach these questions is to familiarize yourself with some general rules about what will happen when various reactants are combined and to get comfortable with the kinds of questions that might be asked about each type of reaction.

I. Look for two uncombined elements

There's only one thing you can do with uncombined elements: combine them. Just make sure that you give each element in the product compound a sensible oxidation state.

(a) Hydrogen gas is burned in air.

$$2 H_2 + O_2 \rightarrow 2 H_2O$$

Q: Describe the enthalpy change in this reaction.
A: The reaction is exothermic. Energy is released.

Notice that air is a synonym for oxygen.

(b) Solid sulfur is burned in oxygen.

$$S + O_2 \rightarrow SO_2$$

Q: What is the oxidation state of solid sulfur before and after the reaction?
A: The oxidation state of an uncombined element is always zero, so sulfur has an oxidation state of 0 before the reaction. After it combines with oxygen, sulfur has an oxidation state of +4.

(c) Solid magnesium is heated in nitrogen gas.

$$3 Mg + N_2 \rightarrow Mg_3N_2$$

Q: Is this an oxidation-reduction reaction? Explain.
A: Yes. Magnesium gives up electrons and is oxidized, while nitrogen gains electrons and is reduced.

(d) A piece of solid zinc is heated in chlorine gas.

$$Zn + Cl_2 \rightarrow ZnCl_2$$

Q: Describe the bonding present in the product(s) of this reaction.
A: Zinc chloride contains a metal and a non-metal, so it is held together by ionic bonds.

(e) A piece of solid barium is placed in oxygen gas.

$$2 Ba + O_2 \rightarrow 2 BaO$$

Q: Is the entropy change for this reaction positive or negative?

A: Negative. The number of moles decreases from reactants to products and a gas phase reactant is incorporated into a solid.

(f) A piece of solid sodium is placed in hydrogen gas.

$$2 \, Na + H_2 \rightarrow 2 \, NaH$$

Q: Describe the bonding present in solid sodium.

A: Solid sodium is held together by metallic bonding.

II. Look for a single reactant

If there is only one reactant, all it can do is break up in a decomposition reaction, so you're guaranteed the reactant partial credit point.

Decomposition reactions usually produce simple salts and oxide gases.

(a) A solution of hydrogen peroxide is placed under a bright light.

$$2 \, H_2O_2 \rightarrow 2 \, H_2O + O_2$$

Q: What are the oxidation states of hydrogen and oxygen in hydrogen peroxide?

A: Hydrogen is +1 and oxygen is −1.

(b) Solid calcium carbonate is heated.

$$CaCO_3 \rightarrow CaO + CO_2$$

Q: Calcium carbonate contains both ionic and covalent bonds. Explain.

A: The carbon and oxygen atoms in the carbonate ion (CO_3^-) are held together by covalent bonds. The calcium ion and the carbonate ion are held together by an ionic bond.

(c) Solid potassium chlorate is heated in the presence of a catalyst.

$$2 \, KClO_3 \rightarrow 2 \, KCl + 3 \, O_2$$

Q: How does the catalyst affect the activation energy of the reaction?

A: The catalyst speeds the reaction by lowering the activation energy.

(d) A sample of solid ammonium carbonate is heated.

$$(NH_4)_2CO_3 \rightarrow 2 \, NH_3 + CO_2 + H_2O$$

Q: Describe the polarity of the products of this reaction.
A: Ammonia (NH_3) and water (H_2O) are polar molecules. Carbon dioxide (CO_2) is nonpolar.

(e) A piece of solid potassium nitrate is heated.

$$2 \, KNO_3 \rightarrow 2 \, KNO_2 + O_2$$

Q: What is the oxidation number of nitrogen before and after the reaction?
A: Nitrogen has an oxidation number of +5 in KNO_3 and and oxidation number of +3 in KNO_2.

III. Look for water as a reactant

When metals and nonmetals react with water they tend to react in predictable ways. In general, metals react with water to form bases and nonmetals react with water to form acids.
Here are three variations.

1. A pure metal or a metal hydride in water will produce a base and hydrogen gas.

(a) Sodium metal is added to distilled water.

$$2 \, Na + 2 \, H_2O \rightarrow 2 \, Na^+ + 2 \, OH^- + H_2$$

Q: Sparks are observed as the reaction progresses. Identify the product of the reaction that undergoes combustion to produce the sparks.
A: Hydrogen gas (H_2) is extremely flammable and can burn in this reaction.

(b) Calcium metal is added to distilled water.

$$Ca + 2 \, H_2O \rightarrow Ca(OH)_2 + H_2$$

Q: The reaction takes place in a flask and the pH is monitored. Does the pH increase or decrease as the reaction proceeds? Explain.
A: The pH increases because calcium hydroxide, a base, is produced in the reaction.

(c) Solid calcium hydride is added to water.

$$CaH_2 + 2 \, H_2O \rightarrow Ca(OH)_2 + 2 \, H_2$$

Q: What is the oxidation state of calcium before the reaction takes place? Is calcium oxidized or reduced during the reaction?
A: The oxidation state is +2 before the reaction and +2 after the reaction, so calcium is neither oxidized nor reduced.

(d) Solid lithium hydride is added to distilled water.

$$LiH + H_2O \rightarrow Li^+ + OH^- + H_2$$

Q: Electrodes attached to a battery and a light bulb are placed in the solution. Will the light bulb shine brighter or dimmer as the reaction progresses? Explain.

A: The bulb will grow brighter because the reaction is producing charged particles which will conduct electricity through the solution.

2. A metal oxide in water will produce a base.

Metal oxides are called basic anhydrides because they produce bases when added to water.

(a) Solid potassium oxide is added to water.

$$K_2O + H_2O \rightarrow 2\,K^+ + 2\,OH^-$$

Q: Will the freezing point of the aqueous solution that results from this reaction be higher or lower than that of pure water?

A: The freezing point will be lower because particles of any kind added to pure water will cause freezing point depression.

(b) Solid barium oxide is added to water.

$$BaO + H_2O \rightarrow Ba^{2+} + 2\,OH^-$$

Q: A splint dipped in the solution resulting from this reaction will produce a green flame. Identify the product of the reaction that burns to produce the flame.

A: Barium ion (Ba^{2+}) produces a green flame in a flame test.

(c) Solid calcium oxide is added to water.

$$CaO + H_2O \rightarrow Ca(OH)_2$$

Q: Describe the change that takes place in the concentration of hydrogen ions in the solution as the reaction progresses? Explain.

A: The hydrogen ion concentration decreases because the product of the reaction is a base.

3. A nonmetal oxide in water will produce an acid.

Nonmetal oxides are called acid anhydrides because they produce acids in water.

(a) Solid dinitrogen pentoxide is added to water.

$$N_2O_5 + H_2O \rightarrow 2\,H^+ + 2\,NO_3^-$$

Q: Describe the change in the pH of the solution as the reaction progresses. Explain.

A: The pH of the solution will decrease because the product of the reaction is an acid.

(b) Carbon dioxide gas is bubbled through water.

$$CO_2 + H_2O \rightarrow H_2CO_3$$

Q: Briefly explain how non-metal oxides cause acid rain.

A: Nonmetal oxides are acid anhydrides. They react with water in the atmosphere to make rain water acidic.

(c) Solid phosphorous pentoxide is added to distilled water.

$$P_2O_5 + 3\ H_2O \rightarrow 2\ H_3PO_4$$

Q: How many moles of products are produced when 4 moles of phosphorous pentoxide react with excess water?

A: From the balanced equation, 4 moles of P_2O_5 produce 8 moles of H_3PO_4.

(d) Sulfur dioxide gas is bubbled through water.

$$SO_2 + H_2O \rightarrow H_2SO_3$$

Q: What is the oxidation state of sulfur before the reaction? What is the oxidation state of sulfur after the reaction? Is sulfur oxidized during the reaction?

A: The oxidation state of sulfur is +4 both before and after the reaction, so sulfur is neither oxidized nor reduced.

(e) Sulfur trioxide gas is bubbled through water.

$$SO_3 + H_2O \rightarrow H^+ + HSO_4^-$$

Q: How many moles of water are required to react with 3 moles of sulfur trioxide gas?

A: From the balanced equation, 3 moles of H_2O will react with 3 moles of SO_3.

IV. Look for an acid-base neutralization

There are a few variations on this, but you can follow all of them through to get the standard proton donor–proton acceptor equations that you are more used to dealing with.

1. An acid and a base.

(a) Equal molar amounts of potassium hydroxide and hydrochloric acid are mixed.

$$H^+ + OH^- \rightarrow H_2O$$

Strong acid and strong base.

Q: What will be the approximate pH of the solution that results from this reaction?

A: The pH will be about 7 because the strong acid and strong base will neutralize each other completely.

(b) A solution of sodium hydroxide is added to a solution of acetic acid.

$$HC_2H_3O_2 + OH^- \rightarrow C_2H_3O_2^- + H_2O$$

Weak acid and strong base.

Q: If a 0.1 M solution of sodium hydroxide is combined with a 0.1 M solution of acetic acid, will the pH of the solution that results from this reaction be greater than, less than, or equal to 7?

A: The pH will be greater than 7 because equimolar solutions of a strong base and a weak acid will react to form a basic solution.

(c) Solutions of ammonia and nitric acid are mixed.

$$NH_3 + H^+ \rightarrow NH_4^+$$

Strong acid and weak base.

Q: Briefly explain the difference between a strong acid and a weak acid.

A: A strong acid dissociates completely in water, producing many hydrogen ions. A weak acid dissociates only slightly, producing far fewer hydrogen ions.

(d) Solutions of ammonia and carbonic acid are mixed.

$$H_2CO_3 + NH_3 \rightarrow HCO_3^- + NH_4^+$$

Weak acid and weak base.

Q: Identify the conjugate acid-base pairs in this reaction.

A: H_2CO_3 is conjugate acid and HCO_3^- is a conjugate base. NH_4^+ is a conjugate acid and NH_3 is a conjugate base.

2. An acid and a basic salt.

This is still a straightforward acid-base reaction.

(a) Solutions of hydrochloric acid and sodium bicarbonate are mixed.

$$H^+ + HCO_3^- \rightarrow H_2CO_3$$

Q: If bubbles are observed at the completion of the reaction, identify the gas.

A: Carbonic acid decomposes into water and carbon dioxide, so the gas is CO_2.

(b) Dilute sulfuric acid is added to a solution of potassium fluoride.

$$H^+ + F^- \rightarrow HF$$

Q: Identify the Lewis base in the reaction and describe its role.

A: Fluorine ion acts as a Lewis base because it donates an electron pair to the hydrogen ion.

(c) Excess hydrochloric acid is added to a solution of potassium sulfide.

$$2H^+ + S^{2-} \rightarrow H_2S$$

Q: Identify the limiting reactant in the reaction.
A: It is stated that there is excess hydrochloric acid, so potassium sulfide is the limiting reactant.

Actually, the word "excess" is there to indicate that there are enough H^+ ions to form H_2S, instead of just HS^-.

3. A base and an acid salt.

(a) Sodium hydroxide solution is added to a solution of ammonium nitrate.

$$NH_4^+ + OH^- \rightarrow NH_3 + H_2O$$

Q: Identify the conjugate acid/base pairs in the reaction.
A: NH_4^+ is conjugate acid and NH_3 is a conjugate base. H_2O is a conjugate acid and OH^- is a conjugate base.

(b) Solutions of barium hydroxide and ammonium chloride are mixed.

$$NH_4^+ + OH^- \rightarrow NH_3 + H_2O$$

Q: If 1.0 L volumes of 1.0 M solutions of barium hydroxide and ammonium chloride are combined, how many moles of barium hydroxide will be consumed? Assume the reaction goes to completion.
A: If 1 mole of NH_4Cl reacts with 1 mole of $Ba(OH)_2$, NH_4Cl will run out when only half of the $Ba(OH)_2$ has been consumed, so only 0.5 moles of barium hydroxide will be consumed.

4. An acid anhydride and a base.

If you look at these in two steps, you'll see that they're just like other acid-base neutralizations.

(a) Sulfur trioxide gas is bubbled through a sodium hydroxide solution.

$$SO_3 + H_2O \rightarrow H^+ + HSO_4^-$$

$$H^+ + OH^- \rightarrow H_2O$$
_____
$$SO_3 + OH^- \rightarrow HSO_4^-$$

Q: Identify the molecular geometry of the product of the reaction and the hybridization of its central atom.
A: HSO_4^- has a tetrahedral geometry and its central sulfur atom has sp^3 hybridization.

(b) Carbon dioxide gas and ammonia gas are bubbled into distilled water.

$$CO_2 + H_2O \rightarrow H_2CO_3$$

$$H_2CO_3 + NH_3 \rightarrow HCO_3^- + NH_4^+$$
_____
$$CO_2 + H_2O + NH_3 \rightarrow HCO_3^- + NH_4^+$$

Q: Describe the polarity of carbon dioxide, ammonia, and water.
A: Carbon dioxide is nonpolar. Ammonia and water are polar.

V. Look for a mixture of two salt solutions

For these questions, you have to predict which salt precipitates. If you are familiar with the solubility rules, these aren't too bad.

(a) Solutions of calcium nitrate and sodium sulfate are mixed.

$$Ca^{2+} + SO_4^{2-} \rightarrow CaSO_4$$

Q: Is the entropy change for this reaction positive or negative? Explain.
A: The entropy change is negative because aqueous particles are combining and precipitating as a solid.

(b) A solution of silver nitrate is added to a solution of potassium iodide.

$$Ag^+ + I^- \rightarrow AgI$$

Q: Identify the spectator ions in this reaction.
A: The spectator ions are potassium (K^+) and nitrate (NO_3^-).

(c) Solutions of lead(II) nitrate and tri-potassium phosphate are mixed.

$$3 Pb^{2+} + 2 PO_4^{3-} \rightarrow Pb_3(PO_4)_2$$

Q: If 0.6 moles of lead(II) nitrate are consumed in this reaction, how many moles of products are produced?
A: From the balanced equation, if 3 moles of $Pb(NO_3)_2$ are consumed, then 1 mole of $Pb_3(PO_4)_2$ is produced. So if 0.6 moles are consumed, then 0.2 moles are produced.

(d) A solution of ammonium sulfide is added to a solution of magnesium iodide.

$$Mg^{2+} + S^{2-} \rightarrow MgS$$

Q: Identify the bonding in the products of this reaction.
A: MgS is held together by ionic bonds.

VI. Look for the combustion of a carbon compound.

Carbon compounds always burn to form oxide gases. Even if you can't get the reactant formula from the name of a hydrocarbon compound, you can probably get the two product points by knowing that it produces carbon dioxide and water when it burns.

(a) Ethane is burned in oxygen.

$$2 C_2H_6 + 7 O_2 \rightarrow 4 CO_2 + 6 H_2O$$

Q: If equal molar quantities of ethane and oxygen are combined, which will be the limiting reactant?
A: Oxygen; 3.5 moles of oxygen are required for each mole of ethane, so oxygen will run out first.

(b) Methanol is burned in air.

$$CH_3OH + O_2 \rightarrow CO_2 + H_2O$$

Q: Is the enthalpy change of the reaction positive or negative? Explain.
A: The enthalpy change is negative. Combustion releases energy, so it is exothermic.

(c) Carbon disulfide is burned in excess oxygen.

$$CS_2 + 3\,O_2 \rightarrow CO_2 + 2\,SO_2$$

Q: Describe the molecular geometry of the products of this reaction.
A: CO_2 is linear and SO_2 is bent.

The word "excess" is there to indicate that there is enough oxygen to form CO_2, rather than CO.

VII. Look for a solid transition metal placed in solution

This will be a redox reaction and the solid metal will always be oxidized. Look for two variations.

1. The solution is a neutral transition metal salt solution.

The metal ion in solution will be reduced.

(a) Solid manganese flakes are placed in a solution of copper(II) sulfate.

$$Mn + Cu^{2+} \rightarrow Mn^{2+} + Cu$$

Q: Identify the oxidizing and reducing agents in the reaction.
A: Mn is oxidized, so it is the reducing agent. Cu^{2+} is reduced, so it is the oxidizing agent.

(b) A piece of solid nickel is placed in a solution of silver nitrate.

$$Ni + 2\,Ag^+ \rightarrow Ni^{2+} + 2\,Ag$$

Q: If the solid nickel was placed in 0.50 L of a 0.40 M silver nitrate solution and the reaction went to completion, how many grams of nickel were consumed?
A: There are (0.5 L)(0.4 M) = 0.2 mole of silver, so from the balanced equation, 0.1 mole of nickel was consumed. (0.1 mole)(59 g/mol) = 5.9 grams of nickel.

(c) A bar of zinc is immersed in a solution of silver nitrate.

$$Zn + 2\,Ag^+ \rightarrow Zn^{2+} + 2\,Ag$$

Q: Describe the oxidation state of zinc before and after the reaction.
A: Before the reaction, Zn has an oxidation state of zero. After the reaction, the oxidation state is +2.

2. The solution is a strong oxoacid solution.

The anion of the oxoacid will be reduced to an oxide gas and water will form.

(a) A piece of copper is immersed in concentrated warm sulfuric acid.

$$Cu + 3 H^+ + HSO_4^- \rightarrow Cu^{2+} + SO_2 + 2 H_2O$$

Q: What is the oxidation state of sulfur before the reaction? What is the oxidation state of sulfur after the reaction?
A: The oxidation state of sulfur before the reaction is +6. The oxidation state after the reaction is +4.

(b) A piece of silver is placed in dilute nitric acid.

$$3 Ag + 4 H^+ + NO_3^- \rightarrow 3 Ag^+ + NO + 2 H_2O$$

Q: If the reaction above took place in a battery and one mole of silver was consumed, how much charge was generated?
A: If one mole of silver was consumed, then one mole of electrons was transferred. The charge on one mole of electrons is one Faraday, or 96,500 C.

VIII. Look for transition metal ions in solution with ammonia, hydroxide, cyanide, or thiocyanate

Transition metal ions form complex ions with those species. It doesn't matter how many of them you place on the transition metal, as long as you get the charge on the complex ion right.

(a) Excess ammonia is added to solution of silver nitrate.

$$Ag^+ + 2\, NH_3 \rightarrow Ag(NH_3)_2^+$$

Q: Identify the Lewis base in the reaction. Explain.
A: Ammonia acts as the Lewis base because it donates an electron pair to the silver ion.

(b) A solution of sodium cyanide is added to a solution of iron(II) chloride.

$$Fe^{2+} + 6\, CN^- \rightarrow Fe(CN)_6^{4-}$$

Q: Identify the spectator ions in the reaction.
A: Sodium ion (Na^+) and chloride ion (Cl^-) are the spectator ions.

(c) A solution of potassium thiocyanate is added to a solution of copper(II) chloride.

$$Cu^{2+} + SCN^- \rightarrow Cu(SCN)^+$$

Q: Briefly explain why solutions containing transition metal ions tend to be brightly colored.
A: The intervals between energy levels of the d-subshell electrons correspond to the energy levels of visible light. So when these electrons make energy transitions, electromagnetic radiation in the visible spectrum is released.

18

The Princeton Review
AP Chemistry
Practice Exam 1

AP® Chemistry Exam

SECTION I: Multiple-Choice Questions

DO NOT OPEN THIS BOOKLET UNTIL YOU ARE TOLD TO DO SO.

At a Glance

Total Time
1 hour and 30 minutes
Number of Questions
75
Percent of Total Grade
50%
Writing Instrument
Pencil required

Instructions

Section I of this examination contains 75 multiple-choice questions. Fill in only the ovals for numbers 1 through 75 on your answer sheet.

CALCULATORS MAY NOT BE USED IN THIS PART OF THE EXAMINATION.

Indicate all of your answers to the multiple-choice questions on the answer sheet. No credit will be given for anything written in this exam booklet, but you may use the booklet for notes or scratch work. After you have decided which of the suggested answers is best, completely fill in the corresponding oval on the answer sheet. Give only one answer to each question. If you change an answer, be sure that the previous mark is erased completely. Here is a sample question and answer.

Sample Question   Sample Answer

Chicago is a   Ⓐ ● Ⓒ Ⓓ Ⓔ
(A) state
(B) city
(C) country
(D) continent
(E) village

Use your time effectively, working as quickly as you can without losing accuracy. Do not spend too much time on any one question. Go on to other questions and come back to the ones you have not answered if you have time. It is not expected that everyone will know the answers to all the multiple-choice questions.

About Guessing

Many candidates wonder whether or not to guess the answers to questions about which they are not certain. Multiple choice scores are based on the number of questions answered correctly. Points are not deducted for incorrect answers, and no points are awarded for unanswered questions. Because points are not deducted for incorrect answers, you are encouraged to answer all multiple-choice questions. On any questions you do not know the answer to, you should eliminate as many choices as you can, and then select the best answer among the remaining choices.

This page intentionally left blank.

GO ON TO THE NEXT PAGE.

CHEMISTRY
SECTION I
Time—1 hour and 30 minutes

Material in the following table may be useful in answering the questions in this section of the examination.

PERIODIC CHART OF THE ELEMENTS

1 **H** 1.0																	2 **He** 4.0
3 **Li** 6.9	4 **Be** 9.0											5 **B** 10.8	6 **C** 12.0	7 **N** 14.0	8 **O** 16.0	9 **F** 19.0	10 **Ne** 20.2
11 **Na** 23.0	12 **Mg** 24.3											13 **Al** 27.0	14 **Si** 28.1	15 **P** 31.0	16 **S** 32.1	17 **Cl** 35.5	18 **Ar** 39.9
19 **K** 39.1	20 **Ca** 40.1	21 **Sc** 45.0	22 **Ti** 47.9	23 **V** 50.9	24 **Cr** 52.0	25 **Mn** 54.9	26 **Fe** 55.8	27 **Co** 58.9	28 **Ni** 58.7	29 **Cu** 63.5	30 **Zn** 65.4	31 **Ga** 69.7	32 **Ge** 72.6	33 **As** 74.9	34 **Se** 79.0	35 **Br** 79.9	36 **Kr** 83.8
37 **Rb** 85.5	38 **Sr** 87.6	39 **Y** 88.9	40 **Zr** 91.2	41 **Nb** 92.9	42 **Mo** 95.9	43 **Tc** (98)	44 **Ru** 101.1	45 **Rh** 102.9	46 **Pd** 106.4	47 **Ag** 107.9	48 **Cd** 112.4	49 **In** 114.8	50 **Sn** 118.7	51 **Sb** 121.8	52 **Te** 127.6	53 **I** 126.9	54 **Xe** 131.3
55 **Cs** 132.9	56 **Ba** 137.3	57 ***La** 138.9	72 **Hf** 178.5	73 **Ta** 180.9	74 **W** 183.9	75 **Re** 186.2	76 **Os** 190.2	77 **Ir** 192.2	78 **Pt** 195.1	79 **Au** 197.0	80 **Hg** 200.6	81 **Tl** 204.4	82 **Pb** 207.2	83 **Bi** 209.0	84 **Po** (209)	85 **At** (210)	86 **Rn** (222)
87 **Fr** (223)	88 **Ra** 226.0	89 **†Ac** 227.0															

*Lanthanum Series

58 **Ce** 140.1	59 **Pr** 140.9	60 **Nd** 144.2	61 **Pm** (145)	62 **Sm** 150.4	63 **Eu** 152.0	64 **Gd** 157.3	65 **Tb** 158.9	66 **Dy** 162.5	67 **Ho** 164.9	68 **Er** 167.3	69 **Tm** 168.9	70 **Yb** 173.0	71 **Lu** 175.0

†Actinium Series

90 **Th** 232.0	91 **Pa** 231.0	92 **U** 238.0	93 **Np** 237.0	94 **Pu** (244)	95 **Am** (243)	96 **Cm** (247)	97 **Bk** (247)	98 **Cf** (251)	99 **Es** (252)	100 **Fm** (258)	101 **Md** (258)	102 **No** (259)	103 **Lr** (260)

DO NOT DETACH FROM BOOK.

GO ON TO THE NEXT PAGE.

Note: For all questions involving solutions and/or chemical equations, assume that the system is in pure water and at room temperature unless otherwise stated.

Part A

Directions: Each set of lettered choices below refers to the numbered questions or statements immediately following it. Select the one lettered choice that best answers each question or best fits each statement and fill in the corresponding oval on the answer sheet. A choice may be used once, more than once, or not at all in each set.

Questions 1–3 are based on the following energy diagrams.

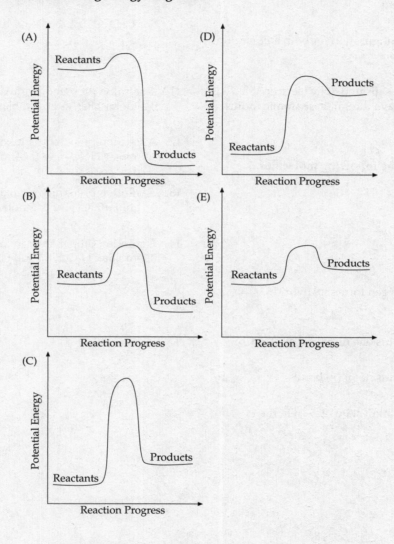

1. This reaction has the largest activation energy.

2. This is the most exothermic reaction.

3. This reaction has the largest positive value for ΔH.

GO ON TO THE NEXT PAGE.

Questions 4–6 refer to the following configurations.

    (A) $1s^2\, 2s^2 2p^6$
    (B) $1s^2\, 2s^2 2p^6\, 3s^2$
    (C) $1s^2\, 2s^2 2p^6\, 3s^2 3p^4$
    (D) $1s^2\, 2s^2 2p^6\, 3s^2 3p^6$
    (E) $1s^2\, 2s^2 2p^6\, 3s^2 3p^6\, 4s^2$

4. The ground state configuration of an atom of a paramagnetic element.

5. The ground state configuration for both a potassium ion and a chloride ion.

6. An atom that has this ground-state electron configuration will have the smallest atomic radius of those listed above.

Questions 7–10 refer to the following molecules.

    (A) CO_2
    (B) H_2O
    (C) SO_2
    (D) NO_2
    (E) O_2

7. In this molecule, oxygen forms sp^3 hybrid orbitals.

8. This molecule contains one unpaired electron.

9. This molecule contains no pi (π) bonds.

10. Oxygen has an oxidation state of zero in the molecule.

Questions 11–14 refer to the following solutions.

    (A)   A solution with a pH of 1
    (B)   A solution with a pH of greater than 1 and less than 7
    (C)   A solution with a pH of 7
    (D)   A solution with a pH of greater than 7 and less than 13
    (E)   A solution with a pH of 13

For CH_3COOH, $K_a = 1.8 \times 10^{-5}$

For NH_3, $K_b = 1.8 \times 10^{-5}$

11. A solution prepared by mixing equal volumes of 0.2-molar HCl and 0.2-molar NH_3.

12. A solution prepared by mixing equal volumes of 0.2-molar HNO_3 and 0.2-molar NaOH.

13. A solution prepared by mixing equal volumes of 0.2-molar HCl and 0.2-molar NaCl.

14. A solution prepared by mixing equal volumes of 0.2-molar CH_3COOH and 0.2-molar NaOH.

GO ON TO THE NEXT PAGE.

Part B

Directions: Each of the questions or incomplete statements below is followed by five suggested answers or completions. Select the one that is best in each case and fill in the corresponding oval on the answer sheet.

15. A pure sample of $KClO_3$ is found to contain 71 grams of chlorine atoms. What is the mass of the sample?

    (A) 122 grams
    (B) 170 grams
    (C) 209 grams
    (D) 245 grams
    (E) 293 grams

16. Which of the following experimental procedures is used to separate two substances by taking advantage of their differing boiling points?

    (A) Titration
    (B) Distillation
    (C) Filtration
    (D) Decantation
    (E) Hydration

17. Which of the following sets of quantum numbers (n, l, m_l, m_s) best describes the highest energy valence electron in a ground-state aluminum atom?

    (A) $2, 0, 0, \dfrac{1}{2}$

    (B) $2, 1, 0, \dfrac{1}{2}$

    (C) $3, 0, 0, \dfrac{1}{2}$

    (D) $3, 0, 1, \dfrac{1}{2}$

    (E) $3, 1, 1, \dfrac{1}{2}$

18. A chemist analyzed the carbon–carbon bond in C_2H_6 and found that it had a bond energy of 350 kJ/mol and a bond length of 1.5 angstroms. If the chemist performed the same analysis on the carbon–carbon bond in C_2H_2, how would the results compare?

    (A) The bond energies and lengths for C_2H_2 would be the same as those of C_2H_6.
    (B) The bond energy for C_2H_2 would be smaller, and the bond length would be shorter.
    (C) The bond energy for C_2H_2 would be greater, and the bond length would be longer.
    (D) The bond energy for C_2H_2 would be smaller, and the bond length would be longer.
    (E) The bond energy for C_2H_2 would be greater, and the bond length would be shorter.

19. $2\,MnO_4^- + 5\,SO_3^{2-} + 6\,H^+ \rightarrow 2\,Mn^{2+} + 5\,SO_4^{2-} + 3\,H_2O$

    Which of the following statements is true regarding the reaction given above?

    (A) MnO_4^- acts as the reducing agent.
    (B) H^+ acts as the oxidizing agent.
    (C) SO_3^{2-} acts as the reducing agent.
    (D) Manganese is oxidized
    (E) Sulfur is reduced.

20. Which of the following can function as both a Brønsted-Lowry acid and Brønsted-Lowry base?

    (A) HCl
    (B) H_2SO_4
    (C) HSO_3^-
    (D) SO_4^{2-}
    (E) H^+

GO ON TO THE NEXT PAGE.

21. Which of the following substances experiences the strongest attractive intermolecular forces?

    (A) H_2
    (B) N_2
    (C) CO_2
    (D) NH_3
    (E) CH_4

22. A mixture of gases at equilibrium over water at 43°C contains 9.0 moles of nitrogen, 2.0 moles of oxygen, and 1.0 mole of water vapor. If the total pressure exerted by the gases is 780 mmHg, what is the vapor pressure of water at 43°C?

    (A) 65 mmHg
    (B) 130 mmHg
    (C) 260 mmHg
    (D) 580 mmHg
    (E) 720 mmHg

23. $...MnO_4^- + ...e^- + ...H^+ \rightarrow ...Mn^{2+} + ...H_2O$

    When the half-reaction above is balanced, what is the coefficient for H^+ if all the coefficients are reduced to the lowest whole number?

    (A) 3
    (B) 4
    (C) 5
    (D) 8
    (E) 10

24. The boiling point of water is known to be lower at high elevations. This is because

    (A) hydrogen bonds are weaker at high elevations.
    (B) the heat of fusion is lower at high elevations.
    (C) the vapor pressure of water is higher at high elevations.
    (D) the atmospheric pressure is lower at high elevations.
    (E) water is denser at high elevations.

25. A student examined 2.0 moles of an unknown carbon compound and found that the sample contained 48 grams of carbon, 64 grams of oxygen, and 8 grams of hydrogen. Which of the following could be the molecular formula of the compound?

    (A) CH_2O
    (B) CH_2OH
    (C) CH_3COOH
    (D) CH_3CO
    (E) C_2H_5OH

26. $S(s) + O_2(g) \rightarrow SO_2(g)$ $\Delta H = x$

    $S(s) + \dfrac{3}{2} O_2(g) \rightarrow SO_3(g)$ $\Delta H = y$

    Based on the information above, what is the standard enthalpy change for the following reaction?

    $$2\,SO_2(g) + O_2(g) \rightarrow 2\,SO_3(g)$$

    (A) $x - y$
    (B) $y - x$
    (C) $2x - y$
    (D) $2x - 2y$
    (E) $2y - 2x$

27. How much water must be added to a 50.0 ml solution of 0.60 M HNO_3 to produce a 0.40 M solution of HNO_3?

    (A) 25 ml
    (B) 33 ml
    (C) 50 ml
    (D) 67 ml
    (E) 75 ml

28. In which of the following equilibria would the concentrations of the products be increased if the volume of the system were decreased at constant temperature?

    (A) $H_2(g) + Cl_2(g) \rightleftharpoons 2\,HCl(g)$
    (B) $2\,CO(g) + O_2(g) \rightleftharpoons 2\,CO_3(g)$
    (C) $NO(g) + O_3(g) \rightleftharpoons NO_2(g) + O_2(g)$
    (D) $2\,HI(g) \rightleftharpoons H_2(g) + I_2(g)$
    (E) $N_2O_4(g) \rightleftharpoons 2\,NO_2(g)$

GO ON TO THE NEXT PAGE.

29. A 100 ml sample of 0.10 molar NaOH solution was added to 100 ml of 0.10 molar $H_3C_6H_5O_7$. After equilibrium was established, which of the ions listed below was present in the greatest concentration?

   (A) $H_2C_6H_5O_7^-$
   (B) $HC_6H_5O_7^{2-}$
   (C) $C_6H_5O_7^{3-}$
   (D) OH^-
   (E) H^+

30. Which of the following can be determined directly from the difference between the boiling point of a pure solvent and the boiling point of a solution of a nonionic solute in the solvent, if k_b for the solvent is known?

   I.   The mass of solute in the solution
   II. The molality of the solution
   III. The volume of the solution

   (A) I only
   (B) II only
   (C) III only
   (D) I and II only
   (E) I and III only

31. The value of the equilibrium constant K_{eq} is greater than 1 for a certain reaction under standard state conditions. Which of the following statements must be true regarding the reaction?

   (A) $\Delta G°$ is negative.
   (B) $\Delta G°$ is positive.
   (C) $\Delta G°$ is equal to zero.
   (D) $\Delta G°$ is negative if the reaction is exothermic and positive if the reaction is endothermic.
   (E) $\Delta G°$ is negative if the reaction is endothermic and positive if the reaction is exothermic.

32. Which of the following aqueous solutions has the highest boiling point?

   (A) $0.1\ m$ NaOH
   (B) $0.1\ m$ HF
   (C) $0.1\ m$ Na_2SO_4
   (D) $0.1\ m$ $KC_2H_3O_2$
   (E) $0.1\ m$ NH_4NO_3

33. The molecular formula for hydrated iron (III) oxide, or rust, is generally written as $Fe_2O_3 \cdot x\ H_2O$ because the water content in rust can vary. If a 1-molar sample of hydrated iron (III) oxide is found to contain 108 g of H_2O, what is the molecular formula for the sample?

   (A) $Fe_2O_3 \cdot H_2O$
   (B) $Fe_2O_3 \cdot 3\ H_2O$
   (C) $Fe_2O_3 \cdot 6\ H_2O$
   (D) $Fe_2O_3 \cdot 10\ H_2O$
   (E) $Fe_2O_3 \cdot 12\ H_2O$

34. In which of the following reactions does the greatest increase in entropy take place?

   (A) $H_2O(l) \rightarrow H_2O(g)$
   (B) $2\ NO(g) + O_2(g) \rightarrow 2\ NO_2(g)$
   (C) $CaH_2(s) + H_2O(l) \rightarrow Ca(OH)_2(s) + H_2O(g)$
   (D) $NH_4Cl(s) \rightarrow NH_3(g) + HCl(g)$
   (E) $2\ HCl(g) \rightarrow H_2(g) + Cl_2(g)$

35. The density of a sample of water decreases as it is heated above a temperature of 4°C. Which of the following will be true of an aqueous solution of $NaC_2H_3O_2$ when it is heated from 10°C to 60°C?

   (A) The molarity will increase.
   (B) The molarity will decrease
   (C) The molality will increase.
   (D) The molality will decrease.
   (E) The molarity and molality will remain unchanged.

36. $... + n \rightarrow {}_3^7Li + {}_2^4He$

   For the nuclear reaction shown above, what is the missing reactant?

   (A) ${}_4^9Be$
   (B) ${}_5^9B$
   (C) ${}_4^{10}Be$
   (D) ${}_5^{10}B$
   (E) ${}_5^{11}B$

GO ON TO THE NEXT PAGE.

37. A boiling-water bath is sometimes used instead of a flame in heating objects. Which of the following could be an advantage of a boiling-water bath over a flame?

    (A) The relatively low heat capacity of water will cause the object to become hot more quickly.
    (B) The relatively high density of water will cause the object to become hot more quickly.
    (C) The volume of boiling water remains constant over time.
    (D) The temperature of boiling water remains constant at 100°C.
    (E) The vapor pressure of boiling water is equal to zero.

38. The addition of a catalyst to a chemical reaction will bring about a change in which of the following characteristics of the reaction?

    I.   The activation energy
    II. The enthalpy change
    III. The value of the equilibrium constant

    (A) I only
    (B) II only
    (C) I and II only
    (D) I and III only
    (E) II and III only

39. $$2\ NO(g) + O_2(g) \rightarrow 2\ NO_2(g)$$

    The reaction above occurs by the following two-step process:

    Step I: $NO(g) + O_2(g) \rightarrow NO_3(g)$

    Step II: $NO_3(g) + NO(g) \rightarrow 2\ NO_2(g)$

    Which of the following is true of Step II if it is the rate-limiting step?

    (A) Step II has a lower activation energy and occurs more slowly than Step I.
    (B) Step II has a higher activation energy and occurs more slowly than Step I.
    (C) Step II has a lower activation energy and occurs more quickly than Step I.
    (D) Step II has a higher activation energy and occurs more quickly than Step I.
    (E) Step II has the same activation energy and occurs at the same speed as Step I.

40. $$C_3H_7OH + ...O_2 \rightarrow ...CO_2 + ...H_2O$$

    One mole of C_3H_7OH underwent combustion as shown in the reaction above. How many moles of oxygen were required for the reaction?

    (A) 2 moles

    (B) 3 moles

    (C) $\dfrac{7}{2}$ moles

    (D) $\dfrac{9}{2}$ moles

    (E) 5 moles

Questions 41–42 refer to the phase diagram below.

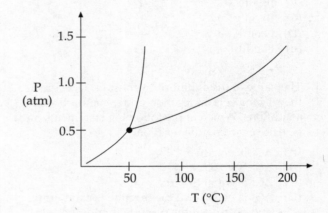

41. If the pressure of the substance shown in the diagram is decreased from 1.0 atmosphere to 0.5 atmosphere at a constant temperature of 100°C, which phase change will occur?

    (A) Freezing
    (B) Vaporization
    (C) Condensation
    (D) Sublimation
    (E) Deposition

GO ON TO THE NEXT PAGE.

42. Under what conditions can all three phases of the substance shown in the diagram exist simultaneously in equilibrium?

    (A) Pressure = 1.0 atm, temperature = 150°C
    (B) Pressure = 1.0 atm, temperature = 100°C
    (C) Pressure = 1.0 atm, temperature = 50°C
    (D) Pressure = 0.5 atm, temperature = 100°C
    (E) Pressure = 0.5 atm, temperature = 50°C

43. Which of the following statements regarding atomic theory is NOT true?

    (A) The Bohr model of the atom was based on Planck's quantum theory.
    (B) Rutherford's experiments with alpha particle scattering led to the conclusion that positive charge was concentrated in an atom's nucleus.
    (C) Heisenberg's uncertainty principle describes the equivalence of mass and energy.
    (D) Millikan's oil drop experiment led to the calculation of the charge on an electron.
    (E) Thomson's cathode ray experiments confirmed the existence of the electron.

44.

Time (min)	$[A]$ M
0	0.50
10	0.36
20	0.25
30	0.18
40	0.13

Reactant A underwent a decomposition reaction. The concentration of A was measured periodically and recorded in the chart above. Based on the data in the chart, which of the following statements is NOT true?

(A) The reaction is first order in $[A]$.
(B) The reaction is first order overall.
(C) The rate of the reaction is constant over time.
(D) The half-life of reactant A is 20 minutes.
(E) The graph of ln $[A]$ will be a straight line.

45. A solution to be used as a reagent for a reaction is to be removed from a bottle marked with its concentration. Which of the following is NOT part of the proper procedure for this process?

    (A) Pouring the solution down a stirring rod into a beaker.
    (B) Inserting a pipette directly into the bottle and drawing out the solution.
    (C) Placing the stopper of the bottle upside down on the table top.
    (D) Pouring the solution down the side of a tilted beaker.
    (E) Touching the stopper of the bottle only on the handle.

46.
$$N_2(g) + 3\,H_2(g) \rightleftharpoons 2\,NH_3(g) + energy$$

Which of the following changes to the equilibrium situation shown above will bring about an increase in the number of moles of NH_3 present at equilibrium?

    I. Adding N_2 gas to the reaction chamber
    II. Increasing the volume of the reaction chamber at constant temperature
    III. Increasing the temperature of the reaction chamber at constant volume

(A) I only
(B) II only
(C) I and II only
(D) I and III only
(E) II and III only

47.
$$CH_4 + 2\,O_2 \rightarrow CO_2 + 2\,H_2O$$

If 16 grams of CH_4 reacts with 16 grams of O_2 in the reaction shown above, which of the following will be true?

(A) The mass of H_2O formed will be twice the mass of CO_2 formed.
(B) Equal masses of H_2O and CO_2 will be formed.
(C) Equal numbers of moles of H_2O and CO_2 will be formed.
(D) The limiting reagent will be CH_4.
(E) The limiting reagent will be O_2.

GO ON TO THE NEXT PAGE.

48. Which of the following sets of gases would be most difficult to separate if the method of gaseous effusion is used?

    (A) O_2 and CO_2
    (B) N_2 and C_2H_4
    (C) H_2 and CH_4
    (D) He and Ne
    (E) O_2 and He

49. Which of the following equilibrium expressions represents the hydrolysis of the CN^- ion?

    (A) $K = \dfrac{[HCN][OH^-]}{[CN^-]}$

    (B) $K = \dfrac{[CN^-][OH^-]}{[HCN]}$

    (C) $K = \dfrac{[CN^-][H_3O^+]}{[HCN]}$

    (D) $K = \dfrac{[HCN][H_3O^+]}{[CN^-]}$

    (E) $K = \dfrac{[HCN]}{[CN^-][OH^-]}$

50. Which of the following is true under any conditions for a reaction that is spontaneous at any temperature?

    (A) ΔG, ΔS, and ΔH are all positive.
    (B) ΔG, ΔS, and ΔH are all negative.
    (C) ΔG and ΔS are negative, and ΔH is positive.
    (D) ΔG and ΔS are positive, and ΔH is negative.
    (E) ΔG and ΔH are negative, and ΔS is positive.

51. Which of the following pairs of compounds are isomers?

    (A) $HCOOH$ and CH_3COOH
    (B) CH_3CH_2CHO and C_3H_7OH
    (C) C_2H_5OH and CH_3OCH_3
    (D) C_2H_4 and C_2H_6
    (E) C_3H_8 and C_4H_{10}

52. A sample of an ideal gas confined in a rigid 5.00 liter container has a pressure of 363 mmHg at a temperature of 25°C. Which of the following expressions will be equal to the pressure of the gas if the temperature of the container is increased to 35°C?

    (A) $\dfrac{(363)(35)}{(25)}$ mmHg

    (B) $\dfrac{(363)(25)}{(35)}$ mmHg

    (C) $\dfrac{(363)(308)}{(298)}$ mmHg

    (D) $\dfrac{(363)(298)}{(308)}$ mmHg

    (E) $\dfrac{(363)(273)}{(308)}$ mmHg

53. If the temperature at which a reaction takes place is increased, the rate of the reaction will

    (A) increase if the reaction is endothermic and decrease if the reaction is exothermic.
    (B) decrease if the reaction is endothermic and increase if the reaction is exothermic.
    (C) increase if the reaction is endothermic and increase if the reaction is exothermic.
    (D) decrease if the reaction is endothermic and decrease if the reaction is exothermic.
    (E) remain the same for both an endothermic and an exothermic reaction.

54. The acid dissociation constant for HClO is 3.0×10^{-8}. What is the hydrogen ion concentration in a 0.12 M solution of HClO?

    (A) $3.6 \times 10^{-9}\ M$
    (B) $3.6 \times 10^{-8}\ M$
    (C) $6.0 \times 10^{-8}\ M$
    (D) $2.0 \times 10^{-5}\ M$
    (E) $6.0 \times 10^{-5}\ M$

GO ON TO THE NEXT PAGE.

55. $$2 NO(g) + 2 H_2(g) \rightarrow N_2(g) + 2 H_2O(g)$$

Which of the following is true regarding the relative molar rates of disappearance of the reactants and appearance of the products?

  I. N_2 appears at the same rate that H_2 disappears.
  II. H_2O appears at the same rate that NO disappears.
  III. NO disappears at the same rate that H_2 disappears.

(A) I only
(B) I and II only
(C) I and III only
(D) II and III only
(E) I, II, and III

56. $$SO_4{}^{2-}, PO_4{}^{3-}, ClO_4{}^{-}$$

The geometries of the polyatomic ions listed above can all be described as

(A) square planar.
(B) square pyramidal.
(C) seesaw-shaped.
(D) tetrahedral.
(E) trigonal bipyramidal.

57. $$2 ZnS(s) + 3 O_2(g) \rightarrow 2 ZnO(s) + 2 SO_2(g)$$

If the reaction above took place at standard temperature and pressure, what was the volume of $O_2(g)$ required to produce 40.0 grams of ZnO(s)?

(A) $\dfrac{(40.0)(2)}{(81.4)(3)(22.4)}$ L

(B) $\dfrac{(40.0)(3)}{(81.4)(2)(22.4)}$ L

(C) $\dfrac{(40.0)(2)(22.4)}{(81.4)(3)}$ L

(D) $\dfrac{(40.0)(3)(22.4)}{(81.4)(2)}$ L

(E) $\dfrac{(81.4)(2)(22.4)}{(40.0)(3)}$ L

58. Which of the following salts will produce a colorless solution when added to water?

(A) $Cu(NO_3)_2$
(B) $NiCl_2$
(C) $KMnO_4$
(D) $ZnSO_4$
(E) $FeCl_3$

59. A beaker contains 300.0 ml of a 0.20 M $Pb(NO_3)_2$ solution. If 200.0 ml of a 0.20 M solution of $MgCl_2$ is added to the beaker, what will be the final concentration of Pb^{2+} ions in the resulting solution?

(A) 0.020 M
(B) 0.040 M
(C) 0.080 M
(D) 0.120 M
(E) 0.150 M

60. Which of the following procedures will produce a buffered solution?

  I. Equal volumes of 1 M NH_3 and 1 M NH_4Cl solutions are mixed.
  II. Equal volumes of 1 M H_2CO_3 and 1 M $NaHCO_3$ solutions are mixed.
  III. Equal volumes of 1 M NH_3 and 1 M H_2CO_3 solutions are mixed.

(A) I only
(B) III only
(C) I and II only
(D) II and III only
(E) I, II, and III

61. $$H_2(g) + I_2(g) \rightarrow 2 HI(g)$$

At 450 C the equilibrium constant, K_c, for the reaction shown above has a value of 50. Which of the following sets of initial conditions at 450 C will cause the reaction above to produce more H_2?

  I. [HI] = 5-molar, [H_2] = 1-molar, [I_2] = 1-molar
  II. [HI] = 10-molar, [H_2] = 1-molar, [I_2] = 1-molar
  III. [HI] = 10-molar, [H_2] = 2-molar, [I_2] = 2-molar

(A) I only
(B) II only
(C) I and II only
(D) II and III only
(E) I, II, and III

GO ON TO THE NEXT PAGE.

62. $Cu^{2+}(aq) + Zn(s) \rightarrow Cu(s) + Zn^{2+}(aq)$

A galvanic cell that uses the reaction shown above has a standard state electromotive force of 1.1 volts. Which of the following changes to the cell will increase the voltage?

I.   An increase in the mass of $Zn(s)$ in the cell.
II. An increase in the concentration of $Cu^{2+}(aq)$ in the cell.
III. An increase in the concentration of $Zn^{2+}(aq)$ in the cell.

(A) I only
(B) II only
(C) III only
(D) I and II only
(E) I and III only

63. The nuclide $_{26}^{61}Fe$ decays through the emission of a single beta (β^-) particle. What is the resulting nuclide?

(A) $_{26}^{60}Fe$
(B) $_{26}^{62}Fe$
(C) $_{27}^{61}Co$
(D) $_{27}^{62}Co$
(E) $_{25}^{61}Mn$

64. Which of the following statements is true regarding sodium and potassium?

(A) Sodium has a larger first ionization energy and a larger atomic radius.
(B) Sodium has a larger first ionization energy and a smaller atomic radius.
(C) Sodium has a smaller first ionization energy and a larger atomic radius.
(D) Sodium has a smaller first ionization energy and a smaller atomic radius.
(E) Sodium and potassium have identical first ionization energies and atomic radii.

65. $HCl(aq) + AgNO_3(aq) \rightarrow AgCl(s) + HNO_3(aq)$

One-half liter of a 0.20-molar HCl solution is mixed with one-half liter of a 0.40-molar solution of $AgNO_3$. A reaction occurs forming a precipitate as shown above. If the reaction goes to completion, what is the mass of AgCl produced?

(A) 14 grams
(B) 28 grams
(C) 42 grams
(D) 70 grams
(E) 84 grams

66. $H_2(g) + Cl_2(g) \rightarrow 2\,HCl(g)$

Based on the information given in the table below, what is ΔH for the above reaction?

Bond	Average Bond Energy (kJ/mol)
H–H	440
Cl–Cl	240
H–Cl	430

(A) −860 kJ
(B) −620 kJ
(C) −440 kJ
(D) −180 kJ
(E) +240 kJ

67. The first ionization energy for magnesium is 730 kJ/mol. The third ionization energy for magnesium is 7700 kJ/mol. What is the most likely value for magnesium's second ionization energy?

(A)   490 kJ/mol
(B)   1,400 kJ/mol
(C)   4,200 kJ/mol
(D)   7,100 kJ/mol
(E)   8,400 kJ/mol

GO ON TO THE NEXT PAGE.

68. Molten NaCl is electrolyzed with a constant current of 1.00 ampere. What is the shortest amount of time, in seconds, that it would take to produce 1.00 mole of solid sodium?
(1 faraday = 96,500 coulombs)

(A) 19,300 seconds
(B) 32,200 seconds
(C) 48,300 seconds
(D) 64,300 seconds
(E) 96,500 seconds

69. How many moles of KCl must be added to 200 milliliters of a 0.5-molar NaCl solution to create a solution in which the concentration of Cl^- ion is 1.0-molar? (Assume the volume of the solution remains constant.)

(A) 0.1 moles
(B) 0.2 moles
(C) 0.3 moles
(D) 0.4 moles
(E) 0.5 moles

70. A student placed solid barium oxalate in a beaker filled with distilled water and allowed it to come to equilibrium with its dissolved ions. The student then added a nickel (II) nitrate solution to the beaker to create the following equilibrium situation.

$$BaC_2O_4 + Ni^{2+} \leftrightarrow Ba^{2+} + NiC_2O_4$$

If the solubility product for BaC_2O_4 is 2×10^{-7} and the solubility product for NiC_2O_4 is 4×10^{-10}, which of the following gives the value of the equilibrium constant for the reaction above?

(A) 2×10^3
(B) 5×10^2
(C) 5×10^{-2}
(D) 2×10^{-3}
(E) 8×10^{-17}

71. A 100-gram sample of pure $^{37}_{18}Ar$ decays by electron capture with a half-life of 35 days. How long will it take for 90 grams of $^{37}_{17}Cl$ to accumulate?

(A) 31 days
(B) 39 days
(C) 78 days
(D) 116 days
(E) 315 days

72. The solubility product, K_{sp}, of CaF_2 is 4×10^{-11}. Which of the following expressions is equal to the solubility of CaF_2?

(A) $\sqrt{4 \times 10^{-11}}$ M
(B) $\sqrt{2 \times 10^{-11}}$ M
(C) $\sqrt[3]{4 \times 10^{-11}}$ M
(D) $\sqrt[3]{2 \times 10^{-11}}$ M
(E) $\sqrt[3]{1 \times 10^{-11}}$ M

73. When excess hydroxide ions were added to 1.0 liter of $CaCl_2$ solution, $Ca(OH)_2$ precipitate was formed. If all of the calcium ions in the solution were precipitated in 7.4 grams of $Ca(OH)_2$, what was the initial concentration of the $CaCl_2$ solution?

(A) 0.05-molar
(B) 0.10-molar
(C) 0.15-molar
(D) 0.20-molar
(E) 0.30-molar

GO ON TO THE NEXT PAGE.

74. When a solution of $KMnO_4$ was mixed with a solution of HCl, Cl_2 gas bubbles formed and Mn^{2+} ions appeared in the solution. Which of the following has occurred?

    (A) K^+ has been oxidized by Cl^-.
    (B) K^+ has been oxidized by H^+.
    (C) Cl^- has been oxidized by K^+.
    (D) Cl^- has been oxidized by MnO_4^-.
    (E) MnO_4^- has been oxidized by Cl^-.

75. $2\,Cu^+(aq) + M(s) \rightarrow 2\,Cu(s) + M^{2+}(aq)$
    $$E^o = +0.92 \text{ V}$$

    $Cu^+(aq) + e^- \rightarrow Cu(s)$
    $$E^o = +0.52 \text{ V}$$

    Based on the reduction potentials given above, what is the standard reduction potential for the following half-reaction?

    $$M^{2+}(aq) + 2\,e^- \rightarrow M(s)$$

    (A) +0.40 V
    (B) +0.12 V
    (C) −0.12 V
    (D) −0.40 V
    (E) −1.44 V

END OF SECTION I

CHEMISTRY
SECTION II
Time—1 hour and 35 minutes

Percent of total grade—50

Parts A: Time—55 minutes

Part B: Time—40 minutes

<u>General Instructions</u>

CALCULATORS MAY NOT BE USED IN PART B.

Calculators, including those with programming and graphing capabilities, may be used in Part A. However, calculators with typewriter-style (QWERTY) keyboards are NOT permitted.

Pages containing a periodic table, the electrochemical series, and equations commonly used in chemistry will be available for your use.

You may write your answers with either a pen or a pencil. Be sure to write CLEARLY and LEGIBLY. If you make an error, you may save time by crossing it out rather than trying to erase it.

Write all your answers in the essay booklet. Number your answers as the questions are numbered in the examination booklet.

GO ON TO THE NEXT PAGE.

MATERIAL IN THE FOLLOWING TABLE AND IN THE TABLES ON THE NEXT 3 PAGES MAY BE USEFUL IN ANSWERING THE QUESTIONS IN THIS SECTION OF THE EXAMINATION.

PERIODIC CHART OF THE ELEMENTS

1 H 1.0																		2 He 4.0
3 Li 6.9	4 Be 9.0											5 B 10.8	6 C 12.0	7 N 14.0	8 O 16.0	9 F 19.0	10 Ne 20.2	
11 Na 23.0	12 Mg 24.3											13 Al 27.0	14 Si 28.1	15 P 31.0	16 S 32.1	17 Cl 35.5	18 Ar 39.9	
19 K 39.1	20 Ca 40.1	21 Sc 45.0	22 Ti 47.9	23 V 50.9	24 Cr 52.0	25 Mn 54.9	26 Fe 55.8	27 Co 58.9	28 Ni 58.7	29 Cu 63.5	30 Zn 65.4	31 Ga 69.7	32 Ge 72.6	33 As 74.9	34 Se 79.0	35 Br 79.9	36 Kr 83.8	
37 Rb 85.5	38 Sr 87.6	39 Y 88.9	40 Zr 91.2	41 Nb 92.9	42 Mo 95.9	43 Tc (98)	44 Ru 101.1	45 Rh 102.9	46 Pd 106.4	47 Ag 107.9	48 Cd 112.4	49 In 114.8	50 Sn 118.7	51 Sb 121.8	52 Te 127.6	53 I 126.9	54 Xe 131.3	
55 Cs 132.9	56 Ba 137.3	57 *La 138.9	72 Hf 178.5	73 Ta 180.9	74 W 183.9	75 Re 186.2	76 Os 190.2	77 Ir 192.2	78 Pt 195.1	79 Au 197.0	80 Hg 200.6	81 Tl 204.4	82 Pb 207.2	83 Bi 209.0	84 Po (209)	85 At (210)	86 Rn (222)	
87 Fr (223)	88 Ra 226.0	89 †Ac 227.0																

*Lanthanum Series

58 Ce 140.1	59 Pr 140.1	60 Nd 144.2	61 Pm (145)	62 Sm 150.4	63 Eu 152.0	64 Gd 157.3	65 Tb 158.9	66 Dy 162.5	67 Ho 164.9	68 Er 167.3	69 Tm 168.9	70 Yb 173.0	71 Lu 175.0

†Actinium Series

90 Th 232.0	91 Pa 231.0	92 U 238.0	93 Np 237.0	94 Pu (244)	95 Am (243)	96 Cm (247)	97 Bk (247)	98 Cf (251)	99 Es (252)	100 Fm (258)	101 Md (258)	102 No (259)	103 Lr (260)

DO NOT DETACH FROM BOOK.

GO ON TO THE NEXT PAGE.

ADVANCED PLACEMENT CHEMISTRY EQUATIONS AND CONSTANTS

ATOMIC STRUCTURE

$$E = h\nu \qquad\qquad c = \lambda\nu$$

$$\lambda = \frac{h}{mv} \qquad\qquad p = mv$$

$$E_n = \frac{-2.178 \times 10^{-18}}{n^2} \text{ joule}$$

EQUILIBRIUM

$$K_a = \frac{[H^-][A^-]}{[HA]}$$

$$K_b = \frac{[OH^-][HB^+]}{[B]}$$

$$K_w = [OH^-][H^+] = 10^{-14} \text{ @ } 25°C$$

$$\quad = K_a \times K_b$$

$$pH = -\log[H^+],\ pOH = -\log[OH^-]$$

$$14 = pH + pOH$$

$$pH = pK_a + \log\frac{[A^-]}{[HA]}$$

$$pOH = pK_b + \log\frac{[HB^+]}{[B]}$$

$$pK_a = -\log K_a,\ pK_b = -\log K_b$$

$$K_p = K_c(RT)^{\Delta n},$$

where Δn = moles product gas – moles reactant gas

THERMOCHEMISTRY / KINETICS

$$\Delta S° = \Sigma S°_{\text{products}} - \Sigma S°_{\text{reactants}}$$

$$\Delta H° = \Sigma \Delta H°_f{}_{\text{products}} - \Sigma \Delta H°_f{}_{\text{reactants}}$$

$$\Delta G° = \Sigma \Delta G°_f{}_{\text{products}} - \Sigma \Delta G°_f{}_{\text{reactants}}$$

$$\Delta G° = \Delta H° - T\Delta S°$$

$$\quad = -RT \ln K = -2.303\ RT \log K$$

$$\quad = -nFE°$$

$$\Delta G = \Delta G° + RT \ln Q = \Delta G° + 2.303\ RT \log Q$$

$$q = mc\Delta T$$

$$C_p = \frac{\Delta H}{\Delta T}$$

$$\ln[A]_t - \ln[A]_o = -kt$$

$$\frac{1}{[A]_t} - \frac{1}{[A]_o} = kt$$

$$\ln k = \frac{-E_a}{R}\left(\frac{1}{T}\right) + \ln A$$

E = energy $\qquad$ v = velocity

ν = frequency $\qquad$ n = principal quantum number

λ = wavelength $\qquad$ m = mass

p = momentum

Speed of light, $c = 3.00 \times 10^8 \text{ m s}^{-1}$

Planck's constant, $h = 6.63 \times 10^{-34} \text{ J s}$

Boltzmann's constant, $k = 1.38 \times 10^{-23} \text{ J K}^{-1}$

Avogadro's number $= 6.022 \times 10^{23} \text{ mol}^{-1}$

Electron charge, $e = -1.602 \times 10^{-19} \text{ coulomb}$

1 electron volt/atom $= 96.5 \text{ kJ mol}^{-1}$

Equilibrium Constants

K_a (weak acid)
K_b (weak base)
K_w (water)
K_p (gas pressure)
K_c (molar concentrations)

$S°$ = standard entropy
$H°$ = standard enthalpy
$G°$ = standard free energy
$E°$ = standard reduction potential
T = temperature
n = moles
m = mass
q = heat
c = specific heat capacity
C_p = molar heat capacity at constant pressure
E_a = activation energy
k = rate constant
A = frequency factor

Faraday's Constant, $F = 96,500$ coulombs per mole of electrons

Gas Constant, $R = 8.31 \text{ J mol}^{-1} \text{ K}^{-1}$
$\qquad = 0.0821 \text{ L atm mol}^{-1} \text{ K}^{-1}$
$\qquad = 8.31 \text{ volt coulomb mol}^{-1} \text{ K}^{-1}$

GO ON TO THE NEXT PAGE.

ADVANCED PLACEMENT CHEMISTRY EQUATIONS AND CONSTANTS

GASES, LIQUIDS, AND SOLUTIONS

$$PV = nRT$$

$$\left(P + \frac{n^2 a}{V^2}\right)(V - nb) = nRT$$

$$P_A = P_{total} \cdot X_A, \text{ where } X_A = \frac{\text{moles A}}{\text{total moles}}$$

$$P_{total} = P_A + P_B + P_C^+ \ldots$$

$$n = \frac{m}{M}$$

$$K = {}^\circ C + 273$$

$$\frac{P_1 V_1}{T_1} = \frac{P_2 V_2}{T_2}$$

$$D = \frac{m}{V}$$

$$U_{rms} = \sqrt{\frac{3kT}{m}} = \sqrt{\frac{3RT}{M}}$$

$$KE \text{ per molecule} = \frac{1}{2}mv^2$$

$$KE \text{ per mole} = \frac{3}{2}RT$$

$$\frac{r_1}{r_2} = \sqrt{\frac{M_2}{M_1}}$$

molarity, M = moles solute per liter solution
molality = moles solute per kilogram solvent

$$\Delta T_f = iK_f \times \text{molality}$$

$$\Delta T_b = iK_b \times \text{molality}$$

$$\pi = MRT$$

$$A = abc$$

P = pressure
V = volume
T = temperature
n = number of moles
D = density
m = mass
v = velocity

U_{rms} = root-mean-square speed
KE = kinetic energy
r = rate of effusion
M = molar mass
π = osmotic pressure
i = van't Hoff factor
K_f = molal freezing-point depression constant
K_b = molal boiling-point elevation constant
A = absorbance
a = molar absorptivity
b = path length
c = concentration
Q = reaction quotient
l = current (amperes)
q = charge (coulombs)
t = time (seconds)
E° = standard reduction potential
K = equilibrium constant

OXIDATION REDUCTION; ELECTROCHEMISTRY

$$Q = \frac{[C]^c [D]^d}{[A]^a [B]^b} \text{ where } a\,A + b\,B \rightarrow c\,C + d\,D$$

$$I = \frac{q}{t}$$

$$E_{cell} = E^\circ_{cell} - \frac{RT}{nF} \ln Q = E^\circ_{cell} - \frac{0.0592}{n} \log Q \text{ @ } 25^\circ C$$

$$\log K = \frac{nE^\circ}{0.0592}$$

Gas constant, R = 8.31 J mol^{-1}K^{-1}
 = 0.0821 L atm mol^{-1} K^{-1}
 = 8.31 volt coulomb mol^{-1} K^{-1}
Boltzmann's constant, k = 1.38×10^{-23} J K^{-1}
K_f for H_2O = 1.86 K kg mol^{-1}
K_b for H_2O = 0.512 K kg mol^{-1}
1 atm = 760 mm Hg
 = 760 torr
STP = 0.000°C and 1.000 atm
1 faraday, F = 96,500 coulombs per mole
of electrons

GO ON TO THE NEXT PAGE.

STANDARD REDUCTION POTENTIALS IN AQUEOUS SOLUTION AT 25 C (in V)			
$F_2(g) + 2\,e^-$	$\rightarrow$	$2\,F^-$	2.87
$Co^{3+} + e^-$	$\rightarrow$	Co^{2+}	1.82
$Au^{3+} + 3\,e^-$	$\rightarrow$	$Au(s)$	1.50
$Cl_2(g) + 2\,e^-$	$\rightarrow$	$2\,Cl^-$	1.36
$O_2(g) + 4\,H^+ + 4\,e^-$	$\rightarrow$	$2\,H_2O$	1.23
$Br_2(l) + 2\,e^-$	$\rightarrow$	$2\,Br^-$	1.07
$2\,Hg^{2+} + 2\,e^-$	$\rightarrow$	Hg_2^{2+}	0.92
$Hg^{2+} + 2\,e^-$	$\rightarrow$	$Hg(l)$	0.85
$Ag^+ + e^-$	$\rightarrow$	$Ag(s)$	0.80
$Hg_2^{2+} + 2\,e^-$	$\rightarrow$	$2\,Hg(l)$	0.79
$Fe^{3+} + e^-$	$\rightarrow$	Fe^{2+}	0.77
$I_2(s) + 2\,e^-$	$\rightarrow$	$2\,I^-$	0.53
$Cu^+ + e^-$	$\rightarrow$	$Cu(s)$	0.52
$Cu^{2+} + 2\,e^-$	$\rightarrow$	$Cu(s)$	0.34
$Cu^{2+} + e^-$	$\rightarrow$	Cu^+	0.15
$Sn^{4+} + 2\,e^-$	$\rightarrow$	Sn^{2+}	0.15
$S(s) + 2\,H^+ + 2\,e^-$	$\rightarrow$	H_2S	0.14
$2\,H^+ + 2\,e^-$	$\rightarrow$	$H_2(g)$	0.00
$Pb^{2+} + 2\,e^-$	$\rightarrow$	$Pb(s)$	−0.13
$Sn^{2+} + 2\,e^-$	$\rightarrow$	$Sn(s)$	−0.14
$Ni^{2+} + 2\,e^-$	$\rightarrow$	$Ni(s)$	−0.25
$Co^{2+} + 2\,e^-$	$\rightarrow$	$Co(s)$	−0.28
$Tl^+ + e^-$	$\rightarrow$	$Tl(s)$	−0.34
$Cd^{2+} + 2\,e^-$	$\rightarrow$	$Cd(s)$	−0.40
$Cr^{3+} + e^-$	$\rightarrow$	Cr^{2+}	−0.41
$Fe^2 + 2\,e^-$	$\rightarrow$	$Fe(s)$	−0.44
$Cr^{3+} + 3\,e^-$	$\rightarrow$	$Cr(s)$	−0.74
$Zn^{2+} + 2\,e^-$	$\rightarrow$	$Zn(s)$	−0.76
$Mn^{2+} + 2\,e^-$	$\rightarrow$	$Mn(s)$	−1.18
$Al^{3+} + 3\,e^-$	$\rightarrow$	$Al(s)$	−1.66
$Be^{2+} + 2\,e^-$	$\rightarrow$	$Be(s)$	−1.70
$Mg^{2+} + 2\,e^-$	$\rightarrow$	$Mg(s)$	−2.37
$Na^+ + e^-$	$\rightarrow$	$Na(s)$	−2.71
$Ca^{2+} + 2\,e^-$	$\rightarrow$	$Ca(s)$	−2.87
$Sr^{2+} + 2\,e^-$	$\rightarrow$	$Sr(s)$	−2.89
$Ba^{2+} + 2\,e^-$	$\rightarrow$	$Ba(s)$	−2.90
$Rb^+ + e^-$	$\rightarrow$	$Rb(s)$	−2.92
$K^+ + e^-$	$\rightarrow$	$K(s)$	−2.92
$Cs^+ + e^-$	$\rightarrow$	$Cs(s)$	−2.92
$Li^+ + e^-$	$\rightarrow$	$Li(s)$	−3.05

GO ON TO THE NEXT PAGE.

CHEMISTRY

Section II

(Total time—95 minutes)

Part A

Time—55 minutes

YOU MAY USE YOUR CALCULATOR FOR PART A

THE METHODS USED AND THE STEPS INVOLVED IN ARRIVING AT YOUR ANSWERS MUST BE SHOWN CLEARLY. It is to your advantage to do this since you may obtain partial credit if you do, and you will receive little or no credit if you do not. Attention should be paid to significant figures.

Be sure to write your answers in the space provided following each question.

Answer Questions 1, 2, and 3. The Section II score for question 1 is 9 points, question 2 is 10 points, and question 3 is 9 points.

1. A 0.20-molar solution of acetic acid, $HC_2H_3O_2$, at a temperature of 25°C, has a pH of 2.73.

    (a) Calculate the hydroxide ion concentration, $[OH^-]$.

    (b) What is the value of the acid ionization constant, K_a, for acetic acid at 25°C?

    (c) How many moles of sodium acetate must be added to 500.0 ml of a 0.200-molar solution of acetic acid to create a buffer with a pH of 4.00? Assume that the volume of the solution is not changed by the addition of sodium acetate.

    (d) In a titration experiment, 100.0 ml of sodium hydroxide solution was added to 200 ml of a 0.400-molar solution of acetic acid until the equivalence point was reached. What was the pH at the equivalence point?

    (e) The pK_a values for several indicators are given in the table below. Which of the indicators on the table is most suitable for this titration? Justify your answer.

Indicator	pK_a
Thymol Blue	2
Bromcresol Purple	6
Phenolphthalein	9

GO ON TO THE NEXT PAGE.

2.

$$2 \, NO(g) + Cl_2(g) \rightarrow 2 \, NOCl(g)$$

The following data were collected for the reaction above. All of the measurements were taken at a temperature of 263 K.

Experiment	Initial [NO] (M)	Initial [Cl₂] (M)	Initial rate of disappearance of Cl₂ (M/min)
1	0.15	0.15	0.60
2	0.15	0.30	1.2
3	0.30	0.15	2.4
4	0.25	0.25	?

(a) Write the expression for the rate law for the reaction above.
(b) Calculate the value of the rate constant for the above reaction, and specify the units.
(c) What is the initial rate of appearance of NOCl in experiment 2?
(d) What is the initial rate of disappearance of Cl_2 in experiment 4?
(e) Each of the experimental trials took place in a closed container. Explain or calculate each of the following:
(i) What was the partial pressure due to NO(g) at the start of experiment 1?
(ii) What was the total pressure at the start of experiment 1? Assume that no NOCl is present.

3.

$$CH_4(g) + 2 \, O_2(g) \rightarrow CO_2(g) + 2 \, H_2O(l)$$

The above reaction for the combustion of methane gas has a standard entropy change, $\Delta S°$, with a value of -242.7 J/mol-K. The following data are also available:

Compound	ΔH_f (kJ/mol)
$CH_4(g)$	-74.8
H_2O (l)	-285.9
CO_2 (g)	-393.5

(a) What are the values of $\Delta H°_f$ and $\Delta G°_f$ for $O_2(g)$?
(b) Calculate the standard change in enthalpy, $\Delta H°$, for the combustion of methane.
(c) Calculate the standard free energy change, $\Delta G°$, for the combustion of methane.
(d) How would the value of $\Delta S°$ for the reaction be affected if the water produced in the combustion remained in the gas phase?
(e) A 20.0-gram sample of $CH_4(g)$ underwent combustion in a bomb calorimeter with excess oxygen gas.
(i) Calculate the mass of carbon dioxide produced.
(ii) Calculate the heat released by the reaction.

GO ON TO THE NEXT PAGE.

Part B

Time—40 minutes

NO CALCULATORS MAY BE USED FOR PART B

Answer Question 4 below. The Section II score for this question is 15 points.

4. You will be given three chemical reactions below. In part (i), write the balanced equation for the reaction, leaving coefficients in terms of lowest whole numbers. Then answer the question pertaining to that reaction in part (ii). For each of the following three reactions, assume that solutions are aqueous unless it says otherwise. Substances in solutions should be represented as ions if these substances are extensively ionized. Omit formulas for ions or molecules that are not affected by the reaction. Only equations inside the answer boxes will be graded.

EXAMPLE:

A piece of solid zinc is placed in a solution of silver(I) acetate

(i) Write the balanced equation in this box:

$$Zn + 2Ag^+ \rightarrow Zn^{2+} + 2Ag$$

(ii) Which substance is reduced in the reaction?

Ag^+ is reduced

(a) Sulfur dioxide gas is bubbled through cold water.

(i) Write the balanced equation in this box:

(ii) As the reaction progresses, will the hydroxide concentration in the solution increase, decrease, or remain the same?

_____

_____

GO ON TO THE NEXT PAGE.

(b) Ethane is burned in air.

(i) Write the balanced equation in this box:

(ii) If the enthalpy change for the reaction was measured to be –3,100 kJ and the total heat of formation for all products formed is –3,300 kJ, what is the approximate heat of formation of ethane measured in kJ/mole?

_____

_____

(c) Chlorine gas is bubbled through a solution of sodium bromide.

(i) Write the balanced equation in this box:

(ii) Which substance is reduced in this reaction?

_____

_____

GO ON TO THE NEXT PAGE.

Answer Question 5 and Question 6. The Section II score for question 5 is 9 points and question 6 is 8 points.

Answering these questions provides an opportunity to demonstrate your ability to present your material in logical, coherent, and convincing English. Your responses will be judged on the basis of accuracy and importance of the detail cited and on the appropriateness of the descriptive material used. Specific answers are preferable to broad, diffuse responses. Illustrative examples and equations may be helpful.

5. Oxygen is found in the atmosphere as a diatomic gas, O_2, and as ozone, O_3. Ozone has a dipole moment of 0.5 debye. Oxygen gas has a dipole moment of zero.

   (a)   Draw the Lewis dot structures for both molecules.

   (b)   Use the principles of bonding and molecular structure to account for the fact that ozone has a higher boiling point than diatomic oxygen.

   (c)   Use the principles of bonding and molecular structure to account for the fact that ozone is more soluble than diatomic oxygen in water.

   (d)   Explain why the two bonds in O_3 are of equal length and are longer than the bond length of the bond in diatomic oxygen.

   (e)   Elemental oxygen is strongly affected by a magnetic field. Explain why.

   (f)   For the equilibrium reaction below at 25°C and 1 atm, $K_{eq} = 10^{-57}$.

   $$3 O_2(g) \leftrightarrow 2 O_3(g)$$

   Which form of oxygen is more abundant under normal conditions? Explain.

GO ON TO THE NEXT PAGE.

6.

$NaC_2H_3O_2$, $Ba(NO_3)_2$, KCl

(a) Aqueous solutions of equal concentration of the three compounds listed above are prepared. What would an experimenter expect to observe when each of the following procedures is performed on each of the solutions?

(i) The pH of each solution is measured.

(ii) SO_4^{2-} ions are introduced into each solution.

(iii) The freezing point of each solution is measured, and the three temperatures are compared.

(iv) Each solution is subjected to a flame test.

(b) At 25°C and 1.0 atmosphere pressure, a balloon contains a mixture of four ideal gases: oxygen, nitrogen, carbon dioxide, and helium. The partial pressure due to each gas is 0.25 atmosphere. Use the ideas of kinetic molecular theory to answer each of the following questions.

(i) Rank the gases in increasing order of average molecular velocity.

(ii) How would the volume of the balloon be affected if the temperature of the gases in the balloon were increased at constant pressure?

(iii) What changes to the temperature and pressure of the gases would cause deviation from ideal behavior, and which gas would be most affected?

STOP

END OF EXAM

19

Answers and Explanations for Practice Exam 1

ANSWER KEY EXAM 1

1. C		39. B	
2. A		40. D	
3. D		41. B	
4. C		42. E	
5. D		43. C	
6. A		44. C	
7. B		45. B	
8. D		46. A	
9. B		47. E	
10. E		48. B	
11. B		49. A	
12. C		50. E	
13. A		51. C	
14. D		52. C	
15. D		53. C	
16. B		54. E	
17. E		55. D	
18. E		56. D	
19. C		57. D	
20. C		58. D	
21. D		59. B	
22. A		60. C	
23. D		61. B	
24. D		62. B	
25. C		63. C	
26. E		64. B	
27. A		65. A	
28. B		66. D	
29. A		67. B	
30. B		68. E	
31. A		69. A	
32. C		70. B	
33. C		71. D	
34. D		72. E	
35. B		73. B	
36. D		74. D	
37. D		75. D	
38. A			

HOW TO SCORE PRACTICE TEST 1

SECTION I: MULTIPLE-CHOICE

_____ × 1.0000 = _____
Number of Correct   Weighted
(out of 75)   Section I Score
                                 (Do not round)

SECTION II: FREE RESPONSE

(See if you can find a teacher or classmate to score your essays using
the guidelines in Chapter 3.)

Question 1   _____ × 1.6666 = _____
         (out of 9)   (Do not round)

Question 2   _____ × 1.5000 = _____
         (out of 10)   (Do not round)

Question 3   _____ × 1.6666 = _____
         (out of 9)   (Do not round)

Question 4   _____ × .5000 = _____
         (out of 15)   (Do not round)

Question 5   _____ × 1.2500 = _____
         (out of 9)   (Do not round)

Question 6   _____ × 1.4062 = _____
         (out of 8)   (Do not round)

AP Score Conversion Chart Chemistry

Composite Score Range	AP Score
100–150	5
81–99	4
62–80	3
49–61	2
0–48	1

Sum   = _____
         Weighted Section II
         Score (Do not round)

COMPOSITE SCORE

_____ + _____ = _____
Weighted   Weighted   Composite Score
Section I Score   Section II Score   (Round to nearest
                                                    whole number)

QUESTIONS	EXPLANATIONS

SECTION I—MULTIPLE CHOICE QUESTIONS

Questions 1–3 are based on the following energy diagrams.

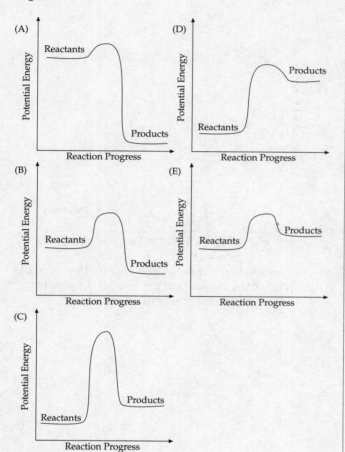

(A)

Potential Energy / Reaction Progress — Reactants, Products

(B)

Potential Energy / Reaction Progress — Reactants, Products

(C)

Potential Energy / Reaction Progress — Reactants, Products

(D)

Potential Energy / Reaction Progress — Reactants, Products

(E)

Potential Energy / Reaction Progress — Reactants, Products

1.

(C)

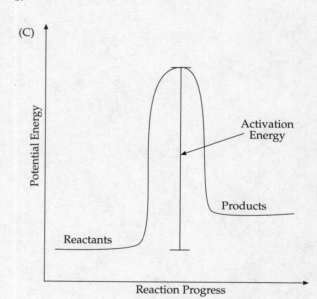

Activation Energy

Potential Energy / Reaction Progress — Reactants, Products

1. This reaction has the largest activation energy.

C   This reaction has the largest rise from the energy level of the reactants to the peak energy that must be overcome in order for the reaction to proceed.

QUESTIONS	EXPLANATIONS
2. This is the most exothermic reaction.	2.

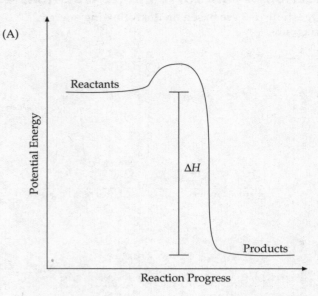

(A)

A This reaction has the largest energy drop from the level of the reactants to the level of the products. That makes it the most exothermic.

3. This reaction has the largest positive value for ΔH.

3.

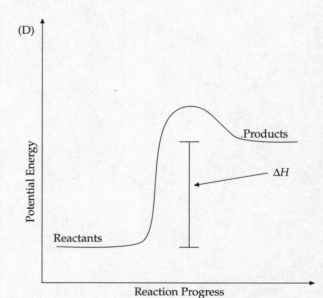

(D)

D This reaction has the largest rise in energy from the level of the reactants to the level of the products. That makes it the most endothermic and gives it the largest positive value for ΔH.

QUESTIONS	EXPLANATIONS

Questions 4–6 refer to the following configurations.

(A) $1s^2 \, 2s^2 2p^6$
(B) $1s^2 \, 2s^2 2p^6 \, 3s^2$
(C) $1s^2 \, 2s^2 2p^6 \, 3s^2 3p^4$
(D) $1s^2 \, 2s^2 2p^6 \, 3s^2 3p^6$
(E) $1s^2 \, 2s^2 2p^6 \, 3s^2 3p^6 \, 4s^2$

4. The ground state configuration of an atom of a paramagnetic element.

4. **C** A paramagnetic element is one whose electrons are not completely spin-paired. For choice (C), sulfur, Hund's rule says that the first three electrons in the $3p$ subshell will each occupy an empty orbital. The fourth electron will pair up, leaving two electrons unpaired. All of the other choices have completed subshells, so all the electrons will be spin-paired.

5. The ground state configuration for both a potassium ion and a chloride ion.

5. **D** A potassium atom loses an electron when it ionizes; a chlorine atom gains an electron. Both end up with the same ground state electron configuration as an argon atom, given in choice (D).

6. An atom that has this ground-state electron configuration will have the smallest atomic radius of those listed above.

6. **A** This atom (neon) has electrons in only two shells. All of the other choices have electrons at higher energy levels, farther away from the nucleus.

QUESTIONS	EXPLANATIONS

Questions 7–10 refer to the following molecules.

(A) CO_2
(B) H_2O
(C) SO_2
(D) NO_2
(E) O_2

7. In this molecule, oxygen forms sp^3 hybrid orbitals.

7. **B** The Lewis dot structure for water is shown below.

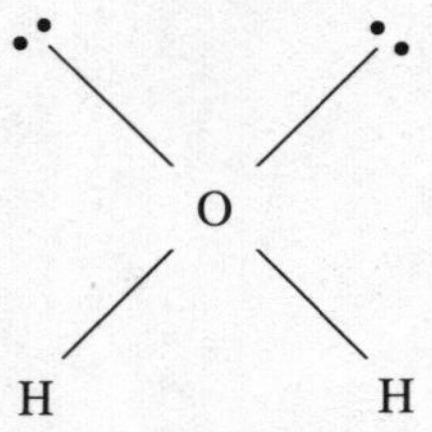

The central oxygen atom forms sp^3 hybrid orbitals, resulting in a tetrahedral structure. The central oxygen atom has two unbonded electron pairs, which gives the molecule a bent shape.

8. This molecule contains one unpaired electron.

8. **D** NO_2 has an odd number (17) of valence electrons, so when we draw the Lewis dot structure, there must be an unpaired electron.

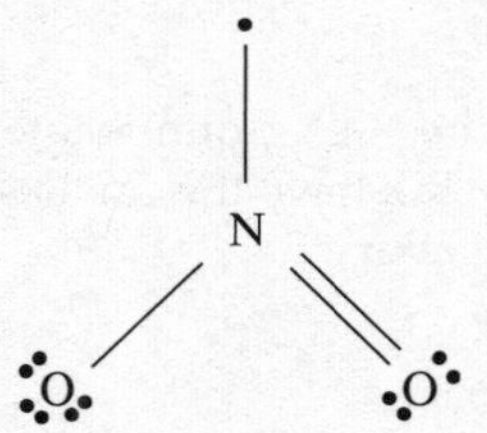

9. This molecule contains no pi (π) bonds.

9. **B** Water has only single bonds or sigma (σ) bonds. All of the other molecules have at least one double bond. The second bond in a double bond is a pi (π) bond.

10. Oxygen has an oxidation state of zero in this molecule.

10. **E** The oxidation state for an uncombined element is zero.

QUESTIONS	EXPLANATIONS

Questions 11–14 refer to the following solutions.

(A) A solution with a pH of 1
(B) A solution with a pH of greater than 1 and less than 7
(C) A solution with a pH of 7
(D) A solution with a pH of greater than 7 and less than 13
(E) A solution with a pH of 13

For CH_3COOH, $K_a = 1.8 \times 10^{-5}$
For NH_3, $K_b = 1.8 \times 10^{-5}$

11. A solution prepared by mixing equal volumes of 0.2-molar HCl and 0.2-molar NH_3.

11. **B** This is a mixture of a strong acid and a weak base, so at the equivalence point, the mixture will be acidic with a pH less than 7.

12. A solution prepared by mixing equal volumes of 0.2-molar HNO_3 and 0.2-molar NaOH.

12. **C** This is a mixture of a strong acid and a strong base, so at the equivalence point, they will have completely neutralized each other and only salt water will be left. Therefore, the solution will be neutral, and the pH will be 7.

13. A solution prepared by mixing equal volumes of 0.2-molar HCl and 0.2-molar NaCl.

13. **A** Both HCl and NaCl dissociate completely, but NaCl will have no effect on the pH of the solution.

Because we are doubling the volume of the HCl solution by adding the salt water, the concentration of HCl will be cut in half, to 0.1-molar. HCl is a strong acid, so [H^+] = 0.1-molar and pH = $-\log$[H^+] = 1.

14. A solution prepared by mixing equal volumes of 0.2-molar CH_3COOH and 0.2-molar NaOH.

14. **D** This is a mixture of a weak acid and a strong base, so at the equivalence point, the mixture will be basic with a pH greater than 7.

QUESTIONS	EXPLANATIONS

15. A pure sample of $KClO_3$ is found to contain 71 grams of chlorine atoms. What is the mass of the sample?

   (A) 122 grams
   (B) 170 grams
   (C) 209 grams
   (D) 245 grams
   (E) 293 grams

15. **D** Remember: moles $= \dfrac{grams}{MW}$

So moles of chlorine $= \dfrac{(71g)}{(35.5 \text{ g / mol})} = 2$ moles

If there are 2 moles of chlorine, there must be 2 moles of $KClO_3$.

The molecular weight of $KClO_3$ is 122.5.

So grams of $KClO_3$ = (moles)(MW)

    $= (2 \text{ mol})(122.5 \text{ g/mol})$
    $= 245$ grams

16. Which of the following experimental procedures is used to separate two substances by taking advantage of their differing boiling points?

   (A) Titration
   (B) Distillation
   (C) Filtration
   (D) Decantation
   (E) Hydration

16. **B** In distillation, two substances are heated until one of them boils. The gaseous substance is separated from the remaining liquid or solid and condensed in a separate container.

About the other answers:

(A) Titration is used to determine the volume of one solution required to react with a given volume of another solution.

(C) Filtration is used to separate a solid from a liquid by passing the solution through a membrane.

(D) Decantation is used to separate a solid from a liquid by letting the solid settle to the bottom of a container and then pouring off the liquid.

(E) Hydration occurs when ions enter into solution with water.

17. Which of the following sets of quantum numbers (n, l, m_l, m_s) best describes the highest energy valence electron in a ground-state aluminum atom?

   (A) $2, 0, 0, \dfrac{1}{2}$

   (B) $2, 1, 0, \dfrac{1}{2}$

   (C) $3, 0, 0, \dfrac{1}{2}$

   (D) $3, 0, 1, \dfrac{1}{2}$

   (E) $3, 1, 1, \dfrac{1}{2}$

17. **E** Aluminum's valence electrons are in the $3p$ subshell.

That means that $n = 3$, $l = 1$, $m_l = -1$, 0, or 1, and

$m_s = \dfrac{1}{2}$ or $-\dfrac{1}{2}$.

QUESTIONS	EXPLANATIONS

18. A chemist analyzed the carbon–carbon bond in C_2H_6 and found that it had a bond energy of 350 kJ/mol and a bond length of 1.5 angstroms. If the chemist performed the same analysis on the carbon–carbon bond in C_2H_2, how would the results compare?

(A) The bond energies and lengths for C_2H_2 would be the same as those of C_2H_6.
(B) The bond energy for C_2H_2 would be smaller, and the bond length would be shorter.
(C) The bond energy for C_2H_2 would be greater, and the bond length would be longer.
(D) The bond energy for C_2H_2 would be smaller, and the bond length would be longer.
(E) The bond energy for C_2H_2 would be greater, and the bond length would be shorter.

18. **E** First, you need to know that C_2H_6 is an alkane, which has a single C–C bond, and C_2H_2 is an alkyne, which has a triple C–C bond. Once you realize that, you know that a triple bond will be stronger and shorter than a single bond.

19. $2\,MnO_4^- + 5\,SO_3^{2-} + 6\,H^+ \rightarrow 2\,Mn^{2+} + 5\,SO_4^{2-} + 3\,H_2O$

Which of the following statements is true regarding the reaction given above?

(A) MnO_4^- acts as the reducing agent.
(B) H^+ acts as the oxidizing agent.
(C) SO_3^{2-} acts as the reducing agent.
(D) Manganese is oxidized.
(E) Sulfur is reduced.

19. **C** SO_3^{2-} is oxidized ($S^{4+} \rightarrow S^{6+} + 2\,e^-$, LEO), so SO_3^{2-} acts as the reducing agent.

MnO_4^- is reduced ($Mn^{7+} + 5\,e^- \rightarrow Mn^{2+}$, GER), so MnO_4^- acts as the oxidizing agent.

20. Which of the following can function as both a Brønsted-Lowry acid and Brønsted-Lowry base?

(A) HCl
(B) H_2SO_4
(C) HSO_3^-
(D) SO_4^{2-}
(E) H^+

20. **C** HSO_3^- is amphoteric.

It can act as a Brønsted-Lowry acid, giving up a proton to become SO_3^{2-}.

It can act as a Brønsted-Lowry base, gaining a proton to become H_2SO_3.

HCl (A) and H_2SO_4 (B) can give up only protons.

SO_4^{2-} (D) can gain only protons.

H^+ (E) *is* a proton.

21. Which of the following substances experiences the strongest attractive intermolecular forces?

(A) H_2
(B) N_2
(C) CO_2
(D) NH_3
(E) CH_4

21. **D** NH_3 is the only molecule listed that undergoes hydrogen bonding. In fact, it is the only polar molecule listed.

QUESTIONS	EXPLANATIONS
22. A mixture of gases at equilibrium over water at 43°C contains 9.0 moles of nitrogen, 2.0 moles of oxygen, and 1.0 mole of water vapor. If the total pressure exerted by the gases is 780 mmHg, what is the vapor pressure of water at 43°C? (A) 65 mmHg (B) 130 mmHg (C) 260 mmHg (D) 580 mmHg (E) 720 mmHg	22. **A** From Dalton's law, partial pressure of a gas in a sample is directly proportional to its molar quantity, so if $\frac{1}{12}$ of the gas in the sample is water vapor, then $\frac{1}{12}$ of the total pressure will be due to water vapor. So the partial pressure of water vapor is (780) = 65 mmHg. The gases are at equilibrium, so the partial pressure of the water vapor will be the same as the vapor pressure of the water.
23. $...MnO_4^- + ...e^- + ...H^+ \rightarrow ...Mn^{2+} + ...H_2O$ When the half-reaction above is balanced, what is the coefficient for H^+ if all the coefficients are reduced to the lowest whole number? (A) 3 (B) 4 (C) 5 (D) 8 (E) 10	23. **D** Backsolve. Start with (C). If there are 5 H^+, there can't be a whole number coefficient for H_2O, so (C) is wrong. You should also be able to see that the answer can't be an odd number, so (A) is also wrong. Try (D). If there are 8 H^+, then there are 4 H_2O. If there are 4 H_2O, then there is 1 MnO_4^-. If there is 1 MnO_4^-, then there is 1 Mn^{2+}. Mn^{7+} (in MnO_4^-) is reduced to Mn^{2+}, so there are 5 e^-. These are the lowest whole number coefficients, so (D) is correct.
24. The boiling point of water is known to be lower at high elevations. This is because (A) hydrogen bonds are weaker at high elevations. (B) the heat of fusion is lower at high elevations. (C) the vapor pressure of water is higher at high elevations. (D) the atmospheric pressure is lower at high elevations. (E) water is more dense at high elevations.	24. **D** Vapor pressure increases with increasing temperature. Water boils when its vapor pressure is equal to the atmospheric pressure. So if the atmospheric pressure is lowered, then water will boil at a lower temperature.

QUESTIONS	EXPLANATIONS

25. A student examined 2.0 moles of an unknown carbon compound and found that the sample contained 48 grams of carbon, 64 grams of oxygen, and 8 grams of hydrogen. Which of the following could be the molecular formula of the compound?

   (A) CH_2O
   (B) CH_2OH
   (C) CH_3COOH
   (D) CH_3CO
   (E) C_2H_5OH

25. **C** You can find the number of moles of C, H, and O.

$$Moles = \frac{grams}{MW}$$

$$Moles\ of\ carbon = \frac{(48g)}{(12g/mol)} = 4\ moles$$

$$Moles\ of\ hydrogen = \frac{(8g)}{(1g/mol)} = 8\ moles$$

$$Moles\ of\ oxygen = \frac{(64g)}{(16g/mol)} = 4\ moles$$

The student had 2 moles of the compound, so 1 mole would contain half as much stuff. That's 2 moles of C, 4 moles of H, and 2 moles of O. That corresponds to CH_3COOH, acetic acid. By the way, choice (A), CH_2O, is wrong because it's the empirical formula, not the molecular formula.

26.

$$S(s) + O_2(g) \rightarrow SO_2(g) \qquad \Delta H = x$$

$$S(s) + \frac{3}{2} O_2(g) \rightarrow SO_3(g) \qquad \Delta H = y$$

Based on the information above, what is the standard enthalpy change for the following reaction?

$$2\ SO_2(g) + O_2(g) \rightarrow 2\ SO_3(g)$$

   (A) $x - y$
   (B) $y - x$
   (C) $2x - y$
   (D) $2x - 2y$
   (E) $2y - 2x$

26. **E** The equations given on top give the heats of formation of all the reactants and products (remember: the heat of formation of O_2, an element in its most stable form, is zero).

$$\Delta H°_{reaction} = \Delta H°_{products} - \Delta H°_{reactants}$$

First, the products.

From SO_3, we get $2y$. That's it for the products.

Now the reactants.

From SO_2, we get $2x$. The heat of formation of O_2 is defined to be zero, so that's it for the reactants.

ΔH for the reaction $= 2y - 2x$

QUESTIONS	EXPLANATIONS

27. How much water must be added to a 50.0 ml solution of 0.60 M HNO_3 to produce a 0.40 M solution of HNO_3?

    (A) 25 ml
    (B) 33 ml
    (C) 50 ml
    (D) 67 ml
    (E) 75 ml

27. **A** Remember: moles = (molarity)(volume)

The number of moles of HNO_3 will remain constant during the dilution.

Moles of HNO_3 = (0.60 M)(50.0 ml)
           = 30 millimoles

Now we can find how much water it will take to create a 0.40 M solution with 30 millimoles of HNO_3.

$$\text{Volume} = \frac{\text{moles}}{\text{molarity}} = \frac{30 \text{ millimoles}}{0.40\ M} = 75 \text{ ml}$$

But 75 ml isn't the answer. We started with 50 ml of solution and ended up with 75 ml of solution, so we must have added 25 ml of water.

28. In which of the following equilibria would the concentrations of the products be increased if the volume of the system were decreased at constant temperature?

    (A) $H_2(g) + Cl_2(g) \rightleftharpoons 2\,HCl(g)$
    (B) $2\,CO(g) + O_2(g) \rightleftharpoons 2\,CO_3(g)$
    (C) $NO(g) + O_3(g) \rightleftharpoons NO_2(g) + O_2(g)$
    (D) $2\,HI(g) \rightleftharpoons H_2(g) + I_2(g)$
    (E) $N_2O_4(g) \rightleftharpoons 2\,NO_2(g)$

28. **B** According to Le Châtelier's law, the equilibrium will shift to counteract any stress that is placed on it.

If the volume is decreased, the equilibrium must shift toward the side with fewer moles of gas. Only choice (B) has fewer moles of gas on the product side (2 moles) than on the reactant side (3 moles).

29. A 100 ml sample of 0.10 molar NaOH solution was added to 100 ml of 0.10 molar $H_3C_6H_5O_7$. After equilibrium was established, which of the ions listed below was present in the greatest concentration?

    (A) $H_2C_6H_5O_7^-$
    (B) $HC_6H_5O_7^{2-}$
    (C) $C_6H_5O_7^{3-}$
    (D) OH^-
    (E) H^+

29. **A** Just enough NaOH has been added to neutralize all of the $H_3C_6H_5O_7$ in the reaction shown below.

$$H_3C_6H_5O_7(aq) \rightarrow H^+(aq) + H_2C_6H_5O_7^-(aq)$$

So the $H_3C_6H_5O_7(aq)$ is gone, and the $H_2C_6H_5O_7^-(aq)$ will be present in the greatest concentration because it won't dissociate much further.

QUESTIONS	EXPLANATIONS

30. Which of the following can be determined directly from the difference between the boiling point of a pure solvent and the boiling point of a solution of a nonionic solute in the solvent, if k_b for the solvent is known?

  I.   The mass of solute in the solution
  II. The molality of the solution
  III. The volume of the solution

(A) I only
(B) II only
(C) III only
(D) I and II only
(E) I and III only

30. B We don't know how much solution we have, so we can't find out (I), the mass of the solute, or (III), the volume of the solution.

We can get (II), the molality, from the boiling point elevation with the expression $\Delta T = k_b m x$. Because the substance is nonionic, it will not dissociate and x will be equal to 1, so we can leave it out of the calculation.

$$m = \frac{\Delta T}{k_b}$$

31. The value of the equilibrium constant K_{eq} is greater than 1 for a certain reaction under standard state conditions. Which of the following statements must be true regarding the reaction?

(A) $\Delta G°$ is negative.
(B) $\Delta G°$ is positive.
(C) $\Delta G°$ is equal to zero.
(D) $\Delta G°$ is negative if the reaction is exothermic and positive if the reaction is endothermic.
(E) $\Delta G°$ is negative if the reaction is endothermic and positive if the reaction is exothermic.

31. A From the relationship $\Delta G° = -RT \ln K$, we can see that if K is greater than 1, then $\ln K$ must be greater than 1, which means that $\Delta G°$ must be less than zero.

32. Which of the following aqueous solutions has the highest boiling point?

(A) 0.1 m NaOH
(B) 0.1 m HF
(C) 0.1 m Na$_2$SO$_4$
(D) 0.1 m KC$_2$H$_3$O$_2$
(E) 0.1 m NH$_4$NO$_3$

32. C The formula for boiling-point elevation is $\Delta T = k_b m x$, where x is the number of particles into which the solute dissociates. So, the more particles into which the solute dissociates, the greater the boiling point elevation.

The salts in all the answer choices except (C) dissociate into two particles. For choice (C), each Na$_2$SO$_4$ dissociates into three particles, two Na$^+$ and one SO$_4^{2-}$.

33. The molecular formula for hydrated iron (III) oxide, or rust, is generally written as Fe$_2$O$_3$ • x H$_2$O because the water content in rust can vary. If a 1-molar sample of hydrated iron (III) oxide is found to contain 108 g of H$_2$O, what is the molecular formula for the sample?

(A) Fe$_2$O$_3$ • H$_2$O
(B) Fe$_2$O$_3$ • 3 H$_2$O
(C) Fe$_2$O$_3$ • 6 H$_2$O
(D) Fe$_2$O$_3$ • 10 H$_2$O
(E) Fe$_2$O$_3$ • 12 H$_2$O

33. C Moles $= \dfrac{\text{grams}}{\text{MW}}$
Moles of H$_2$O $= \dfrac{(108 \text{ g})}{(18 \text{ g / mol})} = 6$ moles

So if 1 mole of hydrate contains 6 moles of H$_2$O, then its formula must be Fe$_2$O$_3$ 6 H$_2$O.

QUESTIONS	EXPLANATIONS

34. In which of the following reactions does the greatest increase in entropy take place?

    (A) $H_2O(l) \rightarrow H_2O(g)$
    (B) $2 NO(g) + O_2(g) \rightarrow 2 NO_2(g)$
    (C) $CaH_2(s) + H_2O(l) \rightarrow Ca(OH)_2(s) + H_2O(g)$
    (D) $NH_4Cl(s) \rightarrow NH_3(g) + HCl(g)$
    (E) $2 HCl(g) \rightarrow H_2(g) + Cl_2(g)$

34. **D** In choice (D), the number of moles goes from 1 to 2, and the phase changes from solid to gas, making this a pretty big entropy change. Entropy is also increasing in (A) and (C), but in both cases, the number of moles stays the same and the phase change is only from liquid to gas. The entropy decreases in (B) and stays about the same in (E).

35. The density of a sample of water decreases as it is heated above a temperature of 4°C. Which of the following will be true of an aqueous solution of $NaC_2H_3O_2$ when it is heated from 10°C to 60°C?

    (A) The molarity will increase.
    (B) The molarity will decrease.
    (C) The molality will increase.
    (D) The molality will decrease.
    (E) The molarity and molality will remain unchanged.

35. **B** Remember the definition of density.

$$\text{Density} = \frac{\text{mass}}{\text{volume}}$$

The mass of water doesn't change as temperature increases, so if the density is decreasing, that must mean that the volume is increasing. Now look at the definition of molarity.

$$\text{Molarity} = \frac{\text{moles}}{\text{liters}}$$

If the volume is increasing, then the molarity must be decreasing.

Molality is moles per kiligram. Since neither of these is changing, molality is unchanged.

36. $$... + n \rightarrow {}^{7}_{3}Li + {}^{4}_{2}He$$

For the nuclear reaction shown above, what is the missing reactant?

    (A) ${}^{9}_{4}Be$

    (B) ${}^{9}_{5}B$

    (C) ${}^{10}_{4}Be$

    (D) ${}^{10}_{5}B$

    (E) ${}^{11}_{5}B$

36. **D** Adding up the atomic numbers in the products, we get $2 + 3 = 5$. So the reactant must be boron.

Adding up the atomic masses and keeping in mind that the neutron has a mass of 1, we get $y + 1 = 7 + 4$.

So the atomic mass must be 10.

QUESTIONS	EXPLANATIONS

37. A boiling-water bath is sometimes used instead of a flame in heating objects. Which of the following could be an advantage of a boiling-water bath over a flame?

   (A) The relatively low heat capacity of water will cause the object to become hot more quickly.

   (B) The relatively high density of water will cause the object to become hot more quickly.

   (C) The volume of boiling water remains constant over time.

   (D) The temperature of boiling water remains constant at 100°C.

   (E) The vapor pressure of boiling water is equal to zero.

37. D The fact that boiling water maintains a constant temperature of 100°C is useful when a relatively low constant temperature is required. None of the other choices is a true statement.

38. The addition of a catalyst to a chemical reaction will bring about a change in which of the following characteristics of the reaction?

    I. The activation energy
   II. The enthalpy change
  III. The value of the equilibrium constant

   (A) I only
   (B) II only
   (C) I and II only
   (D) I and III only
   (E) II and III only

38. A The addition of a catalyst lowers the activation energy of a reaction, making it easier for the reaction to proceed, so (I) is correct.

Adding a catalyst has no effect on the enthalpy change or equilibrium conditions of a reaction, so (II) and (III) are wrong.

39.
$$2\,NO(g) + O_2(g) \rightarrow 2\,NO_2(g)$$

The reaction above occurs by the following two-step process:

Step I: $NO(g) + O_2(g) \rightarrow NO_3(g)$

Step II: $NO_3(g) + NO(g) \rightarrow 2\,NO_2(g)$

Which of the following is true of Step II if it is the rate-limiting step?

   (A) Step II has a lower activation energy and occurs more slowly than Step I.

   (B) Step II has a higher activation energy and occurs more slowly than Step I.

   (C) Step II has a lower activation energy and occurs more quickly than Step I.

   (D) Step II has a higher activation energy and occurs more quickly than Step I.

   (E) Step II has the same activation energy and occurs at the same speed as Step I.

39. B The rate-limiting step in a reaction is the slowest step in the process.

When a reaction occurs slowly, it is because very few collisions among reactant molecules have enough energy to overcome the high activation energy.

QUESTIONS	EXPLANATIONS

40. $C_3H_7OH + ...O_2 \rightarrow ...CO_2 + ...H_2O$

One mole of C_3H_7OH underwent combustion as shown in the reaction above. How many moles of oxygen were required for the reaction?

(A) 2 moles

(B) 3 moles

(C) $\frac{7}{2}$ moles

(D) $\frac{9}{2}$ moles

(E) 5 moles

40. **D** Backsolve. Start at (C).

Instead of using 1 as the coefficient for C_3H_7OH and $\frac{7}{2}$ for O_2, use 2 for C_3H_7OH and 7 for O_2. This won't change your result, and it will make the math easier.

If there are 2 moles of C_3H_7OH, then there must be 6 moles of CO_2 and 8 moles of H_2O.

That gives us 16 O's in the reactants and 20 O's in the products. So (C) is wrong and we should pick a larger number to put more O's on the reactant side. Try (D).

There are still 2 moles of C_3H_7OH, so there must still be 6 moles of CO_2 and 8 moles of H_2O. Now there are 9 moles of O_2. Now we have 20 O's in the reactants and 20 O's in the products, so (D) is the correct answer.

QUESTIONS	EXPLANATIONS

Questions 41–42 refer to the phase diagram below.

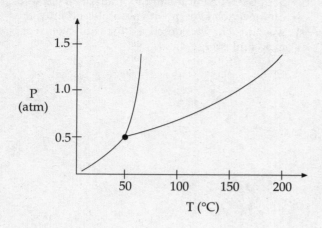

41. If the pressure of the substance shown in the diagram is decreased from 1.0 atmosphere to 0.5 atmosphere at a constant temperature of 100°C, which phase change will occur?

  (A) Freezing
  (B) Vaporization
  (C) Condensation
  (D) Sublimation
  (E) Deposition

41. **B** The phase change will occur as shown in the diagram below.

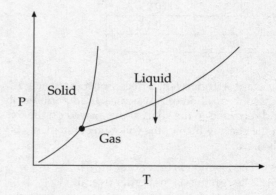

A phase change from liquid to gas is vaporization.

42. Under what conditions can all three phases of the substance shown in the diagram exist simultaneously in equilibrium?

  (A) Pressure = 1.0 atm, temperature = 150°C
  (B) Pressure = 1.0 atm, temperature = 100°C
  (C) Pressure = 1.0 atm, temperature = 50°C
  (D) Pressure = 0.5 atm, temperature = 100°C
  (E) Pressure = 0.5 atm, temperature = 50°C

42. **E** At this point, called the triple point, all the phase change lines converge and all three phases are in equilibrium.

	QUESTIONS		EXPLANATIONS

43. Which of the following statements regarding atomic theory is NOT true?

(A) The Bohr model of the atom was based on Planck's quantum theory.

(B) Rutherford's experiments with alpha particle scattering led to the conclusion that positive charge was concentrated in an atom's nucleus.

(C) Heisenberg's uncertainty principle describes the equivalence of mass and energy.

(D) Millikan's oil drop experiment led to the calculation of the charge on an electron.

(E) Thomson's cathode ray experiments confirmed the existence of the electron.

43. C The Heisenberg uncertainty principle states that both the momentum and location of an electron can never be known with absolute certainty. It was Einstein who described the equivalence of mass and energy.

44.

Time (min)	$[A]$ M
0	0.50
10	0.36
20	0.25
30	0.18
40	0.13

Reactant A underwent a decomposition reaction. The concentration of A was measured periodically and recorded in the chart above. Based on the data in the chart, which of the following statements is NOT true?

(A) The reaction is first order in $[A]$.

(B) The reaction is first order overall.

(C) The rate of the reaction is constant over time.

(D) The half-life of reactant A is 20 minutes.

(E) The graph of ln $[A]$ will be a straight line.

44. C The reaction rate is NOT constant over time, so this is the statement that is not true. In fact, the reaction rate is gradually decreasing as the concentration of reactant A decreases. All of the other statements are true. You can see from the chart that the half-life of A is 20 minutes. The other three statements must be true based on the fact that $[A]$ is decreasing exponentially.

45. A solution to be used as a reagent for a reaction is to be removed from a bottle marked with its concentration. Which of the following is NOT part of the proper procedure for this process?

(A) Pouring the solution down a stirring rod into a beaker.

(B) Inserting a pipette directly into the bottle and drawing out the solution.

(C) Placing the stopper of the bottle upside down on the table top.

(D) Pouring the solution down the side of a tilted beaker.

(E) Touching the stopper of the bottle only on the handle.

45. B The correct procedures listed here are designed to prevent spattering and spilling—choices (A) and (D)—or contamination of the solution in the bottle—choices (C) and (E). Inserting a pipette directly into the bottle could contaminate the solution if the pipette isn't perfectly clean.

46.

$$N_2(g) + 3 H_2(g) \rightleftharpoons 2 NH_3(g) + energy$$

Which of the following changes to the equilibrium situation shown above will bring about an increase in the number of moles of NH_3 present at equilibrium?

   I. Adding N_2 gas to the reaction chamber
  II. Increasing the volume of the reaction chamber at constant temperature
 III. Increasing the temperature of the reaction chamber at constant volume

(A) I only
(B) II only
(C) I and II only
(D) I and III only
(E) II and III only

46. **A** According to Le Châtelier's law, the equilibrium will shift to counteract any stress that is placed on it.

If N_2 is added, the equilibrium will shift to remove the excess N_2. This shift produces more products, and thus, more NH_3. So (I) is correct.

Choice (II) is wrong because increasing the volume will cause the reaction to shift toward the side with more moles of gas. In this case the reactant side has more moles of gas (4 moles) than the product side (2 moles). So increasing the volume will decrease the number of moles of NH_3.

Choice (III) is wrong because increasing the temperature will favor the endothermic direction, which in this case is the reverse reaction. So once again, the number of moles of NH_3 is decreased.

47.

$$CH_4 + 2 O_2 \rightarrow CO_2 + 2 H_2O$$

If 16 grams of CH_4 reacts with 16 grams of O_2 in the reaction shown above, which of the following will be true?

(A) The mass of H_2O formed will be twice the mass of CO_2 formed.
(B) Equal masses of H_2O and CO_2 will be formed.
(C) Equal numbers of moles of H_2O and CO_2 will be formed.
(D) The limiting reagant will be CH_4.
(E) The limiting reagant will be O_2.

47. **E** $Moles = \dfrac{grams}{MW}$

$$Moles\ of\ CH_4 = \frac{(16\ g)}{(16\ g\ /\ mol)} = 1\ mole$$

$$Moles\ of\ O_2 = \frac{(16\ g)}{(32\ g\ /\ mol)} = 0.5\ moles$$

From the balanced equation, 2 moles of O_2 are used up for every mole of CH_4.

When all 0.5 moles of O_2 are used up, only 0.25 moles of CH_4 will be used up, so oxygen is the limiting reagant.

Choices (A), (B), and (C) are wrong because more moles of H_2O will be formed and a greater mass of CO_2 will be formed.

48. Which of the following sets of gases would be most difficult to separate if the method of gaseous effusion is used?

(A) O_2 and CO_2
(B) N_2 and C_2H_4
(C) H_2 and CH_4
(D) He and Ne
(E) O_2 and He

48. **B** From Graham's law, the rate of effusion of a gas depends on its molecular weight. The larger the molecular weight, the slower the rate of effusion.

For separation of gases by effusion to work, the gases must have different molecular weights, which will cause them to effuse at different rates. N_2 and C_2H_4 have the same molecular weight (28 g/mol), so they can't be separated by effusion.

QUESTIONS	EXPLANATIONS

49. Which of the following equilibrium expressions represents the hydrolysis of the CN⁻ ion?

(A) $K = \dfrac{[HCN][OH^-]}{[CN^-]}$

(B) $K = \dfrac{[CN^-][OH^-]}{[HCN]}$

(C) $K = \dfrac{[CN^-][H_3O^+]}{[HCN]}$

(D) $K = \dfrac{[HCN][H_3O^+]}{[CN^-]}$

(E) $K = \dfrac{[HCN]}{[CN^-][OH^-]}$

49. **A** The hydrolysis of the CN⁻ ion is shown by the reaction below.

$$CN^- + H_2O \leftrightarrow HCN + OH^-$$

Putting products over the reactants in the equilibrium expression and omitting water because it is a pure liquid, we get

$$K = \dfrac{[HCN][OH^-]}{[CN^-]}$$

50. Which of the following is true under any conditions for a reaction that is spontaneous at any temperature?

(A) ΔG, ΔS, and ΔH are all positive.
(B) ΔG, ΔS, and ΔH are all negative.
(C) ΔG and ΔS are negative, and ΔH is positive.
(D) ΔG and ΔS are positive, and ΔH is negative.
(E) ΔG and ΔH are negative, and ΔS is positive.

50. **E** For a spontaneous reaction, ΔG is always negative.

From the equation $\Delta G = \Delta H - T\,\Delta S$, we can see that the conditions that will make ΔG always negative are when ΔH is negative and ΔS is positive.

51. Which of the following pairs of compounds are isomers?

(A) HCOOH and CH₃COOH
(B) CH₃CH₃CHO and C₃H₇OH
(C) C₂H₅OH and CH₃OCH₃
(D) C₂H₄ and C₂H₆
(E) C₃H₈ and C₄H₁₀

51. **C** Isomers are different molecules that have the same collection of atoms arranged in different ways. Both of the molecules in choice (C) have 2 carbons, 6 hydrogens, and 1 oxygen.

QUESTIONS	EXPLANATIONS

52. A sample of an ideal gas confined in a rigid 5.00 liter container has a pressure of 363 mmHg at a temperature of 25°C. Which of the following expressions will be equal to the pressure of the gas if the temperature of the container is increased to 35°C?

(A) $\dfrac{(363)(35)}{(25)}$ mmHg

(B) $\dfrac{(363)(25)}{(35)}$ mmHg

(C) $\dfrac{(363)(308)}{(298)}$ mmHg

(D) $\dfrac{(363)(298)}{(308)}$ mmHg

(E) $\dfrac{(363)(273)}{(308)}$ mmHg

52. **C** From the ideal gas laws, we know that with volume constant

$$\frac{P_1}{T_1} = \frac{P_2}{T_2}$$

Solving for P_2, we get

$$P_2 = \frac{P_1 T_2}{T_1} = \frac{(363)(308)}{(298)} \text{ mmHg}$$

53. If the temperature at which a reaction takes place is increased, the rate of the reaction will

(A) increase if the reaction is endothermic and decrease if the reaction is exothermic.
(B) decrease if the reaction is endothermic and increase if the reaction is exothermic.
(C) increase if the reaction is endothermic and increase if the reaction is exothermic.
(D) decrease if the reaction is endothermic and decrease if the reaction is exothermic.
(E) remain the same for both an endothermic and an exothermic reaction.

53. **C** An increase in temperature always increases the rate of a reaction, regardless of the change in enthalpy of the reaction.

Temperature change and enthalpy come into play in establishing whether reactants or products are favored at equilibrium according to Le Châtelier's law, but increasing the temperature will bring the reaction to equilibrium more quickly, regardless of whether the equilibrium favors reactants or products.

54. The acid dissociation constant for HClO is 3.0×10^{-8}. What is the hydrogen ion concentration in a 0.12 M solution of HClO?

(A) $3.6 \times 10^{-9} M$
(B) $3.6 \times 10^{-8} M$
(C) $6.0 \times 10^{-8} M$
(D) $2.0 \times 10^{-5} M$
(E) $6.0 \times 10^{-5} M$

54. **E** Use the formula for K_a.

$$K_a = \frac{\left[H^+ \right]\left[ClO^- \right]}{[HClO]}$$

We're trying to figure out $[H^+]$. Since for every HClO that dissociates, you get 1 H^+ and 1 ClO^-, we can rewrite the expression with $[H^+] = [ClO^-] = x$.

$$3.0 \times 10^{-8} = \frac{x^2}{0.12}$$
$$x^2 = (0.12)(3.0 \times 10^{-8}) = 0.36 \times 10^{-8}$$
$$x = 0.60 \times 10^{-4} = 6.0 \times 10^{-5}$$

So the hydrogen ion concentration is $6.0 \times 10^{-5} M$.

QUESTIONS	EXPLANATIONS

55.

$$2\,NO(g) + 2\,H_2(g) \rightarrow N_2(g) + 2\,H_2O(g)$$

Which of the following is true regarding the relative molar rates of disappearance of the reactants and appearance of the products?

I. N_2 appears at the same rate that H_2 disappears.

II. H_2O appears at the same rate that NO disappears.

III. NO disappears at the same rate that H_2 disappears.

(A) I only
(B) I and II only
(C) I and III only
(D) II and III only
(E) I, II, and III

55. D For every mole of N_2 that appears, 2 moles of H_2 must disappear, so N_2 appears at half the rate that H_2 disappears, and (I) is wrong.

For every 2 moles of H_2O that appear, 2 moles of NO must disappear, so H_2O appears at the same rate that NO disappears, and (II) is correct.

For every 2 moles of NO that disappear, 2 moles of H_2 must disappear, so NO disappears at the same rate that H_2 disappears, and (III) is correct.

56.

$$SO_4^{2-},\ PO_4^{3-},\ ClO_4^{-}$$

The geometries of the polyatomic ions listed above can all be described as

(A) square planar.
(B) square pyramidal.
(C) seesaw-shaped.
(D) tetrahedral.
(E) trigonal bipyramidal.

56. D All of these polyatomic ions have 32 valence electrons distributed in the Lewis dot structure shown below for ClO_4^{-}.

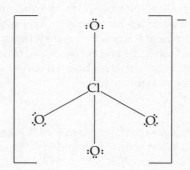

In these polyatomic ions, the central atom forms sp^3 hybrid orbitals, which have a tetrahedral structure. There are no unshared electron pairs on the central atom, so the geometry is tetrahedral.

57. $2\,ZnS(s) + 3\,O_2(g) \rightarrow 2\,ZnO(s) + 2\,SO_2(g)$

If the reaction above took place at standard temperature and pressure, what was the volume of $O_2(g)$ required to produce 40.0 grams of $ZnO(s)$?

(A) $\dfrac{(40.0)(2)}{(81.4)(3)(22.4)}\ L$

(B) $\dfrac{(40.0)(3)}{(81.4)(2)(22.4)}\ L$

(C) $\dfrac{(40.0)(2)(22.4)}{(81.4)(3)}\ L$

(D) $\dfrac{(40.0)(3)(22.4)}{(81.4)(2)}\ L$

(E) $\dfrac{(81.4)(2)(22.4)}{(40.0)(3)}\ L$

57. **D** $Moles = \dfrac{grams}{MW}$

$Moles\ of\ ZnO = \dfrac{(40.0\ g)}{(81.4\ g\,/\,mol)}$

For every 2 moles of ZnO produced, 3 moles of O_2 are consumed.

So moles of $O_2 = \left(\dfrac{3}{2}\right)$ (moles of ZnO)

$= \left(\dfrac{3}{2}\right)\dfrac{(40.0)}{(81.4)}$

At STP, volume of gas = (moles)(22.4 L)

So volume of $O_2 = \left(\dfrac{3}{2}\right)\dfrac{(40.0)}{(81.4)}\ (22.4)\ L$

58. Which of the following salts will produce a colorless solution when added to water?

(A) $Cu(NO_3)_2$
(B) $NiCl_2$
(C) $KMnO_4$
(D) $ZnSO_4$
(E) $FeCl_3$

58. **D** Zn^{2+} and SO_4^{2-} are both colorless in solution.
About the other answers

(A) Cu^{2+} ions are blue in solution.
(B) Ni^{2+} ions are green in solution.
(C) MnO_4^- ions are purple in solution.
(E) Fe^{3+} ions are yellow in solution.

Most salts of transition metals produce colored solutions. That's because energy is released and absorbed when d subshell electrons change energy levels. This energy is manifested as visible light.

Zn^{2+} does not produce a colored solution because it has a full $3d$ subshell. The full subshell means that there are no empty orbitals for the d electrons to jump to. If there are no transitions, there is no light.

QUESTIONS	EXPLANATIONS

59. A beaker contains 300.0 ml of a 0.20 M Pb(NO$_3$)$_2$ solution. If 200.0 ml of a 0.20 M solution of MgCl$_2$ is added to the beaker, what will be the final concentration of Pb^{2+} ions in the resulting solution?

   (A) 0.020 M
   (B) 0.040 M
   (C) 0.080 M
   (D) 0.120 M
   (E) 0.150 M

59. **B** The Pb^{2+} ions and the Cl$^-$ ions will combine and precipitate out of the solution, so let's find out how many of each we have. Remember that each mole of MgCl$_2$ produces 2 moles of Cl$^-$.

Moles = (molarity)(volume)

Moles of Pb^{2+} = (0.20 M)(0.300 L) = 0.060 mole

Moles of Cl$^-$ = (2)(0.20 M)(0.200 L) = 0.080 mole

Since two Cl$^-$ ions are required for each Pb^{2+} ion, subtract half the moles of Cl$^-$ (0.040 mole) from the moles of Pb^{2+} to find the number of moles of Pb^{2+} left in the solution

0.060 mol − 0.040 mol = 0.020 mol

Now use the formula for molarity to find the concentration of Pb^{2+} ions. Don't forget to add the volumes of the two solutions.

$$\text{Molarity} = \frac{\text{moles}}{\text{liters}} = \frac{(0.020 \text{ mol})}{(0.500 \text{ L})} = 0.040 \ M$$

60. Which of the following procedures will produce a buffered solution?

   I. Equal volumes of 1 M NH$_3$ and 1 M NH$_4$Cl solutions are mixed.
   II. Equal volumes of 1 M H$_2$CO$_3$ and 1 M NaHCO$_3$ solutions are mixed.
   III. Equal volumes of 1 M NH$_3$ and 1 M H$_2$CO$_3$ solutions are mixed.

   (A) I only
   (B) III only
   (C) I and II only
   (D) II and III only
   (E) I, II, and III

60. **C** A buffered solution can be prepared by mixing a weak acid or base with an equal amount of its conjugate.

In (I), equal amounts of NH$_3$ (a weak base) and NH$_4$$^+$ (its conjugate acid) are mixed, so (I) creates a buffer.

In (II), equal amounts of H$_2$CO$_3$ (a weak acid) and HCO$_3$$^-$ (its conjugate base) are mixed, so (II) also creates a buffer.

In (III), equal amounts of a weak acid and base that are not conjugates are mixed. These two will neutralize each other and will not create a buffered solution.

QUESTIONS	EXPLANATIONS

61.
$$H_2(g) + I_2(g) \rightarrow 2\,HI(g)$$

At 450°C the equilibrium constant, K_c, for the reaction shown above has a value of 50. Which of the following sets of initial conditions at 450°C will cause the reaction above to produce more H_2?

I. $[HI]$ = 5-molar, $[H_2]$ = 1-molar, $[I_2]$ = 1-molar
II. $[HI]$ = 10-molar, $[H_2]$ = 1-molar, $[I_2]$ = 1-molar
III. $[HI]$ = 10-molar, $[H_2]$ = 2-molar, $[I_2]$ = 2-molar

(A) I only
(B) II only
(C) I and II only
(D) II and III only
(E) I, II, and III

61. B The reaction will proceed in the reverse direction and produce more H_2 when the reaction quotient Q is greater than K_c.

Q takes the form of the equilibrium constant.

$$Q = \frac{[HI]^2}{[H_2][I_2]}$$

For I: $Q = \dfrac{(5)^2}{(1)(1)} = 25$ Q is less than K_c.

For II: $Q = \dfrac{(10)^2}{(1)(1)} = 100$ Q is greater than K_c.

For III: $Q = \dfrac{(10)^2}{(2)(2)} = 25$ Q is less than K_c.

So only (II) will produce more H_2.

62.
$$Cu^{2+}(aq) + Zn(s) \rightarrow Cu(s) + Zn^{2+}(aq)$$

A galvanic cell that uses the reaction shown above has a standard state electromotive force of 1.1 volts. Which of the following changes to the cell will increase the voltage?

I. An increase in the mass of $Zn(s)$ in the cell.
II. An increase in the concentration of $Cu^{2+}(aq)$ in the cell.
III. An increase in the concentration of $Zn^{2+}(aq)$ in the cell.

(A) I only
(B) II only
(C) III only
(D) I and II only
(E) I and III only

62. B Let's look at the Nernst equation, which relates cell potential to concentration.

$$E = E - \frac{0.059}{2} \log Q \text{ at } 25°C$$

The smaller Q becomes, the larger E will become.

Remember that Q is the reaction quotient, which takes the form of K_{eq}, except with initial conditions instead of equilibrium conditions.

In this case $Q = \dfrac{[Zn^{2+}]}{[Cu^{2+}]}$, so increasing $[Cu^{2+}]$ decreases Q, which increases E.

So (II) is right and (III) is wrong.

The amount of solid has no effect, so (I) is wrong.

You can also use Le Châtelier's law for this one.

QUESTIONS	EXPLANATIONS

63. The nuclide $^{61}_{26}$Fe decays through the emission of a single beta (β^μ) particle. What is the resulting nuclide?

   (A) $^{60}_{26}$Fe

   (B) $^{62}_{26}$Fe

   (C) $^{61}_{27}$Co

   (D) $^{62}_{27}$Co

   (E) $^{61}_{25}$Mn

63. **C** The number of nucleons doesn't change in beta decay, so the mass number must remain at 61.

In beta decay, a neutron is converted to a proton, so the atomic number increases by one.

The balanced nuclear reaction is as follows:

$$^{61}_{26}\text{Fe} \rightarrow ^{0}_{-1}\beta + ^{61}_{27}\text{Co}$$

64. Which of the following statements is true regarding sodium and potassium?

   (A) Sodium has a larger first ionization energy and a larger atomic radius.
   (B) Sodium has a larger first ionization energy and a smaller atomic radius.
   (C) Sodium has a smaller first ionization energy and a larger atomic radius.
   (D) Sodium has a smaller first ionization energy and a smaller atomic radius.
   (E) Sodium and potassium have identical first ionization energies and atomic radii.

64. **B** Sodium's valence electron is in the third energy level and potassium's valence electron is in the fourth energy level, so sodium's valence electron is closer to the nucleus than potassium's, so sodium must have a smaller atomic radius.

Also, because sodium's valence electron is closer to the nucleus, it is more difficult to remove, so sodium will have a higher first ionization energy.

65. $\text{HCl}(aq) + \text{AgNO}_3(aq) \rightarrow \text{AgCl}(s) + \text{HNO}_3(aq)$

One-half liter of a 0.20-molar HCl solution is mixed with one-half liter of a 0.40-molar solution of AgNO_3. A reaction occurs forming a precipitate as shown above. If the reaction goes to completion, what is the mass of AgCl produced?

   (A) 14 grams
   (B) 28 grams
   (C) 42 grams
   (D) 70 grams
   (E) 84 grams

65. **A** First we have to find the limiting reagent.

Moles = (molarity)(liters)

Moles of HCl = (0.20 M)(0.50 L) = 0.10 moles

Moles of AgNO_3 = (0.40 M)(0.50 L) = 0.20 moles

From the balanced equation, the two reactants are used up at equal rates. There is twice as much AgNO_3, so when the 0.10 moles of HCl have been used up, there will still be 0.10 moles of AgNO_3. So HCl is the limiting reagent.

From the balanced equation, for every mole of HCl consumed, 1 mole of AgCl is produced. So 0.10 moles of AgCl will be produced.

Grams = (moles)(MW)

Grams of AgCl = (0.10 moles)(143 g/mol)
            = 14 grams

QUESTIONS	EXPLANATIONS

66.

$$H_2(g) + Cl_2(g) \rightarrow 2\,HCl(g)$$

Based on the information given in the table below, what is ΔH for the above reaction?

Bond	Average Bond Energy (kJ/mol)
H–H	440
Cl–Cl	240
H–Cl	430

(A) −860 kJ
(B) −620 kJ
(C) −440 kJ
(D) −180 kJ
(E) +240 kJ

66. D The bond energy is the energy that must be put into a bond to break it.

First let's figure out how much energy must be put in to the reactants to break their bonds.

To break 1 mole of H–H bonds, it takes 440 kJ.

To break 1 mole of Cl–Cl bonds, it takes 240 kJ.

So to break up the reactants, it takes +680 kJ.

Energy is given off when a bond is formed; that's the negative of the bond energy.

Now let's see how much energy is given off when 2 moles of HCl are formed.

2 moles of HCl molecules contain 2 moles of HCl bonds, so (2)(−430) kJ = −860 kJ are given off.

So the value of ΔH for the reaction is

(−860, E given off) + (680, E put in) = −180 kJ.

67. The first ionization energy for magnesium is 730 kJ/mol. The third ionization energy for magnesium is 7,700 kJ/mol. What is the most likely value for magnesium's second ionization energy?

(A)   490 kJ/mol
(B)   1,400 kJ/mol
(C)   4,200 kJ/mol
(D)   7,100 kJ/mol
(E)   8,400 kJ/mol

67. B Magnesium has two valence electrons in the third shell, so we would expect to see a small jump between the first and second ionization energies.

The third electron must be removed from the second shell, so we would expect to see a much larger jump between the second and third ionization energies.

Choice (B) is the only answer that shows this relationship.

QUESTIONS	EXPLANATIONS

68. Molten NaCl is electrolyzed with a constant current of 1.00 ampere. What is the shortest amount of time, in seconds, that it would take to produce 1.00 mole of solid sodium? (1 faraday = 96,500 coulombs)

(A) 19,300 seconds
(B) 32,200 seconds
(C) 48,300 seconds
(D) 64,300 seconds
(E) 96,500 seconds

68. E First let's find out how many moles of electrons we need.

The half-reaction that reduces Na^+ to $Na(s)$ is as follows:

$Na^+ + e^- \rightarrow Na(s)$

So it takes 1 mole of electrons to produce 1 mole of $Na(s)$.

Now let's find out how many coulombs we need.

$$\text{Moles of electrons} = \frac{\text{coulombs}}{96,500}$$

So coulombs = (moles of electrons)(96,500)

$= (1)(96,500) = 96,500$ coulombs.

Now we can find how many seconds it takes

$$\text{Amperes} = \frac{\text{coulombs}}{\text{second}}$$

So seconds $= \dfrac{\text{coulombs}}{\text{amperes}} = \dfrac{(96,500)}{(1)}$

$= 96,500$ seconds

69. How many moles of KCl must be added to 200 milliliters of a 0.5-molar NaCl solution to create a solution in which the concentration of Cl^- ion is 1.0-molar? (Assume the volume of the solution remains constant.)

(A) 0.1 moles
(B) 0.2 moles
(C) 0.3 moles
(D) 0.4 moles
(E) 0.5 moles

69. A For every NaCl in solution, there's one Cl^- ion; and for every KCl we add, we get one Cl^- ion.

Let's find out how many moles of Cl^- ions are already in the solution.

Moles = (molarity)(volume)

Moles of Cl^- = (0.5 M)(0.2 L) = 0.1 moles

We're not changing the volume, so to double the concentration of Cl^- ions from 0.5-molar to 1.0-molar, we just double the number of moles of Cl^- ions. We do that by adding 0.1 moles of KCl.

QUESTIONS	EXPLANATIONS

70. A student placed solid barium oxalate in a beaker filled with distilled water and allowed it to come to equilibrium with its dissolved ions. The student then added a nickel (II) nitrate solution to the beaker to create the following equilibrium situation.

$$BaC_2O_4 + Ni^{2+} \leftrightarrow Ba^{2+} + NiC_2O_4$$

If the solubility product for BaC_2O_4 is 2×10^{-7} and the solubility product for NiC_2O_4 is 4×10^{-10}, which of the following gives the value of the equilibrium constant for the reaction above?

(A) 2×10^3
(B) 5×10^2
(C) 5×10^{-2}
(D) 2×10^{-3}
(E) 8×10^{-17}

70. B We can think of the reaction given in the question as the sum of two other reactions.

$$BaC_2O_4 \leftrightarrow Ba^{2+} + C_2O_4^{2-} \quad K_1 = 2 \times 10^{-7}$$

$$Ni^{2+} + C_2O_4^{2-} \leftrightarrow NiC_2O_4 \quad K_2 = \frac{1}{\left(4 \times 10^{-10}\right)}$$

Notice that we are using the reverse reaction for the solvation of NiC_2O_4, so the reactants and products are reversed, and we must take the reciprocal of the solubility product.

When reactions can be added to get another reaction, their equilibrium constants can be multiplied to get the equilibrium constant of the resulting reaction.

So $K_{eq} = (K_1)(K_2) = (2 \times 10^{-7}) \dfrac{1}{\left(4 \times 10^{-10}\right)}$

$\qquad = 0.5 \times 10^3 = 5 \times 10^2$

71. A 100-gram sample of pure $^{37}_{18}Ar$ decays by electron capture with a half-life of 35 days. How long will it take for 90 grams of $^{37}_{17}Cl$ to accumulate?

(A) 31 days
(B) 39 days
(C) 78 days
(D) 116 days
(E) 315 days

71. D Because they have the same mass number, the mass of $^{37}_{17}Cl$ will accumulate at the same rate that the mass of $^{37}_{18}Ar$ disappears.

We're looking for the moment when 10 grams of $^{37}_{18}Ar$ remains.

Make a chart. Start at time = 0.

Half-Lives	Time	Stuff
0	0 days	100 g
1	35 days	50 g
2	70 days	25 g
3	105 days	12.5 g
4	140 days	6.25 g

It takes between 3 and 4 half-lives for the amount of $^{37}_{18}Ar$ to decrease to 10 grams.

116 days is the only answer choice between 105 days and 140 days.

QUESTIONS	EXPLANATIONS

72. The solubility product, K_{sp}, of CaF_2 is 4×10^{-11}. Which of the following expressions is equal to the solubility of CaF_2?

(A) $\sqrt{4 \times 10^{-11}}\ M$

(B) $\sqrt{2 \times 10^{-11}}\ M$

(C) $\sqrt[3]{4 \times 10^{-11}}\ M$

(D) $\sqrt[3]{2 \times 10^{-11}}\ M$

(E) $\sqrt[3]{1 \times 10^{-11}}\ M$

72. E The solubility of a substance is equal to its maximum concentration in solution.

For every CaF_2 in solution, we get one Ca^{2+} and two F^-, so the solubility of CaF_2 (let's call it x) will be the same as $[Ca^{2+}]$.

So for CaF_2
$$K_{sp} = [Ca^{2+}][F^-]^2$$
$$4 \times 10^{-11} = (x)(2x)^2 = 4x^3$$
$$x^3 = 1 \times 10^{-11}$$
$$x = \sqrt[3]{1 \times 10^{-11}}$$

73. When excess hydroxide ions were added to 1.0 liter of $CaCl_2$ solution, $Ca(OH)_2$ precipitate was formed. If all of the calcium ions in the solution were precipitated in 7.4 grams of $Ca(OH)_2$, what was the initial concentration of the $CaCl_2$ solution?

(A) 0.05-molar
(B) 0.10-molar
(C) 0.15-molar
(D) 0.20-molar
(E) 0.30-molar

73. B
$$\text{Moles} = \frac{\text{grams}}{\text{MW}}$$

$$\text{Moles of } Ca(OH)_2 = \frac{(7.4\ \text{g})}{(74\ \text{g / mol})} = 0.10\ \text{moles}$$

One mole of $CaCl_2$ must have been consumed for every mole of $Ca(OH)_2$ produced, so there must have been 0.10 moles of $CaCl_2$ in the original solution.

$$\text{Molarity} = \frac{\text{moles}}{\text{liters}}$$

$$\text{Molarity of } CaCl_2 = \frac{(0.10\ \text{mol})}{(1.0\ \text{L})} = 0.10\text{-molar}$$

74. When a solution of $KMnO_4$ was mixed with a solution of HCl, Cl_2 gas bubbles formed and Mn^{2+} ions appeared in the solution. Which of the following has occurred?

(A) K^+ has been oxidized by Cl^-.
(B) K^+ has been oxidized by H^+.
(C) Cl^- has been oxidized by K^+.
(D) Cl^- has been oxidized by MnO_4^-.
(E) MnO_4^- has been oxidized by Cl^-.

74. D The oxidation and reduction half-reactions are as follows:

Cl^- is oxidized: $2\ Cl^- \rightarrow Cl_2 + 2e^-$

MnO_4^- is reduced: $Mn^{7+} + 5e^- \rightarrow Mn^{2+}$

So Cl^- is oxidized and MnO_4^- is the oxidizing agent.

QUESTIONS

75. $2\ Cu^{+}(aq) + M(s) \rightarrow 2\ Cu(s) + M^{2+}(aq)$

$E^{o} = +0.92$ V

$Cu^{+}(aq) + e^{-} \rightarrow Cu(s)$

$E^{o} = +0.52$ V

Based on the reduction potentials given above, what is the standard reduction potential for the following half-reaction?

$M^{2+}(aq) + 2\ e^{-} \rightarrow M(s)$

(A) +0.40 V
(B) +0.12 V
(C) –0.12 V
(D) –0.40 V
(E) –1.44 V

EXPLANATIONS

75. **D** We subtract the reduction potential for Cu^{+} (+0.52 V) from the reaction potential for the full reaction (+0.92 V) to get +0.40 V, the *oxidation* potential for M(*s*). Remember: Ignore the coefficients in the reaction when you're calculating reaction potentials. But you're not done. You want the *reduction* potential for $M^{2+}(aq)$, the reverse reaction, so you need to change the sign you got for the oxidation potential. So the answer is –0.40 V.

QUESTIONS	EXPLANATIONS

SECTION II—FREE RESPONSE

1. A 0.20-molar solution of acetic acid, $HC_2H_3O_2$, at a temperature of 25°C, has a pH of 2.73.

   (a) Calculate the hydroxide ion concentration, $[OH^-]$.

   (b) What is the value of the acid ionization constant, K_a, for acetic acid at 25°C?

   (c) How many moles of sodium acetate must be added to 500.0 ml of a 0.200-molar solution of acetic acid to create a buffer with a pH of 4.00? Assume that the volume of the solution is not changed by the addition of sodium acetate.

(a) Knowing the pH, we can calculate the pOH.

$pH + pOH = 14$

$2.73 + pOH = 14$

$pOH = 11.27$

Knowing the pOH, we can calculate $[OH^-]$

$[OH^-] = 10^{-pOH} = 10^{-11.27} = 5.4 \times 10^{-12}$

(b) $K_a = \dfrac{[H^+][C_2H_3O_2^-]}{[HC_2H_3O_2]} = \dfrac{x^2}{0.200 - x}$

Knowing the pH, we can find $[H^+]$. We also know that every $HC_2H_3O_2$ molecule that dissociates will put 1 H^+ ion and 1 $C_2H_3O_2$ ion in solution.

$x = [H^+] = [C_2H_3O_2^-] = 10^{-pH} = 10^{-2.73} = 1.86 \times 10^{-3}$

$[HC_2H_3O_2] = 0.200 - x$

x is very small, so $[HC_2H_3O_2] = 0.200$

Now we can solve for K_a.

$$K_a = \frac{[H^+][C_2H_3O_2^-]}{[HC_2H_3O_2]} = \frac{x^2}{0.200} = \frac{(1.86 \times 10^{-3})^2}{(0.200)}$$

$$= 1.7 \times 10^{-5}$$

(c) We can use the Henderson-Hasselbalch expression to find out what value of $[C_2H_3O_2]$ will create a buffer with a pH of 4.

$$pH = pK_a + \log \frac{[A^-]}{[HA]}$$

$$pH = pK_a + \log \frac{[C_2H_3O_2^-]}{[HC_2H_3O_2]}$$

$pH = 4.00$

$pK_a = -\log(1.73 \times 10^{-5}) = 4.76$

$$\log \frac{[C_2H_3O_2^-]}{[HC_2H_3O_2]} = pH - pK_a = 4.00 - 4.76 = -0.76$$

$$\frac{[C_2H_3O_2^-]}{[HC_2H_3O_2]} = 10^{-0.76} = 0.174$$

$[HC_2H_3O_2] = 0.200\ M$

$[C_2H_3O_2^-] = (0.174)(0.200) = 0.035\ M$

Moles = (molarity)(volume) = (0.035)(0.500)
$\quad\quad = 0.018$ moles

QUESTIONS	EXPLANATIONS

(d) In a titration experiment, 100.0 ml of sodium hydroxide solution was added to 200 ml of a 0.400-molar solution of acetic acid until the equivalence point was reached. What was the pH at the equivalence point?

(d) Because $HC_2H_3O_2$ dissociates to such a small extent, we can assume that all of the $C_2H_3O_2^-$ in the solution came from the NaC_2H_3O.

Use the base ionization constant for $C_2H_3O_2^-$.

$$K_b = \frac{[HC_2H_3O_2][OH^-]}{[C_2H_3O_2^-]}$$

$$K_b = \frac{10^{-14}}{K_a} = \frac{10^{-14}}{1.73\times10^{-5}} = 5.78 \times 10^{-10}$$

At the equivalence point, all of the acetic acid initially present has been converted to acetate ion. So the initial $[HC_2H_3O_2]$ is equal to $[C_2H_3O_2]$ at the equivalence point.

Moles = (molarity)(volume)

Moles of $C_2H_3O_2^-$ = (0.400 M)(0.200 L)
  = 0.080 moles

$$Molarity = \frac{moles}{volume}$$

$$[C_2H_3O_2^-] = \frac{(0.080\ mol)}{(0.200\,L + 0.100\,L)}$$

$$= \frac{(0.080\ mol)}{(0.300\,L)} = 0.267\ M$$

$[HC_2H_3O_2] = [OH^-] = x$

Now we can use the K_b equation to find x, the OH^- concentration.

We'll assume that x is much smaller than 0.267 M.

$$K_b = \frac{[HC_2H_3O_2][OH^-]}{[C_2H_3O_2^-]} = \frac{x^2}{0.267-x} = \frac{x^2}{0.267}$$

$$K_b = 5.78 \times 10^{-10} = \frac{x^2}{0.267}$$

$x = [OH^-] = 1.24 \times 10^{-5}\ M$

Knowing $[OH^-]$, we can calculate pOH, and then pH.

pOH = $-\log[OH^-]$ = $-\log(1.24 \times 10^{-5})$ = 4.91

pH = 14 – pOH

pH = 14 – 4.91 = 9.09

(e) The pK_a values for several indicators are given in the table below. Which of the indicators on the table is most suitable for this titration? Justify your answer.

Indicator	pK_a
Thymol Blue	2
Bromcresol Purple	6
Phenolphthalein	9

(e) Phenolphthalein, with a pK_a of 9, is the best choice. The pH at the equivalence point is about 9, and the indicator should have a pK_a that is close to the equivalence point for the titration.

| | QUESTIONS | | EXPLANATIONS |

QUESTIONS

2. $$2 NO(g) + Cl_2(g) \rightarrow 2 NOCl(g)$$

The following data were collected for the reaction above. All of the measurements were taken at a temperature of 263 K.

Experiment	Initial [NO] (M)	Initial [Cl₂] (M)	Initial rate of disappearance of Cl₂ (M/min)
1	0.15	0.15	0.60
2	0.15	0.30	1.2
3	0.30	0.15	2.4
4	0.25	0.25	?

(a) Write the expression for the rate law for the reaction above.

(b) Calculate the value of the rate constant for the above reaction, and specify the units.

(c) What is the initial rate of appearance of NOCl in experiment 2?

(d) What is the initial rate of disappearance of Cl₂ in experiment 4?

(e) Each of the experimental trials took place in a closed container. Explain or calculate each of the following:

   (i) What was the partial pressure due to NO(g) at the start of experiment 1?

   (ii) What was the total pressure at the start of experiment 1? Assume that no NOCl is present.

EXPLANATIONS

(a) From experiments 1 and 2, we can see that when [Cl₂] doubles, the rate doubles, so the rate law is first order with respect to Cl₂. That is, $[Cl_2]^1$.

From experiments 1 and 3, we can see that when [NO] doubles, the rate quadruples, so the rate law is second order with respect to NO. That is, $[NO]^2$.

Rate = $k[NO]^2[Cl_2]$

(b) We'll use experiment 1 for our calculation.

$$k = \frac{Rate}{[NO]^2[Cl_2]} = \frac{(0.60\,M/min)}{(0.15\,M)^2(0.15\,M)} = 180\,M^{-2}\,min^{-1}$$

(c) From the balanced equation we can see that for every molecule of Cl₂ that disappears, 2 molecules of NOCl appear, so the rate of appearance of NOCl will be twice the rate of disappearance of Cl₂.

In experiment 2, the initial rate of disappearance of Cl₂ is 1.2 M/min, so the initial rate of appearance of NOCl will be 2.4 M/min.

(d) Use the rate law.

Rate = $k[NO]^2[Cl_2]$

Rate = $(180\,M^{-2}\,min^{-1})(0.25\,M)^2(0.25\,M)$
     = 2.8 M/min

(e) (i) Use the gas law

$$P = \frac{nRT}{V} = MRT$$

P = (0.15 mol/L)(0.0821 L-atm/mol-K)(263K) = 3.2 atm

(ii) Since the initial concentrations of NO and Cl₂ are the same for experiment 1, the initial partial pressure due to Cl₂ will also be 3.2 atm. From Dalton's law, the total pressure is equal to the sum of the partial pressures. So 3.2 atm + 3.2 atm = 6.4 atm.

3.
$$CH_4(g) + 2\,O_2(g) \rightarrow CO_2(g) + 2\,H_2O(l)$$

The above reaction for the combustion of methane gas has a standard entropy change, $\Delta S°$, with a value of –242.7 J/mol-K. The following data are also available:

Compound	ΔH_f (kJ/mol)
CH_4 (g)	–74.8
H_2O (l)	–285.9
CO_2 (g)	–393.5

(a) What are the values of $\Delta H°_f$ and $\Delta G°_f$ for $O_2(g)$?

(a) $\Delta H°_f$ and $\Delta G°_f$ for $O_2(g)$ are both equal to zero. The enthalpy and free energy of formation of any element in its standard state are equal to zero.

(b) Calculate the standard change in enthalpy, $\Delta H°$, for the combustion of methane.

(b) $\Delta H = \Sigma\Delta H°_f$ (products) $- \Sigma\Delta H°_f$ (reactants)

$\Delta H = [(-393.5) + (2)(-285.9)]$ kJ $- [-74.8]$ kJ

$\Delta H = [-965.3]$ kJ $- [-74.8]$ kJ

$\Delta H = -890.5$ kJ

(c) Calculate the standard free energy change, $\Delta G°$, for the combustion of methane.

(c) $\Delta G° = \Delta H° - T\,\Delta S°$

$\Delta G° = (-890,500 \text{ J}) - (298 \text{ K})(-242.7 \text{ J/K})$

$\Delta G° = -818,200 \text{ J} = -818.2$ kJ

(d) How would the value of ΔS for the reaction be affected if the water produced in the combustion remained in the gas phase?

(d) $\Delta S°$ would become less negative. $H_2O(g)$ has more entropy than $H_2O(l)$, so the entropy of the products would be increased and the entropy change of the reaction would become more positive (less negative, that is).

(e) A 20.0-gram sample of $CH_4(g)$ underwent combustion in a bomb calorimeter with excess oxygen gas.

(i) Calculate the mass of carbon dioxide produced.

(ii) Calculate the heat released by the reaction.

(e) (i) First find moles of CH_4.

$$\text{Moles} = \frac{\text{grams}}{\text{MW}} = \frac{20.0 \text{ grams}}{16.0 \text{ g/mol}} = 1.25 \text{ mol}$$

There is a one to one ratio in the balance equation between CH_4 and CO_2, so 1.25 moles of CO_2 are produced.

Now find the grams of CO_2.

Grams = (moles)(MW) = (1.25 mol)(44.0 g/mol) = 55 grams

(ii) From (c), the change in enthalpy when one mole of CH_4 is consumed is 890.5 kJ. So when 1.25 moles are consumed, ΔH = (890.5 kJ/mol)(1.25 mol) = 1,110 kJ. So about 1,110 kJ were released.

QUESTIONS	EXPLANATIONS

4.

(a) Sulfur dioxide gas is bubbled through cold water.

  (i) Balanced equation:

    (i) $SO_2 + H_2O \rightarrow H_2SO_3$

  (ii) As the reaction progresses, will the hydroxide concentration in the solution increase, decrease, or remain the same?

    (ii) Sulfur dioxide is an acid anhydride, so the hydroxide ion concentration will decrease as the solution becomes more acidic.

(b) Ethane is burned in air.

  (i) Balanced equation:

    (i) $2\ C_2H_6 + 7\ O_2 \rightarrow 4\ CO_2 + 6\ H_2O$

  (ii) If the enthalpy change for the reaction was measured to be –3,100 kJ and the total heat of formation for a,ll products formed is –3,300 kJ, what is the approximate heat of formation of ethane measured in kJ/mole?

    (ii) The enthalpy change for the reaction comes from subtracting the heat of formation of the reactants from the heat of formation of the products, so the heat of formation of the reactants is –3,300 kJ – (–3,100 kJ) = –200 kJ. Because the heat of formation for oxygen gas is zero and there are 2 moles of ethane in the reaction, the heat of formation of ethane is approximately –100 kJ mole.

(c) Chlorine gas is bubbled through a solution of sodium bromide.

  (i) Balanced equation:

    (i) $Cl_2 + 2\ Br^- \rightarrow 2\ Cl^- + Br_2$

  (ii) Which substance is reduced in this reaction?

    (ii) Chlorine gas gains electrons, so it is reduced.

QUESTIONS	EXPLANATIONS

5. Oxygen is found in the atmosphere as a diatomic gas, O_2, and as ozone, O_3. Ozone has a dipole of 0.5 debye. Oxygen has a dipole moment of zero.

(a) Draw the Lewis dot structures for both molecules.

(a) Lewis dot structure for O_2

Lewis dot structure for O_3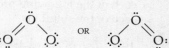

(b) Use the principles of bonding and molecular structure to account for the fact that ozone has a higher boiling point than diatomic oxygen.

(b) O_3 is a polar molecule, while O_2 is nonpolar, so the dipole–dipole attractions between O_3 molecules are stronger than the van der Waals forces between O_2 molecules.

If you didn't recognize the polarity of O_3, you may have gotten some partial credit for noting that O_3 has more electrons than O_2, so O_3 will have stronger van der Waals forces between its molecules.

(c) Use the principles of bonding and molecular structure to account for the fact that ozone is more soluble than diatomic oxygen in water.

(c) Water molecules are polar. O_3 is a polar molecule, while O_2 is nonpolar. So water molecules will be more strongly attracted to O_3 molecules than they are to the nonpolar O_2 molecules.

(d) Explain why the two bonds in O_3 are of equal length and are longer than the bond length of the bond in diatomic oxygen.

(d) The bond in O_2 is a double bond. Ozone, however, has two resonance forms, each with a single and a double bond, so the two bonds in O_3 are each somewhere between a single and a double bond on average. The double bond in O_2 is stronger and shorter than the single/double resonance bonds in O_3.

(e) Elemental oxygen is strongly affected by a magnetic field. Explain why.

(e) Oxygen has unpaired electrons in the $2p$ subshell. This makes oxygen paramagnetic.

(f) For the equilibrium reaction below at 25°C and 1 atm, $K_{eq} = 10^{-57}$.

$3\,O_2(g) \leftrightarrow 2\,O_3(g)$

Which form of oxygen is more abundant under normal conditions? Explain.

(f) The value of the equilibrium constant is very small. This means that the reactants will be far more abundant than the products.

6. $NaC_2H_3O_2$, $Ba(NO_3)_2$, KCl

(a) Aqueous solutions of equal concentration of the three compounds listed above are prepared. What would an experimenter expect to observe when each of the following procedures is performed on each of the solutions?

(i)   The pH of each solution is measured.

(i)   The $NaC_2H_3O_2$ solution will be slightly basic. The other two solutions will be neutral.

$NaC_2H_3O_2$ is a salt composed of the conjugate of a strong base and the conjugate of a weak acid, so it will create a basic solution.

$Ba(NO_3)_2$ and KCl are salts composed of conjugates of strong acids and bases, so they will create neutral solutions.

(ii) SO_4^{2-} ions are introduced into each solution.

(ii) A precipitate will form in the $Ba(NO_3)_2$ solution. The other two solutions will show no change.

$BaSO_4$ is insoluble. Na_2SO_4 and K_2SO_4 are both soluble.

(iii) The freezing point of each solution is measured and the three temperatures are compared.

(iii) The freezing points of the $NaC_2H_3O_2$ and KCl solutions will be less than 0 C and about the same. The freezing point of the $Ba(NO_3)_2$ solution will be lower than the other two.

Freezing-point depression is a colligative property, so it will depend only on the number of particles in solution, not on their identity.

$NaC_2H_3O_2$ and KCl each dissociate into two ions per molecule, while $Ba(NO_3)_2$ dissociates into three ions per unit.

Since the concentrations of all three solutions are the same, the $Ba(NO_3)_2$ solution will have the greatest freezing-point depression because it dissociates into the greatest number of particles. By the same reasoning, the freezing points of the $NaC_2H_3O_2$ and KCl solutions will be about the same.

(iv) Each solution is subjected to a flame test.

(iv) Each flame will have a distinct color: Na^+ will be yellow, Ba^{2+} will be green, and K^+ will be purple.

QUESTIONS	EXPLANATIONS

(b) At 25°C and 1.0 atmosphere pressure, a balloon contains a mixture of four ideal gases: oxygen, nitrogen, carbon dioxide, and helium. The partial pressure due to each gas is 0.25 atmosphere. Use the ideas of kinetic molecular theory to answer each of the following questions:

(i) Rank the gases in increasing order of average molecular velocity and explain.

(i) CO_2, O_2, N_2, He

All of the gases have the same average kinetic energy, and $KE = \frac{1}{2}mv^2$, so the larger the molecular weight, the smaller the average velocity. This relationship is given directly by the expression

$$u_{rms} = \sqrt{\frac{3RT}{(MW)}}.$$

(ii) How would the volume of the balloon be affected if the temperature of the gases in the balloon were increased at constant pressure?

(ii) Volume increases.

$PV = nRT$. From the ideal gas law, we can see that when T increases and P and n are held constant, V must increase to maintain equality.

(iii) What changes to the temperature and pressure of the gases would cause deviation from ideal behavior, and which gas would be most affected?

(iii) Decreased temperature and increased pressure would eventually cause deviation from ideal behavior. Carbon dioxide would be most affected.

Deviation from ideal behavior occurs when gas molecules are brought very close together. When gas molecules are packed close together, the weak attractive forces between them become important. Also, in this situation, the volumes occupied by the individual gas molecules can no longer be ignored. Gas molecules are brought close together by low temperatures and high pressures.

Carbon dioxide would be the most affected of the four gases because it is the largest molecule and it has the most electrons. Because of this, the van der Waals forces among CO_2 molecules will be stronger than for the other gases.

20

The Princeton Review
AP Chemistry
Practice Exam 2

AP® Chemistry Exam

DO NOT OPEN THIS BOOKLET UNTIL YOU ARE TOLD TO DO SO.

At a Glance

Total Time
1 hour and 30 minutes
Number of Questions
75
Percent of Total Grade
50%
Writing Instrument
Pencil required

Instructions

Section I of this examination contains 75 multiple-choice questions. Fill in only the ovals for numbers 1 through 75 on your answer sheet.

CALCULATORS MAY NOT BE USED IN THIS PART OF THE EXAMINATION.

Indicate all of your answers to the multiple-choice questions on the answer sheet. No credit will be given for anything written in this exam booklet, but you may use the booklet for notes or scratch work. After you have decided which of the suggested answers is best, completely fill in the corresponding oval on the answer sheet. Give only one answer to each question. If you change an answer, be sure that the previous mark is erased completely. Here is a sample question and answer.

Sample Question   Sample Answer

Chicago is a   Ⓐ ● Ⓒ Ⓓ Ⓔ
(A) state
(B) city
(C) country
(D) continent
(E) village

Use your time effectively, working as quickly as you can without losing accuracy. Do not spend too much time on any one question. Go on to other questions and come back to the ones you have not answered if you have time. It is not expected that everyone will know the answers to all the multiple-choice questions.

About Guessing

Many candidates wonder whether or not to guess the answers to questions about which they are not certain. Multiple choice scores are based on the number of questions answered correctly. Points are not deducted for incorrect answers, and no points are awarded for unanswered questions. Because points are not deducted for incorrect answers, you are encouraged to answer all multiple-choice questions. On any questions you do not know the answer to, you should eliminate as many choices as you can, and then select the best answer among the remaining choices.

GO ON TO THE NEXT PAGE.

This page intentionally left blank.

GO ON TO THE NEXT PAGE.

CHEMISTRY
SECTION I

Time—1 hour and 30 minutes

Material in the following table may be useful in answering the questions in this section of the examination.

PERIODIC CHART OF THE ELEMENTS

1 H 1.0																		2 He 4.0
3 Li 6.9	4 Be 9.0											5 B 10.8	6 C 12.0	7 N 14.0	8 O 16.0	9 F 19.0	10 Ne 20.2	
11 Na 23.0	12 Mg 24.3											13 Al 27.0	14 Si 28.1	15 P 31.0	16 S 32.1	17 Cl 35.5	18 Ar 39.9	
19 K 39.1	20 Ca 40.1	21 Sc 45.0	22 Ti 47.9	23 V 50.9	24 Cr 52.0	25 Mn 54.9	26 Fe 55.8	27 Co 58.9	28 Ni 58.7	29 Cu 63.5	30 Zn 65.4	31 Ga 69.7	32 Ge 72.6	33 As 74.9	34 Se 79.0	35 Br 79.9	36 Kr 83.8	
37 Rb 85.5	38 Sr 87.6	39 Y 88.9	40 Zr 91.2	41 Nb 92.9	42 Mo 95.9	43 Tc (98)	44 Ru 101.1	45 Rh 102.9	46 Pd 106.4	47 Ag 107.9	48 Cd 112.4	49 In 114.8	50 Sn 118.7	51 Sb 121.8	52 Te 127.6	53 I 126.9	54 Xe 131.3	
55 Cs 132.9	56 Ba 137.3	57 *La 138.9	72 Hf 178.5	73 Ta 180.9	74 W 183.9	75 Re 186.2	76 Os 190.2	77 Ir 192.2	78 Pt 195.1	79 Au 197.0	80 Hg 200.6	81 Tl 204.4	82 Pb 207.2	83 Bi 209.0	84 Po (209)	85 At (210)	86 Rn (222)	
87 Fr (223)	88 Ra 226.0	89 †Ac 227.0																

*Lanthanum Series

58 Ce 140.1	59 Pr 140.9	60 Nd 144.2	61 Pm (145)	62 Sm 150.4	63 Eu 152.0	64 Gd 157.3	65 Tb 158.9	66 Dy 162.5	67 Ho 164.9	68 Er 167.3	69 Tm 168.9	70 Yb 173.0	71 Lu 175.0

†Actinium Series

90 Th 232.0	91 Pa 231.0	92 U 238.0	93 Np 237.0	94 Pu (244)	95 Am (243)	96 Cm (247)	97 Bk (247)	98 Cf (251)	99 Es (252)	100 Fm (258)	101 Md (258)	102 No (259)	103 Lr (260)

DO NOT DETACH FROM BOOK.

GO ON TO THE NEXT PAGE.

Note: For all questions involving solutions and/or chemical equations, assume that the system is in pure water and at room temperature unless otherwise stated.

Part A

Directions: Each set of lettered choices below refers to the numbered questions or statements immediately following it. Select the one lettered choice that best answers each question or best fits each statement and fill in the corresponding oval on the answer sheet. A choice may be used once, more than once, or not at all in each set.

Questions 1–3 refer to the following elements.

    (A) Ne
    (B) Li
    (C) Al
    (D) Cl
    (E) Ca

1. Which of the elements above is most commonly found as a negatively charged ion?

2. Which of the elements above has the same electron structure as Mg^{2+}?

3. Which of the elements above is found as a diatomic gas in its uncombined state?

Questions 4–7 refer to the following substances.

    (A) H_2O
    (B) NH_3
    (C) CH_4
    (D) HF
    (E) CH_3OH

4. This molecule is nonpolar.

5. This substance forms an aqueous solution with a pH that's less than 7.

6. This molecule has a trigonal pyramidal shape.

7. This substance does NOT exhibit hydrogen bonding.

Questions 8–11 refer to the phase diagram shown below for a substance.

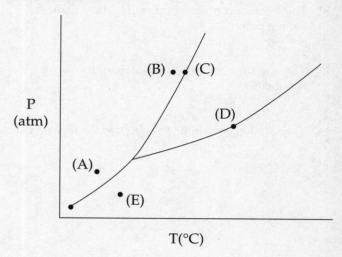

8. This point could be the boiling point for the substance.

9. If pressure is increased from the conditions represented by this point at constant temperature, deposition may occur.

10. At the conditions represented by this point, the substance can exist with liquid and solid phases in equilibrium.

11. If temperature is increased from the conditions represented by this point at constant pressure, no phase change will occur.

GO ON TO THE NEXT PAGE.

Questions 12–14 refer to an experiment in which five individual 1-liter aqueous solutions, each containing a 1-mole sample of one of the salts listed below, were subjected to various tests at room temperature.

(A) $NaC_2H_3O_2$
(B) $NaCl$
(C) $MgBr_2$
(D) $HC_2H_3O_2$
(E) KBr

12. The solution containing this salt had the highest boiling point.

13. The solution containing this salt had the lowest conductivity.

14. The solution containing this salt had the highest pH.

GO ON TO THE NEXT PAGE.

Part B

Directions: Each of the questions or incomplete statements below is followed by five suggested answers or completions. Select the one that is best in each case and then fill in the corresponding oval on the answer sheet.

15. Titanium metal is prepared by heating rutile, an oxide of titanium, along with carbon and chlorine gas. By mass, rutile is 40% oxygen and 60% titanium. What is the empirical formula of rutile?

    (A) TiO
    (B) Ti_2O
    (C) TiO_2
    (D) Ti_2O_3
    (E) Ti_3O_2

16. What is the weight of $NaNO_3$ (molecular weight 85.0) present in 100.0 ml of a 4.00-molar solution?

    (A) 8.50 grams
    (B) 17.0 grams
    (C) 25.5 grams
    (D) 34.0 grams
    (E) 51.0 grams

17. A mixture of gases contains 5.0 moles of oxygen, 12 moles of nitrogen, and 4.0 moles of carbon dioxide. If the partial pressure due to carbon dioxide is 1.6 atmospheres, what is the partial pressure due to oxygen?

    (A) 2.0 atm
    (B) 4.8 atm
    (C) 6.0 atm
    (D) 6.8 atm
    (E) 8.4 atm

18. Which of the diagrams below represents the enthalpy change for an endothermic reaction?

(A)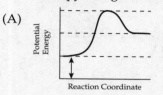

(D)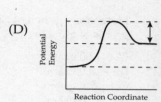

(B)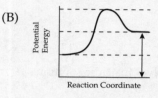

(E)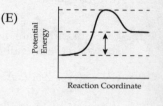

(C)

19. $$2 NO(g) + Br_2(g) \rightarrow 2 NOBr(g)$$

For the reaction above, the experimental rate law is given as follows:

$$Rate = k[NO]^2[Br_2]$$

Which of the statements below is true regarding this reaction?

    (A) The reaction is first-order overall.
    (B) The reaction is first-order with respect to Br_2.
    (C) The reaction is first-order with respect to NO.
    (D) The reaction is second-order overall.
    (E) The reaction is second-order with respect to Br_2.

GO ON TO THE NEXT PAGE.

20. Which of the following is the most likely electron configuration of a sulfur atom in its ground state?

    (A) $1s^2 2s^2 2p^6$
    (B) $1s^2 2s^2 2p^6 3s^2 3p^2$
    (C) $1s^2 2s^2 2p^6 3s^2 3p^4$
    (D) $1s^2 2s^2 2p^6 3s^2 3p^6$
    (E) $1s^2 2s^2 2p^6 3s^2 3p^5 4s^1$

21. How many grams of carbon are present in 270 grams of glucose, $C_6H_{12}O_6$?

    (A) 12.0 grams
    (B) 18.0 grams
    (C) 67.5 grams
    (D) 72.0 grams
    (E) 108 grams

22. Calcium carbonate dissolves in acidic solutions as shown in the equation below.

    $$CaCO_3(s) + 2 H^+(aq) \rightarrow Ca^{2+}(aq) + H_2O(l) + CO_2(g)$$

    If excess $CaCO_3$ is added to 0.250 liters of a 2.00-molar HNO_3 solution, what is the maximum volume of CO_2 gas that could be produced at standard temperature and pressure?

    (A) 5.60 liters
    (B) 11.2 liters
    (C) 16.8 liters
    (D) 33.6 liters
    (E) 39.2 liters

23. A monoprotic acid was titrated with a solution of NaOH. For 55.0 milliliters of the acid, 37.0 milliliters of a 0.450-molar solution of NaOH was required to reach the equivalence point. Which of the following expressions is equal to the initial concentration of the monoprotic acid?

    (A) $\dfrac{(0.450)(0.037)}{(0.055)} M$

    (B) $\dfrac{(0.450)(0.055)}{(0.037)} M$

    (C) $\dfrac{(0.055)}{(0.450)(0.037)} M$

    (D) $\dfrac{(0.037)}{(0.450)(0.055)} M$

    (E) $(0.450)(0.055)(0.037) M$

24. $Zn(s) + NO_3^-(aq) + 10\ H^+(aq) \rightarrow$
    $$Zn^{2+}(aq) + NH_4^+(aq) + 3\ H_2O(l)$$

    Which of the following statements regarding the reaction shown above is correct?

    (A) The oxidation number of hydrogen changes from +1 to 0.
    (B) The oxidation number of hydrogen changes from +1 to −1.
    (C) The oxidation number of nitrogen changes from +5 to −3.
    (D) The oxidation number of nitrogen changes from +5 to +3.
    (E) The oxidation number of nitrogen changes from +6 to +4.

25. $HC_2H_3O_2(aq) + ClO^-(aq) \leftrightarrow HClO(aq) + C_2H_3O_2^-(aq)$

    The standard free energy change for this reaction has a negative value. Based on this information, which of the following statements is true?

    (A) K_a for $HC_2H_3O_2(aq)$ is less than K_a for $HClO(aq)$.
    (B) K_b for $C_2H_3O_2^-(aq)$ is less than K_b for $ClO^-(aq)$.
    (C) K_{eq} for the reaction is less than 1.
    (D) The reaction occurs in the presence of a catalyst.
    (E) $HC_2H_3O_2(aq)$ and $HClO(aq)$ are conjugates.

26. A student added 0.20 mol of NaI and 0.40 mol of KI to 3 liters of water to create an aqueous solution. What is the minimum number of moles of $Pb(C_2H_3O_2)_2$ that the student must add to the solution to precipitate out all of the I^- ions as PbI_2?

    (A) 2.40
    (B) 1.20
    (C) 0.60
    (D) 0.30
    (E) 0.15

GO ON TO THE NEXT PAGE.

27. A molecule with the formula XY_2 whose atoms are arranged linearly could have a central atom with which of the following hybridizations?

    I. sp
    II. sp^3
    III. dsp^3

    (A) I only
    (B) II only
    (C) I and II only
    (D) I and III only
    (E) I, II, and III

28. Which of the following statements regarding fluorine and nitrogen is NOT true?

    (A) Fluorine has greater electronegativity.
    (B) Fluorine has a greater first ionization energy.
    (C) Fluorine has more valence electrons.
    (D) Fluorine has a greater atomic weight.
    (E) Fluorine has a greater atomic radius.

29. Which of the groups below is (are) listed in order from lowest to highest melting point?

    I.   KI, LiF, BeO
    II. F_2, Cl_2, Br_2
    III. K, Na, Li

    (A) I only
    (B) I and II only
    (C) I and III only
    (D) II and III only
    (E) I, II, and III

30. Which of the statements below regarding elemental nitrogen is NOT true?

    (A) It contains one sigma bond.
    (B) It contains two pi bonds.
    (C) It has a bond order of 3.
    (D) It has a large dipole moment.
    (E) It exists as a diatomic gas.

31. The reaction of elemental chlorine with ozone in the atmosphere occurs by the two-step process shown below.

    I. $Cl + O_3 \rightarrow ClO + O_2$
    II. $ClO + O \rightarrow Cl + O_2$

    Which of the statements below is true regarding this process?

    (A) Cl is a catalyst.
    (B) O_3 is a catalyst.
    (C) ClO is a catalyst.
    (D) O_2 is an intermediate.
    (E) O is an intermediate.

32. A student examined the line spectrum of a hydrogen atom and was able to conclude that when the electron in the hydrogen atom made the transition from the $n = 3$ to the $n = 1$ state, the frequency of the ultraviolet radiation emitted was 2.92×10^{15} Hz. When the same transition occurred in a He^+ ion, the frequency emitted was 1.17×10^{16} Hz. Which of the following best accounts for the difference?

    (A) More energy is released by the He^+ transition because of the greater molar mass of the He^+ ion.
    (B) More energy is released by the He^+ transition because of the greater nuclear charge of the He^+ ion.
    (C) More energy is released by the He^+ transition because the ionic radius of He^+ is greater than the atomic radius of H.
    (D) Less energy is released by the He^+ transition because the ionic radius of He^+ is smaller than the atomic radius of H.
    (E) Less energy is released by the He^+ transition because of the greater nuclear charge of the He^+ ion.

GO ON TO THE NEXT PAGE.

33. $CH_3CH_2OH(g) + ...O_2(g) \rightarrow$

$$...CO_2(g) + ...H_2O(g)$$

The reaction above represents the oxidation of ethanol. How many moles of O_2 are required to oxidize 1 mole of CH_3CH_2OH?

(A) $\dfrac{3}{2}$ moles

(B) $\dfrac{5}{2}$ moles

(C) 3 moles

(D) $\dfrac{7}{2}$ moles

(E) 4 moles

34. How many milliliters of water must be added to 10 milliliters of an HCl solution with a pH of 1 to produce a solution with a pH of 2?

(A)   10 ml
(B)   90 ml
(C)   100 ml
(D)   990 ml
(E) 1,000 ml

35. Elemental iodine (I_2) is more soluble in carbon tetrachloride (CCl_4) than it is in water (H_2O). Which of the following statements is the best explanation for this?

(A) I_2 is closer in molecular weight to CCl_4 than it is to H_2O.
(B) The freezing point of I_2 is closer to that of CCl_4 than it is to that of H_2O.
(C) I_2 and CCl_4 are nonpolar molecules, while H_2O is a polar molecule.
(D) The heat of formation of I_2 is closer to that of CCl_4 than it is to that of H_2O.
(E) CCl_4 has a greater molecular weight than does H_2O.

36. A sample of water was electrolyzed to produce hydrogen and oxygen gas, as shown in the reaction below.

$$2\,H_2O(l) \rightarrow 2\,H_2(g) + O_2(g)$$

If 33.6 liters of gas were produced at STP, how many grams of water were consumed in the reaction?

(A) 11.1 grams
(B) 18.0 grams
(C) 24.0 grams
(D) 36.0 grams
(E) 44.8 grams

37. The following data were gathered in an experiment to determine the density of a sample of an unknown substance.

Mass of the sample = 7.50 grams

Volume of the sample = 2.5 milliliters

The density of the sample should be reported as

(A) 3.00 grams per ml.
(B) 3.0 grams per ml.
(C) 3 grams per ml.
(D) 0.3 grams per ml.
(E) 0.33 grams per ml.

38. $\qquad 2\,H_2O_2(g) + S(s) \rightarrow SO_2(g) + 2\,H_2O(g)$

Based on the information given in the table below, what is the enthalpy change in the reaction represented above?

Substance	ΔH_f (kJ/mol)
$H_2O_2(g)$	−150
$SO_2(g)$	−300
$H_2O(g)$	−250
$S(s)$	0

(A) −500 kJ
(B) −200 kJ
(C)   200 kJ
(D)   400 kJ
(E)   600 kJ

GO ON TO THE NEXT PAGE.

39. A chemist creates a buffer solution by mixing equal volumes of a 0.2-molar HOCl solution and a 0.2-molar KOCl solution. Which of the following will occur when a small amount of KOH is added to the solution?

    I. The concentration of undissociated HOCl will increase.

    II. The concentration of OCl⁻ ions will increase.

    III. The concentration of H⁺ ions will increase.

    (A) I only
    (B) II only
    (C) III only
    (D) I and III only
    (E) II and III only

40. A 15-gram sample of neon is placed in a sealed container with a constant volume of 8.5 liters. If the temperature of the container is 27 C, which of the following expressions gives the correct pressure of the gas? The ideal gas constant, R, is 0.08 (L-atm)/(mole-K).

    (A) $\dfrac{(15)(0.08)(300)}{(20)(8.5)}$ atm

    (B) $\dfrac{(20)(0.08)(300)}{(15)(8.5)}$ atm

    (C) $\dfrac{(20)(300)}{(15)(8.5)(0.08)}$ atm

    (D) $\dfrac{(15)(300)}{(20)(8.5)(0.08)}$ atm

    (E) $\dfrac{(15)(0.08)}{(20)(8.5)(300)}$ atm

41. $$C(s) + CO_2(g) + energy \leftrightarrow 2\,CO(g)$$

The system above is currently at equilibrium in a closed container. Which of the following changes to the system would serve to increase the number of moles of CO present at equilibrium?

    I. Raising the temperature
    II. Increasing the volume of the container
    III. Adding more C(s) to the container

    (A) I only
    (B) II only
    (C) I and II only
    (D) I and III only
    (E) I, II, and III

42. In a titration experiment, a 0.10-molar $H_2C_2O_4$ solution was completely neutralized by the addition of a 0.10-molar NaOH solution. Which of the diagrams below illustrates the change in pH that accompanied this process?

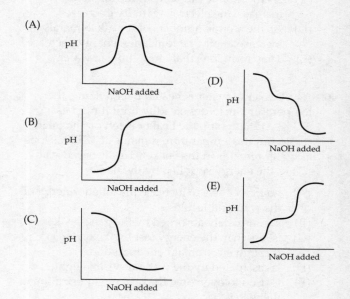

43. If 52 grams of $Ba(NO_3)_2$ (molar mass 260 grams) are completely dissolved in 500 milliliters of distilled water, what are the concentrations of the barium and nitrate ions?

    (A) $[Ba^{2+}] = 0.10\ M$ and $[NO_3^-] = 0.10\ M$
    (B) $[Ba^{2+}] = 0.20\ M$ and $[NO_3^-] = 0.40\ M$
    (C) $[Ba^{2+}] = 0.40\ M$ and $[NO_3^-] = 0.40\ M$
    (D) $[Ba^{2+}] = 0.40\ M$ and $[NO_3^-] = 0.80\ M$
    (E) $[Ba^{2+}] = 0.80\ M$ and $[NO_3^-] = 0.80\ M$

44. An acid solution of unknown concentration is to be titrated with a standardized hydroxide solution that will be released from a buret. The buret should be rinsed with

    (A) hot distilled water.
    (B) distilled water at room temperature.
    (C) a sample of the unknown acid solution.
    (D) a sample of the hydroxide solution.
    (E) a neutral salt solution.

GO ON TO THE NEXT PAGE.

45. All of the following statements concerning the alkali metals are true EXCEPT

    (A) They are strong oxidizing agents.
    (B) They form ions with a +1 oxidation state.
    (C) As the atomic numbers of the alkali metals increase, the electronegativity decreases.
    (D) As the atomic numbers of the alkali metals increase, their first ionization energy decreases.
    (E) They form ions that are soluble in water.

46. As a beaker of water is heated over a flame, the temperature increases steadily until it reaches 373 K. At that point, the beaker is left on the open flame, but the temperature remains at 373 K as long as water remains in the beaker. This is because at 373 K, the energy supplied by the flame

    (A) no longer acts to increase the kinetic energy of the water molecules.
    (B) is completely absorbed by the glass beaker.
    (C) is less than the energy lost by the water through electromagnetic radiation.
    (D) is dissipated by the water as visible light.
    (E) is used to overcome the heat of vaporization of the water.

47. For which of the following reactions will the equilibrium constants K_c and K_p have the same value?

    (A) $2 N_2O_5(g) \leftrightarrow 2 NO_2(g) + O_2(g)$
    (B) $2 CO_2(g) \leftrightarrow 2 CO(g) + O_2(g)$
    (C) $H_2O(g) + CO(g) \leftrightarrow H_2(g) + CO_2(g)$
    (D) $3 O_2(g) \leftrightarrow 2 O_3(g)$
    (E) $CO(g) + Cl_2(g) \leftrightarrow COCl_2(g)$

48. A student added solid potassium iodide to 1 kilogram of distilled water. When the KI had completely dissolved, the freezing point of the solution was measured to be $-3.72°C$. What was the mass of the potassium iodide added? (The freezing point depression constant, k_f, for water is 1.86 K-kg/mol.)

    (A) 334 grams
    (B) 167 grams
    (C) 84 grams
    (D) 42 grams
    (E) 21 grams

49. Each of the following compounds was added to distilled water at 25°C. Which one produced a solution with a pH that was less than 7?

    (A) N_2
    (B) O_2
    (C) NaI
    (D) MgO
    (E) SO_2

50. $...Be_2C + ...H_2O \rightarrow ...Be(OH)_2 + ...CH_4$

    When the equation for the reaction above is balanced with the lowest whole-number coefficients, the coefficient for H_2O will be

    (A) 1
    (B) 2
    (C) 3
    (D) 4
    (E) 5

51. A sample of an ideal gas is placed in a sealed container of constant volume. If the temperature of the gas is increased from 40°C to 70°C, which of the following values for the gas will NOT increase?

    (A) The average speed of the molecules of the gas
    (B) The average kinetic energy of the molecules of the gas
    (C) The pressure exerted by the gas
    (D) The density of the gas
    (E) The frequency of collisions between molecules of the gas

52. A $^{222}_{86}Rn$ nuclide decays through the emission of two beta particles and two alpha particles. The resulting nuclide is

    (A) $^{214}_{84}Po$
    (B) $^{210}_{84}Po$
    (C) $^{214}_{83}Bi$
    (D) $^{210}_{83}Bi$
    (E) $^{214}_{82}Pb$

GO ON TO THE NEXT PAGE.

53. When 80.0 ml of a 0.40 M NaI solution is combined with 20.0 ml of a 0.30 M CaI_2 solution, what will be the molar concentration of I^- ions in the solution?

   (A) 0.70 M
   (B) 0.44 M
   (C) 0.38 M
   (D) 0.35 M
   (E) 0.10 M

54.

Compound	K_{sp} at 25°C
FeS	6.33×10^{-18}
PbS	8.03×10^{-28}
MnS	1.03×10^{-13}

   A solution at 25 C contains Fe^{2+}, Pb^{2+}, and Mn^{2+} ions. Which of the following gives the order in which precipitates will form, from first to last, as Na_2S is steadily added to the solution?

   (A) FeS, PbS, MnS
   (B) MnS, PbS, FeS
   (C) FeS, MnS, PbS
   (D) MnS, FeS, PbS
   (E) PbS, FeS, MnS

55. Hydrogen sulfide is a toxic waste product of some industrial processes. It can be recognized by its distinctive rotten egg smell. One method for elimininating hydrogen sulfide is by reaction with dissolved oxygen, as shown below.

$$2 H_2S + O_2 \rightarrow 2 S + 2H_2O$$

   If 102 grams of H_2S are combined with 64 grams of O_2, what is the maximum mass of elemental sulfur that could be produced by the reaction?

   (A) 16 grams
   (B) 32 grams
   (C) 48 grams
   (D) 64 grams
   (E) 96 grams

56.
$$Mn(s) + Cu^{2+}(aq) \rightarrow Mn^{2+}(aq) + Cu(s)$$

   A voltaic cell based on the reaction above was constructed from manganese and zinc half-cells. The standard cell potential for this reaction was 1.52 volts, but the observed voltage at 25°C was 1.66 volts. Which of the following could explain this observation?

   (A) The Mn^{2+} solution was more concentrated than the Cu^{2+} solution.
   (B) The Cu^{2+} solution was more concentrated than the Mn^{2+} solution.
   (C) The manganese electrode was larger than the copper electrode.
   (D) The copper electrode was larger than the manganese electrode.
   (E) The atomic weight of copper is greater than the atomic weight of manganese.

57. The rate of effusion of helium gas (atomic weight 4.0) at a given temperature and pressure is known to be x. What would be the expected rate of effusion for hydrogen gas (molecular weight 2.0) at the same temperature and pressure?

   (A) $\dfrac{x}{2}$

   (B) $\dfrac{x}{\sqrt{2}}$

   (C) x

   (D) $x\sqrt{2}$

   (E) $2x$

58.
$$H_2O(g) \rightarrow H_2O(l)$$

   Which of the following is true of the values of ΔH, ΔS, and ΔG for the reaction shown above at 25°C?

	ΔH	ΔS	ΔG
(A)	Positive	Positive	Positive
(B)	Positive	Negative	Negative
(C)	Negative	Positive	Negative
(D)	Negative	Negative	Positive
(E)	Negative	Negative	Negative

GO ON TO THE NEXT PAGE.

59. A student added 1 liter of a 1.0 M Na$_2$SO$_4$ solution to 1 liter of a 1.0 M Ag(C$_2$H$_3$O$_2$) solution. A silver sulfate precipitate formed, and nearly all of the silver ions disappeared from the solution. Which of the following lists the ions remaining in the solution in order of decreasing concentration?

    (A) $[SO_4^{2-}] > [C_2H_3O_2^-] > [Na^+]$
    (B) $[C_2H_3O_2^-] > [Na^+] > [SO_4^{2-}]$
    (C) $[C_2H_3O_2^-] > [SO_4^{2-}] > [Na^+]$
    (D) $[Na^+] > [SO_4^{2-}] > [C_2H_3O_2^-]$
    (E) $[Na^+] > [C_2H_3O_2^-] > [SO_4^{2-}]$

60. Iron ore is converted to pure iron by the following reaction, which occurs at extremely high temperatures:

    $$Fe_2O_3 + 3\,CO \rightarrow 2\,Fe + 3\,CO_2$$

    A 1,600 gram sample of iron ore reacted completely to form 558 grams of pure iron. What was the percent of Fe$_2$O$_3$ (molecular weight 160) by mass in the original sample?

    (A) 25%
    (B) 33%
    (C) 50%
    (D) 67%
    (E) 100%

61. Gold and silver have been used throughout history to make coins. Which of the following statements could account for the popularity of these metals for use in coins?

    (A) Network bonds make these metals especially durable.
    (B) Negative oxidation potentials for these metals make them especially unreactive with their surroundings.
    (C) These metals are among the lightest of the elements.
    (D) These metals are commonly found in nature.
    (E) These metals form ions that are extremely soluble in water.

62. A student added a salt to an aqueous solution of sodium sulfate. After the salt was added, a sulfate compound was observed as a precipitate. Which of the following could have been the salt added in this experiment?

    I. BaCl$_2$
    II. Pb(NO$_3$)$_2$
    III. AgC$_2$H$_3$O$_2$

    (A) I only
    (B) III only
    (C) I and II only
    (D) I and III only
    (E) I, II, and III

63. $$2\,HI(g) \rightarrow H_2(g) + I_2(g)$$

    For the reaction given above, $\Delta H°$ is –50 kJ. Based on the information given in the table below, what is the average bond energy of the H–I bond?

Bond	Average Bond Energy (kJ/mol)
H–H	440
I–I	150

    (A) 270 kJ/mol
    (B) 540 kJ/mol
    (C) 590 kJ/mol
    (D) 640 kJ/mol
    (E) 1,180 kJ/mol

64. A solid piece of barium hydroxide is immersed in water and allowed to come to equilibrium with its dissolved ions. The addition of which of the following substances to the solution would cause more solid barium hydroxide to dissolve into the solution?

    (A) NaOH
    (B) HCl
    (C) NaCl
    (D) BaCl$_2$
    (E) NH$_3$

65. All of the elements listed below are gases at room temperature. Which gas would be expected to show the greatest deviation from ideal behavior?

    (A) He
    (B) Ne
    (C) Ar
    (D) Kr
    (E) Xe

GO ON TO THE NEXT PAGE.

Questions 66 and 67 refer to the information given below.

$$F_2(g) + 2\,ClO_2(g) \rightarrow 2\,FClO_2(g)$$

Experiment	$[F_2]$ (M)	$[ClO_2]$ (M)	Initial rate of disappearance of F_2 (M/ sec)
1	0.10	0.010	1.2×10^{-3}
2	0.20	0.010	2.4×10^{-3}
3	0.40	0.020	9.6×10^{-3}

66. Based on the data given in the table, which of the following expressions is equal to the rate law for the reaction given above?

   (A) Rate = $k[F_2]$
   (B) Rate = $k[ClO_2]$
   (C) Rate = $k[F_2][ClO_2]$
   (D) Rate = $k[F_2]^2[ClO_2]$
   (E) Rate = $k[F_2][ClO_2]^2$

67. What is the initial rate of disappearance of ClO_2 in experiment 2?

   (A) 1.2×10^{-3} M/sec
   (B) 2.4×10^{-3} M/sec
   (C) 4.8×10^{-3} M/sec
   (D) 7.0×10^{-3} M/sec
   (E) 9.6×10^{-3} M/sec

68. Hypobromous acid, HBrO, is added to distilled water. If the acid dissociation constant for HBrO is equal to 2×10^{-9}, what is the concentration of HBrO when the pH of the solution is equal to 5?

   (A) 5-molar
   (B) 1-molar
   (C) 0.1-molar
   (D) 0.05-molar
   (E) 0.01-molar

69. $N_2(g) + 3\,Cl_2(g) \rightarrow 2\,NCl_2(g)$ $\qquad \Delta H = 460$ kJ

   Which of the following statements is true regarding the reaction shown above?

   (A) It is not spontaneous at any temperatures.
   (B) It is spontaneous only at very high temperatures.
   (C) It is spontaneous only at very low temperatures.
   (D) It is spontaneous only at very high concentrations.
   (E) It is spontaneous only at very low concentrations.

70. At 25°C, the vapor pressure of water is 24 mmHg. Which of the following expressions gives the vapor pressure of a solution created by adding 2.0 moles of sucrose to 55 moles of water?

   (A) $\dfrac{(24)(2.0)}{(55)}$ mmHg

   (B) $\dfrac{(2.0)}{(24)(55)}$ mmHg

   (C) $\dfrac{(24)(55)}{(57)}$ mmHg

   (D) $\dfrac{(55)}{(24)(57)}$ mmHg

   (E) $\dfrac{(24)(57)}{(55)}$ mmHg

71. $$PCl_5(g) \leftrightarrow PCl_3(g) + Cl_2(g)$$

   When PCl_5 is placed in an evacuated container at 250°C, the above reaction takes place. The pressure in the container before the reaction takes place is 6 atm and is entirely due to the PCl_5 gas. After the reaction comes to equilibrium, the partial pressure due to Cl_2 is found to be 2 atm. What is the value of the equilibrium constant, K_p, for this reaction?

   (A) 0.5
   (B) 1
   (C) 2
   (D) 5
   (E) 10

GO ON TO THE NEXT PAGE.

72. $Co^{2+} + 2\,e^- \rightarrow Co$ $E^o = -0.28$ V

    $Cr^{3+} + e^- \rightarrow Cr^{2+}$ $E^o = -0.41$ V

According to the information shown above, which of the following statements is true regarding the reaction below?

$$2\,Co + Cr^{3+} \rightarrow 2\,Co^{2+} + Cr^{2+}$$

(A) Cr^{3+} acts as the reducing agent.
(B) The reaction would take place in a galvanic cell.
(C) K_{eq} for the reaction is less than 1.
(D) Cobalt metal is reduced in the reaction.
(E) Chromium metal will plate out in the reaction.

73. What is the pH of a solution made by mixing 200 milliliters of a 0.20-molar solution of NH_3 with 200 milliliters of a 0.20-molar of an NH_4Cl solution? (The base dissociation constant, K_b, for NH_3 is 1.8×10^{-5}.)

(A) Between 3 and 4
(B) Between 4 and 5
(C) Between 5 and 6
(D) Between 8 and 9
(E) Between 9 and 10

74. $$POCl_3(l) \rightarrow POCl_3(g)$$

For the reaction above, ΔH is 50 kilojoules per mole and ΔS is 100 joules per mole. What is the boiling point of $POCl_3$? (Assume that ΔH and ΔS remain constant with changing temperature.)

(A) 0.5 K
(B) 2 K
(C) 50 K
(D) 200 K
(E) 500 K

75. When fluorine gas is bubbled through a concentrated aqueous solution of potassium chloride at room temperature, which of the following would be expected to occur?

(A) A reaction will occur, and chlorine gas will be produced.
(B) A reaction will occur, and potassium fluoride will precipitate.
(C) A reaction will occur, and solid chlorine will precipitate.
(D) A reaction will occur, and solid potassium will precipitate.
(E) No reaction will occur, and the fluorine gas will bubble through.

END OF SECTION I

CHEMISTRY
SECTION II
Time—1 hour and 35 minutes
Percent of total grade—50
Parts A: Time—55 minutes
Part B: Time—40 minutes

General Instructions

CALCULATORS MAY NOT BE USED IN PART B.

Calculators, including those with programming and graphing capabilities, may be used in Part A. However, calculators with typewriter-style (QWERTY) keyboards are NOT permitted.

Pages containing a periodic table, the electrochemical series, and equations commonly used in chemistry will be available for your use.

You may write your answers with either a pen or a pencil. Be sure to write CLEARLY and LEGIBLY. If you make an error, you may save time by crossing it out rather than trying to erase it.

Write all your answers in the essay booklet. Number your answers as the questions are numbered in the examination booklet.

GO ON TO THE NEXT PAGE.

MATERIAL IN THE FOLLOWING TABLE AND IN THE TABLES ON THE NEXT 3 PAGES MAY BE USEFUL IN ANSWERING THE QUESTIONS IN THIS SECTION OF THE EXAMINATION.

PERIODIC CHART OF THE ELEMENTS

1 **H** 1.0																	2 **He** 4.0
3 **Li** 6.9	4 **Be** 9.0											5 **B** 10.8	6 **C** 12.0	7 **N** 14.0	8 **O** 16.0	9 **F** 19.0	10 **Ne** 20.2
11 **Na** 23.0	12 **Mg** 24.3											13 **Al** 27.0	14 **Si** 28.1	15 **P** 31.0	16 **S** 32.1	17 **Cl** 35.5	18 **Ar** 39.9
19 **K** 39.1	20 **Ca** 40.1	21 **Sc** 45.0	22 **Ti** 47.9	23 **V** 50.9	24 **Cr** 52.0	25 **Mn** 54.9	26 **Fe** 55.8	27 **Co** 58.9	28 **Ni** 58.7	29 **Cu** 63.5	30 **Zn** 65.4	31 **Ga** 69.7	32 **Ge** 72.6	33 **As** 74.9	34 **Se** 79.0	35 **Br** 79.9	36 **Kr** 83.8
37 **Rb** 85.5	38 **Sr** 87.6	39 **Y** 88.9	40 **Zr** 91.2	41 **Nb** 92.9	42 **Mo** 95.9	43 **Tc** (98)	44 **Ru** 101.1	45 **Rh** 102.9	46 **Pd** 106.4	47 **Ag** 107.9	48 **Cd** 112.4	49 **In** 114.8	50 **Sn** 118.7	51 **Sb** 121.8	52 **Te** 127.6	53 **I** 126.9	54 **Xe** 131.3
55 **Cs** 132.9	56 **Ba** 137.3	57 ***La** 138.9	72 **Hf** 178.5	73 **Ta** 180.9	74 **W** 183.9	75 **Re** 186.2	76 **Os** 190.2	77 **Ir** 192.2	78 **Pt** 195.1	79 **Au** 197.0	80 **Hg** 200.6	81 **Tl** 204.4	82 **Pb** 207.2	83 **Bi** 209.0	84 **Po** (209)	85 **At** (210)	86 **Rn** (222)
87 **Fr** (223)	88 **Ra** 226.0	89 ***Ac** 227.0															

*Lanthanum Series

58 **Ce** 140.1	59 **Pr** 140.9	60 **Nd** 144.2	61 **Pm** (145)	62 **Sm** 150.4	63 **Eu** 152.0	64 **Gd** 157.3	65 **Tb** 158.9	66 **Dy** 162.5	67 **Ho** 164.9	68 **Er** 167.3	69 **Tm** 168.9	70 **Yb** 173.0	71 **Lu** 175.0

†Actinium Series

90 **Th** 232.0	91 **Pa** 231.0	92 **U** 238.0	93 **Np** 237.0	94 **Pu** (244)	95 **Am** (243)	96 **Cm** (247)	97 **Bk** (247)	98 **Cf** (251)	99 **Es** (252)	100 **Fm** (258)	101 **Md** (258)	102 **No** (259)	103 **Lr** (260)

DO NOT DETACH FROM BOOK.

GO ON TO THE NEXT PAGE.

ADVANCED PLACEMENT CHEMISTRY EQUATIONS AND CONSTANTS

ATOMIC STRUCTURE

$$E = h\nu \qquad\qquad c = \lambda\nu$$

$$\lambda = \frac{h}{mv} \qquad\qquad p = mv$$

$$E_n = \frac{-2.178 \times 10^{-18}}{n^2} \text{ joule}$$

EQUILIBRIUM

$$K_a = \frac{[H^-][A^-]}{[HA]}$$

$$K_b = \frac{[OH^-][HB^+]}{[B]}$$

$$K_w = [OH^-][H^+] = 10^{-14} @ 25°C$$
$$\qquad = K_a \times K_b$$

$$pH = -\log[H^+], \ pOH = -\log[OH^-]$$

$$14 = pH + pOH$$

$$pH = pK_a + \log\frac{[A^-]}{[HA]}$$

$$pOH = pK_b + \log\frac{[HB^+]}{[B]}$$

$$pK_a = -\log K_a, \ pK_b = -\log K_b$$
$$K_p = K_c(RT)^{\Delta n},$$
where Δn = moles product gas – moles reactant gas

THERMOCHEMISTRY / KINETICS

$$\Delta S° = \Sigma S°_{\text{products}} - \Sigma S°_{\text{reactants}}$$

$$\Delta H° = \Sigma\Delta H°_{f\ \text{products}} - \Sigma\Delta H°_{f\ \text{reactants}}$$

$$\Delta G° = \Sigma\Delta G°_{f\ \text{products}} - \Sigma\Delta G°_{f\ \text{reactants}}$$

$$\Delta G° = \Delta H° - T\Delta S°$$
$$\qquad = -RT\ln K = -2.303\ RT\log K$$
$$\qquad = -nFE°$$

$$\Delta G = \Delta G° + RT\ln Q = \Delta G° + 2.303\ RT\log Q$$
$$q = mc\Delta T$$
$$C_p = \frac{\Delta H}{\Delta T}$$

$$\ln[A]_t - \ln[A]_o = -kt$$

$$\frac{1}{[A]_t} - \frac{1}{[A]_o} = kt$$

$$\ln k = \frac{-E_a}{R}\left(\frac{1}{T}\right) + \ln A$$

E = energy $\qquad\qquad v$ = velocity

ν = frequency $\qquad\quad n$ = principal quantum number

λ = wavelength $\qquad m$ = mass

p = momentum

Speed of light, $c = 3.00 \times 10^8$ m s^{-1}

Planck's constant, $h = 6.63 \times 10^{-34}$ J s

Boltzmann's constant, $k = 1.38 \times 10^{-23}$ J K^{-1}

Avogadro's number = 6.022×10^{23} mol^{-1}

Electron charge, $e = -1.602 \times 10^{-19}$ coulomb

1 electron volt/atom = 96.5 kJ mol^{-1}

Equilibrium Constants

K_a (weak acid)
K_b (weak base)
K_w (water)
K_p (gas pressure)
K_c (molar concentrations)

$S°$ = standard entropy
$H°$ = standard enthalpy
$G°$ = standard free energy
$E°$ = standard reduction potential
T = temperature
n = moles
m = mass
q = heat
c = specific heat capacity
C_p = molar heat capacity at constant pressure
E_a = activation energy
k = rate constant
A = frequency factor

Faraday's Constant, F = 96,500 coulombs per mole
of electrons

Gas Constant, R = 8.31 J mol^{-1} K^{-1}
$\qquad\qquad\qquad$ = 0.0821 L atm mol^{-1} K^{-1}
$\qquad\qquad\qquad$ = 8.31 volt coulomb mol^{-1} K^{-1}

GO ON TO THE NEXT PAGE.

ADVANCED PLACEMENT CHEMISTRY EQUATIONS AND CONSTANTS

GASES, LIQUIDS, AND SOLUTIONS

$$PV = nRT$$

$$\left(P + \frac{n^2 a}{V^2}\right)(V - nb) = nRT$$

$$P_A = P_{total} \cdot X_A, \text{ where } X_A = \frac{\text{moles A}}{\text{total moles}}$$

$$P_{total} = P_A + P_B + P_{C^+}...$$

$$n = \frac{m}{M}$$

$$K = {}^\circ C + 273$$

$$\frac{P_1 V_1}{T_1} = \frac{P_2 V_2}{T_2}$$

$$D = \frac{m}{V}$$

$$U_{rms} = \sqrt{\frac{3kT}{m}} = \sqrt{\frac{3RT}{M}}$$

$$KE \text{ per molecule} = \frac{1}{2} mv^2$$

$$KE \text{ per mole} = \frac{3}{2} RT$$

$$\frac{r_1}{r_2} = \sqrt{\frac{M_2}{M_1}}$$

molarity, M = moles solute per liter solution

molality = moles solute per kilogram solvent

$$\Delta T_f = iK_f \times \text{molality}$$

$$\Delta T_b = iK_b \times \text{molality}$$

$$\pi = MRT$$

$$A = abc$$

OXIDATION REDUCTION; ELECTROCHEMISTRY

$$Q = \frac{[C]^c [D]^d}{[A]^a [B]^b} \text{ where } a\,A + b\,B \rightarrow c\,C + d\,D$$

$$I = \frac{q}{t}$$

$$E_{cell} = E^\circ_{cell} - \frac{RT}{nF} \ln Q = E^\circ_{cell} - \frac{0.0592}{n} \log Q \text{ @ } 25^\circ C$$

$$\log K = \frac{nE^\circ}{0.0592}$$

P = pressure
V = volume
T = temperature
n = number of moles
D = density
m = mass
v = velocity

U_{rms} = root-mean-square speed
KE = kinetic energy
r = rate of effusion
M = molar mass
π = osmotic pressure
i = van't Hoff factor
K_f = molal freezing-point depression constant
K_b = molal boiling-point elevation constant
A = absorbance
a = molar absorptivity
b = path length
c = concentration
Q = reaction quotient
l = current (amperes)
q = charge (coulombs)
t = time (seconds)
E° = standard reduction potential
K = equilibrium constant

Gas constant, R = 8.31 J mol^{-1}K^{-1}
  = 0.0821 L atm mol^{-1} K^{-1}
  = 8.31 volt coulomb mol^{-1}K^{-1}
Boltzmann's constant, k = 1.38×10^{-23} J K^{-1}
K_f for H_2O = 1.86 K kg mol^{-1}
K_b for H_2O = 0.512 K kg mol^{-1}
1 atm = 760 mm Hg
  = 760 torr
STP = 0.000°C and 1.000 atm
1 faraday, F = 96,500 coulumbs per mole
  of electrons

GO ON TO THE NEXT PAGE.

STANDARD REDUCTION POTENTIALS IN AQUEOUS SOLUTION AT 25 C (in V)		
$F_2(g) + 2\ e^-$	$\rightarrow$	$2\ F^-$	2.87
$Co^{3+} + e^-$	$\rightarrow$	Co^{2+}	1.82
$Au^{3+} + 3e^-$	$\rightarrow$	$Au(s)$	1.50
$Cl_2(g) + 2\ e^-$	$\rightarrow$	$2\ Cl^-$	1.36
$O_2(g) + 4H^+ + 4\ e^-$	$\rightarrow$	$2\ H_2O$	1.23
$Br_2(l) + 2\ e^-$	$\rightarrow$	$2\ Br^-$	1.07
$2\ Hg^{2+} + 2\ e^-$	$\rightarrow$	Hg_2^{2+}	0.92
$Hg^{2+} + 2\ e^-$	$\rightarrow$	$Hg(l)$	0.85
$Ag^+ + e^-$	$\rightarrow$	$Ag(s)$	0.80
$Hg_2^{2+} + 2\ e^-$	$\rightarrow$	$2\ Hg(l)$	0.79
$Fe^{3+} + e^-$	$\rightarrow$	Fe^{2+}	0.77
$I_2(s) + 2\ e^-$	$\rightarrow$	$2\ I^-$	0.53
$Cu^+ + e^-$	$\rightarrow$	$Cu(s)$	0.52
$Cu^{2+} + 2\ e^-$	$\rightarrow$	$Cu(s)$	0.34
$Cu^{2+} + e^-$	$\rightarrow$	Cu^+	0.15
$Sn^{4+} + 2\ e^-$	$\rightarrow$	Sn^{2+}	0.15
$S(s) + 2\ H^+ + 2\ e^-$	$\rightarrow$	H_2S	0.14
$2\ H^+ + 2\ e^-$	$\rightarrow$	$H_2(g)$	0.00
$Pb^{2+} + 2\ e^-$	$\rightarrow$	$Pb(s)$	−0.13
$Sn^{2+} + 2\ e^-$	$\rightarrow$	$Sn(s)$	−0.14
$Ni^{2+} + 2\ e^-$	$\rightarrow$	$Ni(s)$	−0.25
$Co^{2+} + 2\ e^-$	$\rightarrow$	$Co(s)$	−0.28
$Tl^+ + e^-$	$\rightarrow$	$Tl(s)$	−0.34
$Cd^{2+} + 2\ e^-$	$\rightarrow$	$Cd(s)$	−0.40
$Cr^{3+} + e^-$	$\rightarrow$	Cr^{2+}	−0.41
$Fe^2 + 2\ e^-$	$\rightarrow$	$Fe(s)$	−0.44
$Cr^{3+} + 3\ e^-$	$\rightarrow$	$Cr(s)$	−0.74
$Zn^{2+} + 2\ e^-$	$\rightarrow$	$Zn(s)$	−0.76
$Mn^{2+} + 2\ e^-$	$\rightarrow$	$Mn(s)$	−1.18
$Al^{3+} + 3\ e^-$	$\rightarrow$	$Al(s)$	−1.66
$Be^{2+} + 2\ e^-$	$\rightarrow$	$Be(s)$	−1.70
$Mg^{2+} + 2\ e^-$	$\rightarrow$	$Mg(s)$	−2.37
$Na^+ + e^-$	$\rightarrow$	$Na(s)$	−2.71
$Ca^{2+} + 2\ e^-$	$\rightarrow$	$Ca(s)$	−2.87
$Sr^{2+} + 2\ e^-$	$\rightarrow$	$Sr(s)$	−2.89
$Ba^{2+} + 2\ e^-$	$\rightarrow$	$Ba(s)$	−2.90
$Rb^+ + e^-$	$\rightarrow$	$Rb(s)$	−2.92
$K^+ + e^-$	$\rightarrow$	$K(s)$	−2.92
$Cs^+ + e^-$	$\rightarrow$	$Cs(s)$	−2.92
$Li^+ + e^-$	$\rightarrow$	$Li(s)$	−3.05

GO ON TO THE NEXT PAGE.

CHEMISTRY
Section II
(Total time—95 minutes)

Part A
Time—55 minutes
YOU MAY USE YOUR CALCULATOR FOR PART A.

THE METHODS USED AND THE STEPS INVOLVED IN ARRIVING AT YOUR ANSWERS MUST BE SHOWN CLEARLY. It is to your advantage to do this since you may obtain partial credit if you do, and you will receive little or no credit if you do not. Attention should be paid to significant figures.

Be sure to write your answers in the space provided following each question.

Answer Questions 1, 2, and 3. The Section II score for question 1 is 9 points, question 2 is 10 points, and question 3 is 9 points.

1. At 25°C, the solubility product consant, K_{sp}, for nickel hydroxide, $Ni(OH)_2$, is 1.6×10^{-14}.

   (a) Write a balanced equation for the solubility equilibrium for $Ni(OH)_2$.

   (b) What is the molar solubility of $Ni(OH)_2$ in pure water at 25°C?

   (c) A 1.0-molar NaOH solution is slowly added to a saturated solution of $Ni(OH)_2$ at 25°C. If excess solid $Ni(OH)_2$ remains in the solution throughout the procedure, what is the concentration of Ni^{2+} ions in the solution at the moment when the pH is equal to 11?

   (d) Predict whether a precipitate will form when 200.0 milliliters of a 5.0×10^{-5}-molar KOH solution is mixed with 300.0 milliliters of a 2.0×10^{-4}-molar $Ni(NO_3)_2$ solution at 25°C. Show calculations to support your prediction.

   (e) At 25°C, 100 ml of a saturated $Ni(OH)_2$ solution was prepared.

   (i) Calculate the mass of $Ni(OH)_2$ present in the solution.

   (ii) Calculate the pH of the solution.

   (iii) If the solution is allowed to evaporate to a final volume of 50 ml, what will be the pH? Justify your answer.

GO ON TO THE NEXT PAGE.

2. In two separate experiments, a sample of an unknown hydrocarbon was burned in air, and a sample of the same hydrocarbon was placed into an organic solvent.

   (a) When the hydrocarbon sample was burned in a reaction that went to completion, 2.2 grams of water and 3.6 liters of carbon dioxide were produced under standard conditions. What is the empirical formula of the hydrocarbon?

   (b) When 4.05 grams of the unknown hydrocarbon was placed in 100.0 grams of benzene, C_6H_6, the freezing point of the solution was measured to be 1.66°C. The normal freezing point of benzene is 5.50°C and the freezing-point depression constant for benzene is 5.12°C/m. What is the molecular weight of the unknown hydrocarbon?

   (c) What is the molecular formula and name of the hydrocarbon?

   (d) Write the balanced equation for the combustion reaction that took place in (a).

   (e) Draw two isomers for the hydrocarbon.

3. A strip of Ni metal is placed in a 1-molar solution of $Ni(NO_3)_2$ and a strip of Ag metal is placed in a 1-molar solution of $AgNO_3$. An electrochemical cell is created when the two solutions are connected by a salt bridge and the two metal strips are connected by wires to a voltmeter.

   (a) Write the balanced chemical equation for the overall reaction that occurs in the cell, and calculate the cell potential, $E°$.

   (b) Calculate how many grams of metal will be deposited on the cathode if the cell is allowed to run at a constant current of 1.5 amperes for 8.00 minutes.

   (c) Calculate the value of the standard free energy change, $\Delta G°$, for the cell reaction.

   (d) Calculate the cell potential, E, at 25°C for the cell shown above if the initial concentration of $Ni(NO_3)_2$ is 0.100-molar and the initial concentration of $AgNO_3$ is 1.20-molar.

   (e) Is the reaction in the cell spontaneous under conditions described in part (d)? Justify your answer.

GO ON TO THE NEXT PAGE.

Part B
Time—40 minutes
NO CALCULATORS MAY BE USED FOR PART B

Answer Question 4 below. The Section II score for this question is 15 points.

4. You will be given three chemical reactions below. In part (i), write the balanced equation for the reaction, leaving coefficients in terms of lowest whole numbers. Then answer the question pertaining to that reaction in part (ii). For each of the following three reactions, assume that solutions are aqueous unless it says otherwise. Substances in solutions should be represented as ions if these substances are extensively ionized. Omit formulas for ions or molecules that are not affected by the reaction. Only equations inside the answer boxes will be graded.

EXAMPLE:

A piece of solid zinc is placed in a solution of silver(I) acetate

(i) Write the balanced equation in this box:

$$Zn + 2Ag^+ \rightarrow Zn^{2+} + 2Ag$$

(ii) Which substance is reduced in the reaction?

Ag^+ is reduced

(a) Solutions of sodium acetate and nitric acid are mixed.

(i) Write the balanced equation in this box:

(ii) Identify the spectator ion in this reaction.

(b) A piece of solid calcium is heated in oxygen gas.

(i) Write the balanced equation in this box:

(ii) Is the entropy change for this reaction positive or negative? Briefly explain.

GO ON TO THE NEXT PAGE.

(c) Ammonia and boron trichloride gases are mixed.

(i) Write the balanced equation in this box:

(ii) Identify the Lewis acid in this reaction.

_____

_____

GO ON TO THE NEXT PAGE.

Answer Question 5 and Question 6. The Section II score for question 5 is 9 points and question 6 is 8 points.

Answering these questions provides an opportunity to demonstrate your ability to present your material in logical, coherent, and convincing English. Your responses will be judged on the basis of accuracy and importance of the detail cited and on the appropriateness of the descriptive material used. Specific answers are preferable to broad, diffuse responses. Illustrative examples and equations may be helpful.

5. Use your knowledge of chemical principles to answer the following questions.

	First ionization energy (kJ/mol)	Second ionization energy (kJ/mol)
Na	490	4,560
K	420	3,050

(a) The table above shows the first and second ionization energies for potassium and sodium.

(i) Explain the difference between the first ionization energies of potassium and sodium.

(ii) Explain why the second ionization energy for potassium is so much larger than the first ionization energy.

	Atomic radius (pm)
Na	190
Na$^+$	120
K	230
K$^+$	150
Cl	100
Cl$^-$	170
Br	110
Br$^-$	190

(b) The table above shows the atomic radii of several atoms and ions.

(i) The atomic radius of bromine is smaller than the atomic radius of potassium. Explain.

(ii) The potassium and chlorine ions have the same electronic structure. Explain the difference in their atomic radii.

(iii) When potassium bromide and sodium chloride are compared, which will have the higher melting point? Explain.

GO ON TO THE NEXT PAGE.

Name	Formula	Boiling point (°C)
Methane	CH_4	−164
Methoxymethane	CH_3OCH_3	−23
Ethanol	CH_3CH_2OH	78

(c) The table above shows the boiling points for several organic compounds. Explain the trend in increasing boiling point.

GO ON TO THE NEXT PAGE.

6. An experiment is to be performed involving the titration of a solution of benzoic acid, $HC_7H_5O_2$, of unknown concentration with a standardized 1.00-molar NaOH solution. The pK_a of benzoic acid is 4.2.

   (a) What measurements and calculations must be made if the concentration of the acid is to be determined?

   (b) An indicator must be chosen for the titration. If methyl red (pH interval for color change: 4.2–6.3) and thymol blue (pH interval for color change: 8.0–9.6) are the only indicators available, which one should be used? Explain your choice.

   (c) Describe how you would use the titration apparatus and benzoic acid and NaOH solutions to create a buffer solution with a pH of 4.2 after you had measured the volume of NaOH solution required to reach the equivalence point for a given volume of benzoic acid solution.

   (d) Describe a way to test whether the neutralization of benzoic acid by a strong base is an exothermic process, using readily available school laboratory equipment.

STOP

END OF EXAM

21

Answers and Explanations for Practice Exam 2

ANSWER KEY EXAM 2

1.	D	39.	B
2.	A	40.	A
3.	D	41.	C
4.	C	42.	E
5.	D	43.	D
6.	B	44.	D
7.	C	45.	A
8.	D	46.	E
9.	E	47.	C
10.	C	48.	B
11.	E	49.	E
12.	C	50.	D
13.	D	51.	D
14.	A	52.	A
15.	C	53.	B
16.	D	54.	E
17.	A	55.	E
18.	E	56.	B
19.	B	57.	D
20.	C	58.	E
21.	E	59.	E
22.	A	60.	C
23.	A	61.	B
24.	C	62.	E
25.	B	63.	A
26.	D	64.	B
27.	D	65.	E
28.	E	66.	C
29.	E	67.	C
30.	D	68.	D
31.	A	69.	A
32.	B	70.	C
33.	C	71.	B
34.	B	72.	C
35.	C	73.	E
36.	B	74.	E
37.	B	75.	A
38.	A		

HOW TO SCORE PRACTICE TEST 2

SECTION I: MULTIPLE-CHOICE

_____ × 1.0000 = _____
Number of Correct   Weighted
(out of 75)   Section I Score
                                 (Do not round)

SECTION II: FREE RESPONSE

(See if you can find a teacher or classmate to score your essays using the guidelines in Chapter 3.)

Question 1 _____ × 1.6666 = _____
            (out of 9)   (Do not round)

Question 2 _____ × 1.5000 = _____
            (out of 10)   (Do not round)

Question 3 _____ × 1.6666 = _____
            (out of 9)   (Do not round)

Question 4 _____ × .5000 = _____
            (out of 15)   (Do not round)

Question 5 _____ × 1.2500 = _____
            (out of 9)   (Do not round)

Question 6 _____ × 1.4062 = _____
            (out of 8)   (Do not round)

AP Score Conversion Chart Chemistry

Composite Score Range	AP Score
100–150	5
81–99	4
62–80	3
49–61	2
0–48	1

Sum = _____
       Weighted Section II
       Score (Do not round)

COMPOSITE SCORE

_____ + _____ = _____
Weighted   Weighted   Composite Score
Section I Score   Section II Score   (Round to nearest
                                        whole number)

QUESTIONS	EXPLANATIONS

Section I—Multiple Choice

Questions 1–3 refer to the following elements.

    (A) Ne
    (B) Li
    (C) Al
    (D) Cl
    (E) Ca

1. Which of the elements above is most commonly found as a negatively charged ion?

1. **D** Chlorine (Cl) forms a negatively charged ion with an oxidation state of –1. None of the other elements listed normally forms negative ions.

2. Which of the elements above has the same electron structure as Mg^{2+}?

2. **A** Mg^{2+} has the following electron structure: $1s^2\ 2s^2 2p^6$. That's the electron structure that neutral neon (Ne) has.

3. Which of the elements above is found as a diatomic gas in its uncombined state?

3. **D** Chlorine is found as a diatomic gas, Cl_2, in its uncombined state.

Questions 4–7 refer to the following substances.

    (A) H_2O
    (B) NH_3
    (C) CH_4
    (D) HF
    (E) CH_3OH

4. This molecule is nonpolar.

4. **C** CH_4 has a symmetrical tetrahedral shape and is nonpolar.

All of the other choices are polar, as shown on the following page.

(A)

(−)

O

H   H

(+)

(B)

(−)

N

H   H

(+)

(E)

(−)

N

H   H

(+)

(D)   (+) H–F (−)

5. This substance forms an aqueous solution with a pH that's less than 7.

5. **D** HF is the only acid on the list.

6. This molecule has a trigonal pyramidal shape.

6. **B** NH₃ is the only choice with a trigonal pyramidal shape, and it has one lone electron pair. For the shapes of the other molecules, see the answer to Question 4.

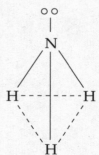

QUESTIONS	EXPLANATIONS

7. This substance does NOT exhibit hydrogen bonding.

7. **C** Remember FON. Only hydrogen compounds that contain fluorine, oxygen, or nitrogen exhibit hydrogen bonding.

Questions 8–11 refer to the phase diagram shown below for a substance.

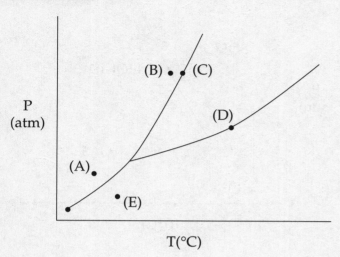

8. This point could be the boiling point for the substance.

8. **D** At the boiling point, the liquid and gas phases are in equilibrium. Only point (D) is on the line between gas and liquid phase (condensation and vaporization), as shown in the diagram below.

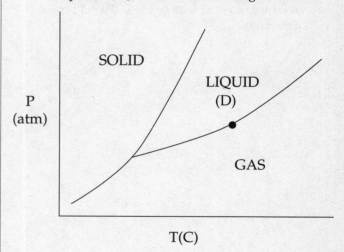

QUESTIONS	EXPLANATIONS

9. If pressure is increased from the conditions represented by this point at constant temperature, deposition may occur.

9. **E** Deposition is the phase change from gas to solid, as shown in the diagram below. As you can see from the arrow below, when temperature is kept constant and pressure is increased, the substance moves from gaseous to solid phase.

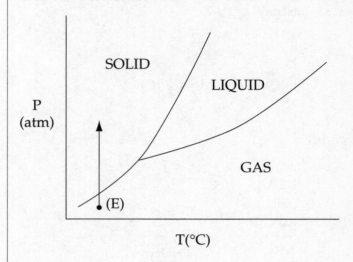

10. At the conditions represented by this point, the substance can exist with liquid and solid phases in equilibrium.

10. **C** Point (C) is on the line between liquid and solid (freezing or melting) as shown in the diagram below.

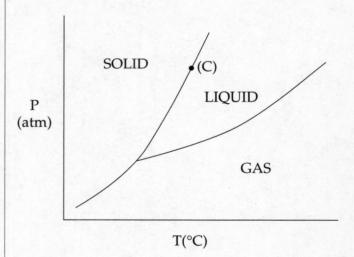

QUESTIONS	EXPLANATIONS

11. If temperature is increased from the conditions represented by this point at constant pressure, no phase change will occur.

11. **E** If the temperature is increased from point (E) at constant pressure, no phase change lines will be crossed, as is shown in the diagram below.

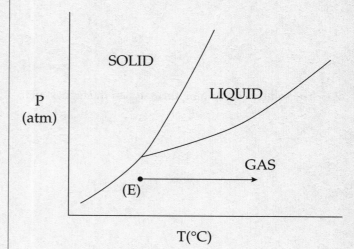

Questions 12–14 refer to an experiment in which five individual 1-liter aqueous solutions, each containing a 1-mole sample of one of the salts listed below, were subjected to various tests at room temperature.

(A) $NaC_2H_3O_2$
(B) $NaCl$
(C) $MgBr_2$
(D) $HC_2H_3O_2$
(E) KBr

12. The solution containing this salt had the highest boiling point.

12. **C** Boiling point elevation is a colligative property; it depends only on the number of particles in solution. Each unit of $MgBr_2$ produces three units of particles in solution, one Mg^{2+} and two Br^-s. Each of the other choices listed produces only two particles in solution. The greater the number of particles a molecule dissociates into in solution, the larger the boiling point elevation.

QUESTIONS	EXPLANATIONS
13. The solution containing this salt had the lowest conductivity.	13. **D** The conductivity of a solution depends on the number of charged particles it contains. $HC_2H_3O_2$ is a weak acid, so very little of the one-molar sample will dissociate in solution, producing very few charged particles. All of the other choices are soluble salts, which dissociate completely.
14. The solution containing this salt had the highest pH.	14. **A** $NaC_2H_3O_2$ is composed of the conjugates of a strong base (NaOH) and a weak acid ($HC_2H_3O_2$). It will produce a basic solution that will have a pH greater than 7. Choices (B), (C), and (E) will produce neutral solutions with pHs of 7, and choice (D) will produce an acidic solution, with a pH of less than 7.

15. Titanium metal is prepared by heating rutile, an oxide of titanium, along with carbon and chlorine gas. By mass, rutile is 40% oxygen and 60% titanium. What is the empirical formula of rutile?

(A) TiO
(B) Ti_2O
(C) TiO_2
(D) Ti_2O_3
(E) Ti_3O_2

15. **C** Let's say you have 100 grams of the rutile. So you have 40 grams of oxygen and 60 grams of titanium. Now convert to moles.

$$\text{Moles} = \frac{\text{grams}}{\text{MW}}$$

$$\text{Moles of oxygen} = \frac{(40\text{ g})}{(16\text{ g/mol})} = 2.5$$

$$\text{Moles of titanium} = \frac{(60\text{ g})}{(48\text{ g/mol})} = 1.25$$

The molar ratio of O to Ti is 2 to 1, so the

empirical formula must be TiO_2.

16. What is the weight of $NaNO_3$ (molecular weight 85.0) present in 100.0 ml of a 4.00-molar solution?

(A) 8.50 grams
(B) 17.0 grams
(C) 25.5 grams
(D) 34.0 grams
(E) 51.0 grams

16. **D** Moles = (molarity)(volume)

Moles of $NaNO_3$ = (4.00 M)(0.100 L)
                              = 0.400 moles

Grams = (moles)(MW)

Grams of $NaNO_3$ = (0.400 moles)(85.0 g/mol)
                               = 34.0 grams

17. A mixture of gases contains 5.0 moles of oxygen, 12 moles of nitrogen, and 4.0 moles of carbon dioxide. If the partial pressure due to carbon dioxide is 1.6 atmospheres, what is the partial pressure due to oxygen?

   (A) 2.0 atm
   (B) 4.8 atm
   (C) 6.0 atm
   (D) 6.8 atm
   (E) 8.4 atm

17. **A** From Dalton's law, the partial pressure of a gas in a sample is directly proportional to its molar quantity. So if 4 moles of carbon dioxide have a partial pressure of 1.6 atm, then 5 moles of oxygen will have a partial pressure of 2.0 atm.

18. Which of the diagrams below represents the enthalpy change for an endothermic reaction?

18. **E** The change in enthalpy, ΔH, is the difference in potential energy between the final and initial states of a reaction.

(A)
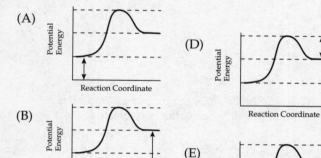

(D)
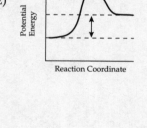

(B)
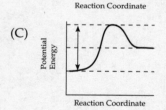

(E)

(C)

QUESTIONS	EXPLANATIONS
19. $2 NO(g) + Br_2(g) \rightarrow 2 NOBr(g)$	19. **B** The exponent for Br_2 in the rate law is equal to 1, so the reaction is first-order with respect to Br_2. By the way, the reaction is second-order with respect to NO and third-order overall.

19. $2 NO(g) + Br_2(g) \rightarrow 2 NOBr(g)$

For the reaction above, the experimental rate law is given as follows:

$$Rate = k[NO]^2[Br_2]$$

Which of the statements below is true regarding this reaction?

(A) The reaction is first-order overall.
(B) The reaction is first-order with respect to Br_2.
(C) The reaction is first-order with respect to NO.
(D) The reaction is second-order overall.
(E) The reaction is second-order with respect to Br_2.

19. **B** The exponent for Br_2 in the rate law is equal to 1, so the reaction is first-order with respect to Br_2. By the way, the reaction is second-order with respect to NO and third-order overall.

20. Which of the following is the most likely electron configuration of a sulfur atom in its ground state?

(A) $1s^2\, 2s^2 2p^6$
(B) $1s^2\, 2s^2 2p^6 3s^2 3p^2$
(C) $1s^2\, 2s^2 2p^6 3s^2 3p^4$
(D) $1s^2\, 2s^2 2p^6 3s^2 3p^6$
(E) $1s^2\, 2s^2 2p^6 3s^2 3p^5 4s^1$

20. **C** $1s^2\, 2s^2 2p^6\, 3s^2 3p^4$ is the proper ground-state configuration for sulfur's 16 electrons.

21. How many grams of carbon are present in 270 grams of glucose, $C_6H_{12}O_6$?

(A) 12.0 grams
(B) 18.0 grams
(C) 67.5 grams
(D) 72.0 grams
(E) 108 grams

21. **E** Moles $= \dfrac{grams}{MW}$

Moles of glucose $= \dfrac{(270\ g)}{(180\ g/mol)} = 1.5$ moles

1.50 moles of $C_6H_{12}O_6$ contains 9.00 moles of carbon.

Grams = (moles)(MW)

Grams of carbon = (9.00 moles)(12.0 grams/mol)
$= 108$ grams

You could also set up a ratio; if 180 g of glucose contains 72 g of carbon, then 270 g glucose contains 108 g of carbon.

QUESTIONS	EXPLANATIONS

22. Calcium carbonate dissolves in acidic solutions as shown in the equation below.

$$CaCO_3(s) + 2 H^+(aq) \rightarrow Ca^{2+}(aq) + H_2O(l) + CO_2(g)$$

If excess $CaCO_3$ is added to 0.250 liters of a 2.00-molar HNO_3 solution, what is the maximum volume of CO_2 gas that could be produced at standard temperature and pressure?

(A) 5.60 liters
(B) 11.2 liters
(C) 16.8 liters
(D) 33.6 liters
(E) 39.2 liters

22. A H^+ is the limiting reagent, so we need to find out how many moles of H^+ are consumed.

Moles = (molarity)(volume)

Moles of H^+ = (2.00 M)(0.250 L) = 0.500 moles

We can see from the balanced equation that for every 2 moles of H^+ consumed, 1 mole of CO_2 is produced, so if 0.500 moles of H^+ is consumed, 0.250 moles of CO_2 is produced.

Volume of gas at STP = (Moles of gas)(22.4 liters/mol)

Volume of CO_2 = (0.250 moles)(22.4 L/mol)
= 5.60 liters

23. A monoprotic acid was titrated with a solution of NaOH. For 55.0 milliliters of the acid, 37.0 milliliters of a 0.450-molar solution of NaOH was required to reach the equivalence point. Which of the following expressions is equal to the initial concentration of the monoprotic acid?

(A) $\dfrac{(0.450)(0.037)}{(0.055)} M$

(B) $\dfrac{(0.450)(0.055)}{(0.037)} M$

(C) $\dfrac{(0.055)}{(0.450)(0.037)} M$

(D) $\dfrac{(0.037)}{(0.450)(0.055)} M$

(E) $(0.450)(0.055)(0.037) M$

23. A First let's find the number of moles of NaOH added.

Moles = (molarity)(volume)

Moles of NaOH = (0.450)(0.037) moles

At the equivalence point, the moles of NaOH added are equal to the moles of acid originally present. Now we can find the original molar concentration of the acid.

Molarity = $\dfrac{\text{moles}}{\text{liters}}$

Molarity of the acid = $\dfrac{(0.450)(0.037)}{(0.055)} M$

24. $Zn(s) + NO_3^-(aq) + 10\ H^+(aq) \rightarrow$
$$Zn^{2+}(aq) + NH_4^+(aq) + 3\ H_2O(l)$$

Which of the following statements regarding the reaction shown above is correct?

(A) The oxidation number of hydrogen changes from +1 to 0.

(B) The oxidation number of hydrogen changes from +1 to −1.

(C) The oxidation number of nitrogen changes from +5 to −3.

(D) The oxidation number of nitrogen changes from +5 to +3.

(E) The oxidation number of nitrogen changes from +6 to +4.

24. **C** Remember that the oxidation state of oxygen is almost always −2. This means that the oxidation number of nitrogen in NO_3^- is +5. [Nitrogen + (−2)(3) = −1]

The oxidation state for hydrogen is almost always +1, so the oxidation state of nitrogen in NH_4^+ is −3. [Nitrogen + (1)(4) = 1] So the oxidation state of nitrogen changes from +5 to −3.

25. $HC_2H_3O_2(aq) + ClO^-(aq) \leftrightarrow HClO(aq) + C_2H_3O_2^-(aq)$

The standard free energy change for this reaction has a negative value. Based on this information, which of the following statements is true?

(A) K_a for $HC_2H_3O_2(aq)$ is less than K_a for $HClO(aq)$

(B) K_b for $C_2H_3O_2^-(aq)$ is less than K_b for $ClO^-(aq)$

(C) K_{eq} for the reaction is less than 1.

(D) The reaction occurs in the presence of a catalyst.

(E) $HC_2H_3O_2(aq)$ and $HClO(aq)$ are conjugates.

25. **B** If the standard free energy change is negative, then the forward reaction is favored under standard conditions, which means that K_{eq} will be greater than 1. For the forward reaction to be favored, $HC_2H_3O_2(aq)$ must be a stronger acid than $HClO(aq)$. If $HC_2H_3O_2(aq)$ is the stronger acid, then its conjugate, $C_2H_3O_2^-(aq)$, must be the weaker base. Thus, the base dissociation constant, K_b, is smaller for $C_2H_3O_2^-(aq)$ than for $ClO^-(aq)$.

26. A student added 0.20 mol of NaI and 0.40 mol of KI to 3 liters of water to create an aqueous solution. What is the minimum number of moles of $Pb(C_2H_3O_2)_2$ that the student must add to the solution to precipitate out all of the I⁻ ions as PbI_2?

(A) 2.40

(B) 1.20

(C) 0.60

(D) 0.30

(E) 0.15

26. **D** The solution contains 0.60 moles of I⁻ ions, 0.20 from NaI and 0.40 from KI. Each Pb^{2+} ion will remove 2 I⁻ ions, so the student needs to add only half as much $Pb(C_2H_3O_2)_2$ or 0.30 moles.

QUESTIONS	EXPLANATIONS

27. A molecule with the formula XY_2, whose atoms are arranged linearly could have a central atom with which of the following hybridizations?

    I. sp
    II. sp^3
    III. dsp^3

    (A) I only
    (B) II only
    (C) I and II only
    (D) I and III only
    (E) I, II, and III

27. **D** If the central atom has sp (choice I) hybridization, the molecule can be linear, as shown below.

$$B - A - B$$

Choice II is not correct because sp^3 hybridization will bring about a basic tetrahedral shape for the molecule. This molecule will not be linear no matter how many lone electron pairs are present.

Choice III is correct because a central atom with dsp^3 hybridization can bring about a molocule with a linear shape if there are three lone electron pairs on the central atom, as shown below.

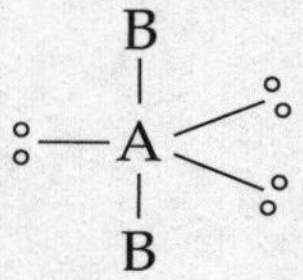

28. Which of the following statements regarding fluorine and nitrogen is NOT true?

    (A) Fluorine has greater electronegativity.
    (B) Fluorine has a greater first ionization energy.
    (C) Fluorine has more valence electrons.
    (D) Fluorine has a greater atomic weight.
    (E) Fluorine has a greater atomic radius.

28. **E** Fluorine has a smaller atomic radius than nitrogen. That's because fluorine has the same number of electron shells as nitrogen, but it has more protons in its nucleus. So each of fluorine's negatively charged electrons feels a stronger pull toward its more positively charged nucleus. This makes its radius smaller. All of the other answer choices are true.

QUESTIONS	EXPLANATIONS

29. Which of the groups below is (are) listed in order from lowest to highest melting point?

   I.   KI, LiF, BeO
   II. F_2, Cl_2, Br_2
   III. K, Na, Li

   (A) I only
   (B) I and II only
   (C) I and III only
   (D) II and III only
   (E) I, II, and III

29. E All three groups are listed in order of increasing melting point. Remember: Stronger bonds or intermolecular forces means higher melting point.

I. These are ionic compounds. LiF has stronger bonds than KI because Li^+ and F^- are smaller than K^+ and I^-. BeO has stronger bonds than LiF because Be^{2+} and O^{2-} are more highly charged than Li^+ and F^-. So BeO melts at the highest temperature, and KI melts at the lowest.

II. These are nonpolar covalent compounds, so their solids are held together by weak London dispersion forces. Since London dispersion forces depend on the random movement of electrons, the more electrons in a compound the stronger the London dispersion forces. So F_2 has the weakest intermolecular forces and the lowest melting point, and Br_2 has the strongest forces and the highest melting point.

III. These three are held together by metallic bonding. In general, the smaller the nucleus of a metal, the stronger the metallic bonds, so K has the weakest bonds and the lowest melting point, and Li has the strongest bonds and the highest melting point.

30. Which of the statements below regarding elemental nitrogen is NOT true?

   (A) It contains one sigma bond.
   (B) It contains two pi bonds.
   (C) It has a bond order of 3.
   (D) It has a large dipole moment.
   (E) It exists as a diatomic gas.

30. D Elemental nitrogen exists as a diatomic gas, N_2. N_2 is nonpolar, so it has a dipole moment of zero. N_2 has a triple bond, so it has a bond order of 3, with one sigma bond and two pi bonds.

31. The reaction of elemental chlorine with ozone in the atmosphere occurs by the two-step process shown below.

   I. $Cl + O_3 \rightarrow ClO + O_2$
   II. $ClO + O \rightarrow Cl + O_2$

Which of the statements below is true regarding this process?

   (A) Cl is a catalyst.
   (B) O_3 is a catalyst.
   (C) ClO is a catalyst.
   (D) O_2 is an intermediate.
   (E) O is an intermediate.

31. A The overall reaction is as follows:

$$O_3 + O \rightarrow 2 O_2$$

Cl acts as a catalyst in the reaction because it must be there for the reaction to occur, but it does not appear in the overall reaction. By the way, ClO is an intermediate in the process because it is the only component that's produced during the process and then completely consumed.

32. A student examined the line spectrum of a hydrogen atom and was able to conclude that when the electron in the hydrogen atom made the transition from the $n = 3$ to the $n = 1$ state, the frequency of the ultraviolet radiation emitted was 2.92×10^{15} Hz. When the same transition occurred in a He$^+$ ion, the frequency emitted was 1.17×10^{16} Hz. Which of the following best accounts for the difference?

(A) More energy is released by the He$^+$ transition because of the greater molar mass of the He$^+$ ion.

(B) More energy is released by the He$^+$ transition because of the greater nuclear charge of the He$^+$ ion.

(C) More energy is released by the He$^+$ transition because the ionic radius of He$^+$ is greater than the atomic radius of H.

(D) Less energy is released by the He$^+$ transition because the ionic radius of He$^+$ is smaller than the atomic radius of H.

(E) Less energy is released by the He$^+$ transition because of the greater nuclear charge of the He$^+$ ion.

32. B The frequency of radiation emitted by the He$^+$ is greater than the frequency emitted by the H atom for the same transition, so the energy emitted by the He$^+$ ion is greater (Remember: $E = hf$). That rules out (D) and (E).

The He$^+$ ion has 2 protons and the H atom has only one, so the electron is more strongly attracted to the He$^+$ ion and when it moves closer to the nucleus more energy will be released, so (B) is correct.

Choice (A) is wrong because the effect of mass in this situation is insignificant compared to charge. Choice (C) is wrong because He$^+$ has a smaller radius than H.

33. $CH_3CH_2OH(g) + ...O_2(g) \rightarrow$

$...CO_2(g) + ...H_2O(g)$

The reaction above represents the oxidation of ethanol. How many moles of O_2 are required to oxidize 1 mole of CH_3CH_2OH?

(A) $\dfrac{3}{2}$ moles

(B) $\dfrac{5}{2}$ moles

(C) 3 moles

(D) $\dfrac{7}{2}$ moles

(E) 4 moles

33. C Since we're given a coefficient here ($1CH_3CH_2OH$) and the answer choices look kind of ugly for backsolving, let's just put in the coefficient we have and work from there.

If we have 1 CH_3CH_2OH, then we have 2 Cs and thus, 2 CO_2s.

Also, if we have 1 CH_3CH_2OH, then we have 6 Hs, and thus, 3 H_2Os.

That means we have 7 Os on the right, so we must have 7 on the left. There is 1 O in CH_3CH_2OH, so we must get 6 from O_2, which means the coefficient of O_2 must be 3.

QUESTIONS	EXPLANATIONS

34. How many milliliters of water must be added to 10 milliliters of an HCl solution with a pH of 1 to produce a solution with a pH of 2?

(A)   10 ml
(B)   90 ml
(C)   100 ml
(D)   990 ml
(E)   1,000 ml

34. B HCl is a strong acid, so it will completely dissociate into H^+ and Cl^- ions. A pH of 1 means $[H^+] = 0.1$ M. A pH of 2 means $[H^+] = 0.01$ M. So we need to go from a concentration of 0.1 M to 0.01 M. As the solution is diluted, the number of moles of H^+ will remain constant.

Moles = (molarity)(volume)

Moles of H^+ = (0.1 M)(10 ml) = (0.01 M)(x)

x = 100 ml, but that's not the answer.

To get a 0.01-molar solution, we need 100 ml, but we already have 10 ml, so we need to *add* 90 ml of water. That's the answer.

35. Elemental iodine (I_2) is more soluble in carbon tetrachloride (CCl_4) than it is in water (H_2O). Which of the following statements is the best explanation for this?

(A) I_2 is closer in molecular weight to CCl_4 than it is to H_2O.
(B)   The freezing point of I_2 is closer to that of CCl_4 than it is to that of H_2O.
(C) I_2 and CCl_4 are nonpolar molecules, while H_2O is a polar molecule.
(D)   The heat of formation of I_2 is closer to that of CCl_4 than it is to that of H_2O.
(E) CCl_4 has a greater molecular weight than does H_2O.

35. C Like dissolves like. Polar molecules are more soluble in polar solvents, and nonpolar molecules are more soluble in nonpolar solvents.

36. A sample of water was electrolyzed to produce hydrogen and oxygen gas, as shown in the reaction below.

$$2\ H_2O(l) \rightarrow 2\ H_2(g) + O_2(g)$$

If 33.6 liters of gas were produced at STP, how many grams of water were consumed in the reaction?

(A)   11.1 grams
(B)   18.0 grams
(C)   24.0 grams
(D)   36.0 grams
(E)   44.8 grams

36. B Moles of gas at STP = $\dfrac{\text{liters}}{22.4 \text{ moles/liter}}$

Moles of gas produced = $\dfrac{33.6 \text{ L}}{22.4 \text{ moles/L}}$ = 1.5 moles

3 moles of gas (1 oxygen and 2 hydrogen) are produced for every 2 moles of water consumed, so if 1.5 moles of gas were produced, then 1 mole of water was consumed.

Grams = (moles)(MW)

Grams of water consumed = (1 mole)(18.0 grams/mole) = 18.0 grams

37. The following data were gathered in an experiment to determine the density of a sample of an unknown substance.

Mass of the sample = 7.50 grams

Volume of the sample = 2.5 milliliters

The density of the sample should be reported as

(A) 3.00 grams per ml.
(B) 3.0 grams per ml.
(C) 3 grams per ml.
(D) 0.3 grams per ml.
(E) 0.33 grams per ml.

37. **B** Density = $\dfrac{\text{mass}}{\text{volume}}$

Density of the sample = $\dfrac{7.50 \text{ grams}}{2.5 \text{ ml}}$ = 3.0 grams/ml

The volume (2.5 ml) has only two significant digits, so the final result (3.0 grams/ml) is limited to two significant digits.

---

38. $$2 H_2O_2(g) + S(s) \rightarrow SO_2(g) + 2 H_2O(g)$$

Based on the information given in the table below, what is the enthalpy change in the reaction represented above?

Substance	ΔH°_f (kJ/mol)
$H_2O_2(g)$	−150
$SO_2(g)$	−300
$H_2O(g)$	−250
$S(s)$	0

(A) −500 kJ
(B) −200 kJ
(C) 200 kJ
(D) 400 kJ
(E) 600 kJ

38. **A** We are given the heats of formation of all the reactants and products:

ΔH° for a reaction =
$\Sigma(\Delta H^{\circ}_f$ for the products) − $\Sigma(\Delta H^{\circ}_f$ for the reactants)

First the products.

From SO_2, we get (−300 kJ) = −300 kJ

From H_2O, we get (2)(−250 kJ) = −500 kJ

So ΔH° for the products = (−300 kJ) + (−500 kJ)
= −800 kJ

Now the reactants.

From H_2O_2, we get (2)(−150 kJ) = −300 kJ, and from S we get 0.

ΔH° for the reaction = (−800 kJ) − (−300 kJ) = (−800 kJ) + (300 kJ) = −500 kJ

QUESTIONS	EXPLANATIONS

39. A chemist creates a buffer solution by mixing equal volumes of a 0.2-molar HOCl solution and a 0.2-molar KOCl solution. Which of the following will occur when a small amount of KOH is added to the solution?

    I. The concentration of undissociated HOCl will increase.

    II. The concentration of OC1$^-$ ions will increase.

    III. The concentration of H$^+$ ions will increase.

(A) I only
(B) II only
(C) III only
(D) I and III only
(E) II and III only

39. **B** Look at the equilibrium reaction for this buffer solution.

$$HOCl(aq) \leftrightarrow H^+(aq) + OCl^-(aq)$$

When OH$^-$ is added, it will react to neutralize H$^+$. So adding OH$^-$ is the same as removing H$^+$, and when H$^+$ is removed, the equilibrium shifts to the right to replace the lost H$^+$ ions. This causes the concentration of HOCl to decrease and the concentration of OCl$^-$ to increase (that's choice II). The concentration of H$^+$ decreases a little bit but stays about the same, so really only choice II is correct.

40. A 15-gram sample of neon is placed in a sealed container with a constant volume of 8.5 liters. If the temperature of the container is 27°C, which of the following expressions gives the correct pressure of the gas? The ideal gas constant, R, is 0.08 (L-atm)/(mole-K).

A) $\dfrac{(15)(0.08)(300)}{(20)(8.5)}$ atm

(B) $\dfrac{(20)(0.08)(300)}{(15)(8.5)}$ atm

(C) $\dfrac{(20)(300)}{(15)(8.5)(0.08)}$ atm

(D) $\dfrac{(15)(300)}{(20)(8.5)(0.08)}$ atm

(E) $\dfrac{(15)(0.08)}{(20)(8.5)(300)}$ atm

40. **A** First we need to find the number of moles of neon.

$$\text{Moles} = \frac{\text{grams}}{\text{MW}} = \frac{15\,\text{g}}{20\,\text{g/mol}} = n$$

Now we can use the ideal gas equation to find the pressure. Don't forget to convert 27°C to 300 K.

$$P = \frac{nRT}{V} =$$

$$\frac{\left(\dfrac{15}{20}\,\text{mol}\right)\left(0.08\,\dfrac{\text{L-atm}}{\text{mol-K}}\right)(300\,\text{K})}{(8.5\,\text{L})} =$$

$$\frac{(15)(0.08)(300)}{(20)(8.5)}\,\text{atm}$$

QUESTIONS	EXPLANATIONS

41.
$$C(s) + CO_2(g) + energy \leftrightarrow 2\,CO(g)$$

The system above is currently at equilibrium in a closed container. Which of the following changes to the system would serve to increase the number of moles of CO present at equilibrium?

   I. Raising the temperature
  II. Increasing the volume of the container
 III. Adding more C(s) to the container

(A) I only
(B) II only
(C) I and II only
(D) I and III only
(E) I, II, and III

41. C Use Le Châtelier's law.

I. Raising the temperature favors the endothermic, forward reaction, increasing the CO produced at equilibrium, so (I) is correct.

II. Increasing the volume of the container favors the side with more moles of gas. That's the CO side, so (II) is correct.

III. Adding more of a solid will have no effect on an equilibrium situation, so choice (III) is not correct.

42. In a titration experiment, a 0.10-molar $H_2C_2O_4$ solution was completely neutralized by the addition of a 0.10-molar NaOH solution. Which of the diagrams below illustrates the change in pH that accompanied this process?

(A)

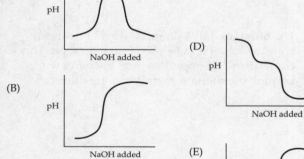

(B)

(D)

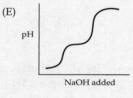

(C)

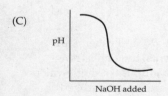

(E)

42. E The diagram for the titration of a diprotic acid (such as $H_2C_2O_4$) by a base shows increasing pH and two endpoints.

QUESTIONS	EXPLANATIONS

43. If 52 grams of $Ba(NO_3)_2$ (molar mass 260 grams) are completely dissolved in 500 milliliters of distilled water, what are the concentrations of the barium and nitrate ions?

    (A) $[Ba^{2+}] = 0.10\ M$ and $[NO_3^-] = 0.10\ M$
    (B) $[Ba^{2+}] = 0.20\ M$ and $[NO_3^-] = 0.40\ M$
    (C) $[Ba^{2+}] = 0.40\ M$ and $[NO_3^-] = 0.40\ M$
    (D) $[Ba^{2+}] = 0.40\ M$ and $[NO_3^-] = 0.80\ M$
    (E) $[Ba^{2+}] = 0.80\ M$ and $[NO_3^-] = 0.80\ M$

43. **D** First let's find the number of moles of $Ba(NO_3)_2$, Ba^{2+}, and NO_3^- in the solution.

$$\text{Moles} = \frac{\text{grams}}{\text{MW}}$$

$$\text{Moles of } Ba(NO_3)_2 = \frac{52\ g}{260\ g/mol} = 0.20\ \text{moles}$$

For every mole of $Ba(NO_3)_2$ in solution, we get 1 mole of Ba^{2+} and 2 moles of NO_3^-, so we have 0.20 moles of Ba^{2+} and 0.04 moles of NO_3^-.

Now we can find the concentrations of Ba^{2+} and NO_3^-.

$$\text{Molarity} = \frac{\text{moles}}{\text{liters}}$$

$$[Ba^{2+}] = \frac{0.20\ \text{moles}}{0.500\ L} = 0.40\ M$$

$$[NO_3^-] = \frac{0.40\ \text{moles}}{0.500\ L} = 0.80\ M$$

44. An acid solution of unknown concentration is to be titrated with a standardized hydroxide solution that will be released from a buret. The buret should be rinsed with

    (A) hot distilled water.
    (B) distilled water at room temperature.
    (C) a sample of the unknown acid solution.
    (D) a sample of the hydroxide solution.
    (E) a neutral salt solution.

44. **D** The buret used in a titration should be rinsed with the solution to be used in the titration. This cleans dust and impurites from the buret while keeping the solution from becoming dilute.

45. All of the following statements concerning the alkali metals are true EXCEPT

    (A) They are strong oxidizing agents.
    (B) They form ions with a +1 oxidation state.
    (C) As the atomic numbers of the alkali metals increase, their electronegativity decreases.
    (D) As the atomic numbers of the alkali metals increase, the first ionization energy decreases.
    (E) They form ions that are soluble in water.

45. **A** The alkali metals (Li, Na, K, and others) are not strong oxidizing agents, they are strong *reducing* agents. They give up their single valence electrons easily, so they are easily oxidized (LEO). If they are easily oxidized, they are strong reducing agents.

QUESTIONS	EXPLANATIONS
46. As a beaker of water is heated over a flame, the temperature increases steadily until it reaches 373 K. At that point, the beaker is left on the open flame, but the temperature remains at 373 K as long as water remains in the beaker. This is because at 373 K, the energy supplied by the flame	46. **E** The boiling point of water is 373 K (100°C). At this point, the energy put into the water is used to overcome the intermolecular forces of attraction of the water molecules as it vaporizes. (That's the heat of vaporization.)

46.
(A) no longer acts to increase the kinetic energy of the water molecules.
(B) is completely absorbed by the glass beaker.
(C) is less than the energy lost by the water through electromagnetic radiation.
(D) is dissipated by the water as visible light.
(E) is used to overcome the heat of vaporization of the water.

47. For which of the following reactions will the equilibrium constants K_c and K_p have the same value?

(A) $2 N_2O_5(g) \leftrightarrow 2 NO_2(g) + O_2(g)$
(B) $2 CO_2(g) \leftrightarrow 2 CO(g) + O_2(g)$
(C) $H_2O(g) + CO(g) \leftrightarrow H_2(g) + CO_2(g)$
(D) $3 O_2(g) \leftrightarrow 2 O_3(g)$
(E) $CO(g) + Cl_2(g) \leftrightarrow COCl_2(g)$

47. **C** The expression that connects K_p and K_c is $K_p = K_c (RT)^{\Delta n}$, where Δn is the change in moles of gas over the course of the reaction. For choice (C), there are 2 moles of reactant gases and 2 moles of product gases, so $\Delta n = 0$ and $K_p = K_c$. None of the other choices has $\Delta n = 0$.

48. A student added solid potassium iodide to 1 kilogram of distilled water. When the KI had completely dissolved, the freezing point of the solution was measured to be –3.72°C. What was the mass of the potassium iodide added? (The freezing point depression constant, k_f, for water is 1.86 K-kg/mol.)

(A) 334 grams
(B) 167 grams
(C) 84 grams
(D) 42 grams
(E) 21 grams

48. **B** Use the formula $\Delta T = k_f m x$ to find the molality of the KI. Potassium iodide dissociates into 2 particles, so $x = 2$.

$3.72 = (1.86)(m)(2)$
$m = 1.00$

Since there was 1 kilogram of water, a 1-molal solution must contain 1 mole of KI. Now you use the molar mass of KI to find the number of grams.

Grams = (moles)(MW) = (1 mol)(167 g/mol)
= 167 grams

49. Each of the following compounds was added to distilled water at 25°C. Which one produced a solution with a pH that was less than 7?

    (A) N_2
    (B) O_2
    (C) NaI
    (D) MgO
    (E) SO_2

49. **E** SO_2 is an acid anhydride; it reacts with water to produce H_2SO_3. So when SO_2 is added to water, the solution will be acidic and its pH will be less than 7. N_2 and O_2 will not react with water, NaI is a neutral salt, and MgO is a basic anhydride that will produce a solution with a pH greater than 7.

50. $...Be_2C + ...H_2O \rightarrow ...Be(OH)_2 + ...CH_4$

    When the equation for the reaction above is balanced with the lowest whole-number coefficients, the coefficient for H_2O will be

    (A) 1
    (B) 2
    (C) 3
    (D) 4
    (E) 5

50. **D** Backsolve, starting with (C).

    If there are 3 H_2Os, there is no way to balance O on the other side of the equation because, with a whole number coefficent for $Be(OH)_2$, there can be only an even number of Os. That means that (A) and (E) are also wrong.

    Try (D).

    If there are 4 H_2Os, then there are 2 $Be(OH)_2$s.

    If there are 2 $Be(OH)_2$s, then there is 1 Be_2C.

    If there is 1 Be_2C, then there is 1 CH_4.

    Now there are 8 Hs on the left and 8 Hs on the right, so the equation is balanced with the lowest whole number coefficients and (D) is correct.

51. A sample of an ideal gas is placed in a sealed container of constant volume. If the temperature of the gas is increased from 40°C to 70°C, which of the following values for the gas will NOT increase?

    (A) The average speed of the molecules of the gas
    (B) The average kinetic energy of the molecules of the gas
    (C) The pressure exerted by the gas
    (D) The density of the gas
    (E) The frequency of collisions between molecules of the gas

51. **D** According to kinetic-molecular theory, as the temperature of a gas is increased, the speed, kinetic energy, and frequency of collision of its molecules all increase. From the ideal gas law, we know that as temperature increases at constant volume, pressure increases. The temperature increase affects neither the volume of the container nor the mass of the gas contained, so the density of the gas is not changed.

52. An $^{222}_{86}$Rn nuclide decays through the emission of two beta particles and two alpha particles. The resulting nuclide is

   (A) $^{214}_{84}$Po

   (B) $^{210}_{84}$Po

   (C) $^{214}_{83}$Bi

   (D) $^{210}_{83}$Bi

   (E) $^{214}_{82}$Pb

52. **A** Let's do the math.

$$^{222}_{86}\text{Rn} - {}_{-1}\beta - {}_{-1}\beta - {}^{4}_{2}\alpha - {}^{4}_{2}\alpha =$$

For the mass number we have

$$222 - 4 - 4 = 214$$

For the proton number we have

$$86 - (-1) - (-1) - 2 - 2 = 86 + 2 - 4 = 84$$

So the answer is $^{214}_{84}$Po.

53. When 80.0 ml of a 0.40 M NaI solution is combined with 20.0 ml of a 0.30 M CaI$_2$ solution, what will be the molar concentration of I$^-$ ions in the solution?

   (A) 0.70 M

   (B) 0.44 M

   (C) 0.38 M

   (D) 0.35 M

   (E) 0.10 M

53. **B** The I$^-$ ions from the two salts will both be present in the solution, so we need to find the number of moles of I$^-$ contributed by each salt.

Moles = (molarity)(volume)

Each NaI produces 1 I$^-$

Moles of I$^-$ from NaI = (0.40 M)(0.080 L)
  = 0.032 mole

Each CaI$_2$ produces 2 I$^-$

Moles of I$^-$ from CaI$_2$ = (2)(0.30 M)(0.020 L)
  = 0.012 mole

To find the number of moles of Cl$^-$ in the solution, add the two together.

0.032 mole + 0.012 mole = 0.044 mole

Now use the formula for molarity to find the concentration of I$^-$ ions. Don't forget to add the volumes of the two solutions.

$$\text{Molarity} = \frac{\text{moles}}{\text{liters}} = \frac{(0.044\text{mol})}{(0.100\text{L})} = 0.44 \ M$$

54.

Compound	K_{sp} at 25°C
FeS	6.33×10^{-18}
PbS	8.03×10^{-28}
MnS	1.03×10^{-13}

A solution at 25°C contains Fe^{2+}, Pb^{2+}, and Mn^{2+} ions. Which of the following gives the order in which precipitates will form, from first to last, as Na_2S is steadily added to the solution?

(A) FeS, PbS, MnS
(B) MnS, PbS, FeS
(C) FeS, MnS, PbS
(D) MnS, FeS, PbS
(E) PbS, FeS, MnS

54. E The lower the value of the K_{sp}, the less soluble the compound and the earlier it will precipitate. The correct answer lists the compounds from least to greatest K_{sp}.

55. Hydrogen sulfide is a toxic waste product of some industrial processes. It can be recognized by its distinctive rotten egg smell. One method for elimininating hydrogen sulfide is by reaction with dissolved oxygen, as shown below.

$$2\,H_2S + O_2 \rightarrow 2\,S + 2\,H_2O$$

If 102 grams of H_2S are combined with 64 grams of O_2, what is the maximum mass of elemental sulfur that could be produced by the reaction?

(A) 16 grams
(B) 32 grams
(C) 48 grams
(D) 64 grams
(E) 96 grams

55. E We need to find the limiting reagent.

$$\text{Moles} = \frac{\text{grams}}{\text{MW}}$$

$$\text{Moles of } H_2S = \frac{102 \text{ grams}}{34 \text{ g/mol}} = 3 \text{ moles}$$

$$\text{Moles of } O_2 = \frac{64 \text{ grams}}{32 \text{ g/mol}} = 2 \text{ moles}$$

We can see from the balanced equation that 2 moles of H_2S are consumed for every 1 mole of O_2. Since the number of moles of H_2S is less than twice the number of moles of O_2, H_2S will run out first and is the limiting reagent.

Also, from the balanced equation, the number of moles of S produced is equal to the number of moles of H_2S consumed, so 3 moles of S are produced.

Grams = (moles)(MW)

Grams of S = (3 moles)(32 g/mol) = 96 grams

QUESTIONS	EXPLANATIONS

56.

$$Mn(s) + Cu^{2+}(aq) \rightarrow Mn^{2+}(aq) + Cu(s)$$

A voltaic cell based on the reaction above was constructed from manganese and zinc half-cells. The standard cell potential for this reaction was 1.52 volts, but the observed voltage at 25°C was 1.66 volts. Which of the following could explain this observation?

(A) The Mn^{2+} solution was more concentrated than the Cu^{2+} solution.
(B) The Cu^{2+} solution was more concentrated than the Mn^{2+} solution.
(C) The manganese electrode was larger than the copper electrode.
(D) The copper electrode was larger than the manganese electrode.
(E) The atomic weight of copper is greater than the atomic weight of manganese.

56. B You can use the Nernst equation.

$$E_{cell} = E^{\circ}_{cell} - \frac{0.0592}{n} \log \frac{\left[Mn^{2+}\right]}{\left[Cu^{2+}\right]}$$

Or you can use Le Châtelier's law. Let's do it. If the reactants (Cu^{2+}) are more concentrated than the products (Mn^{2+}), the reaction becomes more spontaneous. If the reaction becomes more spontaneous, the reaction potential, or voltage, will increase. By the way, the relative sizes of the electrodes are not important because solids are not considered in the equilibrium expression.

57. The rate of effusion of helium gas (atomic weight 4.0) at a given temperature and pressure is known to be x. What would be the expected rate of effusion for hydrogen gas (molecular weight 2.0) at the same temperature and pressure?

(A) $\dfrac{x}{2}$

(B) $\dfrac{x}{\sqrt{2}}$

(C) x

(D) $x\sqrt{2}$

(E) $2x$

57. D Use Graham's law.

$$\frac{v_1}{v_2} = \sqrt{\frac{MW_2}{MW_1}}$$

$$= \sqrt{\frac{2}{1}} v_{helium} = x\sqrt{2}$$

58.

$$H_2O(g) \rightarrow H_2O(l)$$

Which of the following is true of the values of ΔH, ΔS, and ΔG for the reaction shown above at 25°C?

	ΔH	ΔS	ΔG
(A)	Positive	Positive	Positive
(B)	Positive	Negative	Negative
(C)	Negative	Positive	Negative
(D)	Negative	Negative	Positive
(E)	Negative	Negative	Negative

58. E Heat is given off when water condenses, so ΔH is negative. The entropy decreases in a phase change from gas to liquid, so ΔS is negative. The water condenses spontaneously at room temperature, so ΔG is negative.

QUESTIONS	EXPLANATIONS

59. A student added 1 liter of a 1.0 M Na_2SO_4 solution to 1 liter of a 1.0 M $Ag(C_2H_3O_2)$ solution. A silver sulfate precipitate formed, and nearly all of the silver ions disappeared from the solution. Which of the following lists the ions remaining in the solution in order of decreasing concentration?

(A) $[SO_4^{2-}] > [C_2H_3O_2^-] > [Na^+]$
(B) $[C_2H_3O_2^-] > [Na^+] > [SO_4^{2-}]$
(C) $[C_2H_3O_2^-] > [SO_4^{2-}] > [Na^+]$
(D) $[Na^+] > [SO_4^{2-}] > [C_2H_3O_2^-]$
(E) $[Na^+] > [C_2H_3O_2^-] > [SO_4^{2-}]$

59. **E** At the start, the concentrations of the ions are as follows:

$[Na^+] = 2\ M$
$[SO_4^{2-}] = 1\ M$
$[Ag^+] = 1\ M$
$[C_2H_3O_2^-] = 1\ M$

After Ag_2SO_4 forms, the concentrations are

$[Na^+] = 2\ M$
$[SO_4^{2-}] = 0.5\ M$
$[Ag^+] = 0\ M$
$[C_2H_3O_2^-] = 1\ M$

So from greatest to least

$[Na^+] > [C_2H_3O_2^-] > [SO_4^{2-}]$

---

60. Iron ore is converted to pure iron by the following reaction, which occurs at extremely high temperatures:

$$Fe_2O_3 + 3\,CO \rightarrow 2\,Fe + 3\,CO_2$$

A 1,600-gram sample of iron ore reacted completely to form 558 grams of pure iron. What was the percent of Fe_2O_3 (molecular weight 160) by mass in the original sample?

(A) 25%
(B) 33%
(C) 50%
(D) 67%
(E) 100%

60. **C** Moles = $\dfrac{grams}{MW}$

Moles of Fe = $\dfrac{558\ grams}{55.8\ g/mol}$ = 10 moles

2 moles of Fe are produced for every 1 mole

of Fe_2O_3 consumed, so if 10 moles of Fe were produced, 5 moles of Fe_2O_3 were consumed.

Grams = (moles)(MW)

Grams of Fe_2O_3 = (5 moles)(160 g/mol)
        = 800 grams

Percent of Fe_2O_3 by

mass = $\dfrac{grams\ of\ Fe_2O_3}{grams\ of\ sample} = \dfrac{800\ g}{1600\ g}$ = 0.50 = 50%

---

61. Gold and silver have been used throughout history to make coins. Which of the following statements could account for the popularity of these metals for use in coins?

(A) Network bonds make these metals especially durable.
(B) Negative oxidation potentials for these metals make them especially unreactive with their surroundings.
(C) These metals are among the lightest of the elements.
(D) These metals are commonly found in nature.
(E) These metals form ions that are extremely soluble in water.

61. **B** A coin must be stable; it must not react with its surroundings. Elemental silver and gold are among the few metals that have negative oxidation potentials, which makes them unreactive. All of the statements in the other answer choices are false.

QUESTIONS	EXPLANATIONS

62. A student added a salt to an aqueous solution of sodium sulfate. After the salt was added, a sulfate compound was observed as a precipitate. Which of the following could have been the salt added in this experiment?

    I. $BaCl_2$
    II. $Pb(NO_3)_2$
    III. $AgC_2H_3O_2$

(A) I only
(B) III only
(C) I and II only
(D) I and III only
(E) I, II, and III

62. **E** Ba^{2+}, Pb^{2+}, and Ag^+ all form insoluble sulfates. Any of the three salts could have been added to form a precipitate.

63.
$$2\,HI(g) \rightarrow H_2(g) + I_2(g)$$

For the reaction given above, $\Delta H°$ is –50 kJ. Based on the information given in the table below, what is the average bond energy of the H–I bond?

Bond	Average Bond Energy (kJ/mol)
H–H	440
I–I	150

(A) 270 kJ/mol
(B) 540 kJ/mol
(C) 590 kJ/mol
(D) 640 kJ/mol
(E) 1180 kJ/mol

63. **A** Use the following equation to solve the problem.

$\Delta H°$ = [Bond energies of bonds broken] – [Bond energies of the bonds formed]

$\Delta H°$ = [(2)(H–I)] – [(H–H) + (I–I)]

–50kJ = (2)(H–I) – [440 kJ + 150 kJ]

(2)(H–I) = –50 kJ + 590 kJ

(2)(H–I) = 540 kJ

H–I = 270 kJ

64. A solid piece of barium hydroxide is immersed in water and allowed to come to equilibrium with its dissolved ions. The addition of which of the following substances to the solution would cause more solid barium hydroxide to dissolve into the solution?

(A) NaOH
(B) HCl
(C) NaCl
(D) $BaCl_2$
(E) NH_3

64. **B** The addition of H^+ ions to the solution will neutralize the OH^- ions produced by the solid barium hydroxide. According to Le Châtelier's law, this decrease in the concentration of OH^- ions will cause the solubility equilibrium to shift to add more OH^- ions to the solution, thus consuming more solid barium hydroxide.

Adding NaOH or NH_3 will increase the concentration of OH^- ions and will inhibit the dissolution of the solid. Adding $BaCl_2$ will add Ba^{2+} ions to the solution, also inhibiting the dissolution of the solid. Adding NaCl will have no effect on the barium hydroxide solubility equilibrium.

65. All of the elements listed below are gases at room temperature. Which gas would be expected to show the greatest deviation from ideal behavior?

    (A) He
    (B) Ne
    (C) Ar
    (D) Kr
    (E) Xe

65. **E** Deviation from ideal behavior comes about because of attractive forces between gas molecules and because of molecular size. Of the noble gases listed, Xe has the most electrons, so it will have the strongest London dispersion forces, which are the only attractions experienced by the noble gases. Xe is also the largest atom listed.

Questions 66 and 67 refer to the information given below.

$$F_2(g) + 2\,ClO_2(g) \rightarrow 2\,FClO_2(g)$$

Experiment	$[F_2]$ (M)	$[ClO_2]$ (M)	Initial rate of disappearance of F_2 (M/sec)
1	0.10	0.010	1.2×10^{-3}
2	0.20	0.010	2.4×10^{-3}
3	0.40	0.020	9.6×10^{-3}

66. Based on the data given in the table, which of the following expressions is equal to the rate law for the reaction given above?
    (A) Rate = $k[F_2]$
    (B) Rate = $k[ClO_2]$
    (C) Rate = $k[F_2][ClO_2]$
    (D) Rate = $k[F_2]^2[ClO_2]$
    (E) Rate = $k[F_2][ClO_2]^2$

66. **C** Compare experiments 1 and 2. When $[F_2]$ is doubled while $[ClO_2]$ is held constant, the rate doubles, so the reaction is first order with respect to $[F_2]$.

The next part is a little tricky. Compare experiments 1 and 3. When both $[F_2]$ and $[ClO_2]$ are doubled, the rate quadruples. (Quadrupling is like doubling twice.) One of the rate doublings came from $[F_2]$, so the other one must have come from $[ClO_2]$. So doubling $[ClO_2]$ also causes the rate to double, making the reaction first order with respect to $[ClO_2]$.

The reaction is first order with respect to $[F_2]$ and first order with respect to $[ClO_2]$, so the rate law says that Rate = $k[F_2][ClO_2]$.

QUESTIONS	EXPLANATIONS

67. What is the initial rate of disappearance of ClO_2 in experiment 2?

   (A) 1.2×10^{-3} M/sec
   (B) 2.4×10^{-3} M/sec
   (C) 4.8×10^{-3} M/sec
   (D) 7.0×10^{-3} M/sec
   (E) 9.6×10^{-3} M/sec

67. **C** We can see from the balanced equation that for every mole of F_2 consumed, 2 moles of ClO_2 are consumed. So ClO_2 will disappear at twice the rate that F_2 disappears. In experiment 2, F_2 disappears at the rate of 2.4×10^{-3} M/sec, so ClO_2 will disappear at twice that rate, or 4.8×10^{-3} M/sec.

68. Hypobromous acid, HBrO, is added to distilled water. If the acid dissociation constant for HBrO is equal to 2×10^{-9}, what is the concentration of HBrO when the pH of the solution is equal to 5?

   (A) 5-molar
   (B) 1-molar
   (C) 0.1-molar
   (D) 0.05-molar
   (E) 0.01-molar

68. **D** Use the K_a expression for HBrO. In the denominator, we can ignore the amount of HBrO that dissociates because it is such a small number.

$$K_a = \frac{\left[H^+\right]\left[BrO^-\right]}{\left[HBrO\right]}$$

If the pH is 5, then $[H^+] = [BrO^-] = 1 \times 10^{-5} M$

$$2 \times 10^{-9} = \frac{\left(1\times10^{-5}\right)\left(1\times10^{-5}\right)}{\left[HBrO\right]}$$

$$[HBrO] = \frac{\left(1\times10^{-10}\right)}{\left(2\times10^{-9}\right)} M = 0.5 \times 10^{-1} M = 0.05\ M$$

69. $N_2(g) + 3\ Cl_2(g) \rightarrow 2\ NCl_2(g)$ $\qquad \Delta H = 460$ kJ

   Which of the following statements is true regarding the reaction shown above?

   (A) It is not spontaneous at any temperature.
   (B) It is spontaneous only at very high temperatures.
   (C) It is spontaneous only at very low temperatures.
   (D) It is spontaneous only at very high concentrations.
   (E) It is spontaneous only at very low concentrations.

69. **A** This reaction goes from 4 moles of gas in the reactants to 2 moles in the products, so ΔS is negative. You can see that ΔH is positive. When ΔS is negative and ΔH is positive, the equation $\Delta G = \Delta H - T\Delta S$ tells us that ΔG will always be positive. If ΔG is always positive, then the reaction is never spontaneous, no matter what the temperature. As for (D) and (E), concentration doesn't affect the spontaneity of a reaction, only the speed.

QUESTIONS	EXPLANATIONS

70. At 25°C, the vapor pressure of water is 24 mmHg. Which of the following expressions gives the vapor pressure of a solution created by adding 2.0 moles of sucrose to 55 moles of water?

(A) $\dfrac{(24)(2.0)}{(55)}$ mmHg

(B) $\dfrac{(2.0)}{(24)(55)}$ mmHg

(C) $\dfrac{(24)(55)}{(57)}$ mmHg

(D) $\dfrac{(55)}{(24)(57)}$ mmHg

(E) $\dfrac{(24)(57)}{(55)}$ mmHg

70. C Use Raoult's law.

$$VP_{SOLUTION} = (VP_{SOLVENT})(\text{mole fraction of solvent})$$

$$VP_{SOLUTION} = (24 \text{ mmHg})$$

$$\left(\frac{55 \text{ moles of water}}{(55+2.0) \text{ moles of solution}}\right) = (24)\left(\frac{55}{57}\right) = \text{mmHg}$$

$$VP_{SOLUTION} = \frac{(24)(55)}{(57)} \text{ mmHg}$$

71.
$$PCl_5(g) \leftrightarrow PCl_3(g) + Cl_2(g)$$

When PCl_5 is placed in an evacuated container at 250°C, the above reaction takes place. The pressure in the container before the reaction takes place is 6 atm and is entirely due to the PCl_5 gas. After the reaction comes to equilibrium, the partial pressure due to Cl_2 is found to be 2 atm. What is the value of the equilibrium constant, K_p, for this reaction?

(A) 0.5
(B) 1
(C) 2
(D) 5
(E) 10

71. B First we need to find out the equilibrium partial pressures of each of the gases. The question says that the partial pressure of Cl_2 at equilibrium is 2 atm.

From the balanced equation, for every mole of Cl_2 generated by the reaction, 1 mole of PCl_3 is also generated. There was no Cl_2 or PCl_3 to start with, so if there are 2 atm of Cl_2 at equilibrium, there must also be 2 atm of PCl_3.

Also from the balanced equation, for every mole of Cl_2 generated, 1 mole of PCl_5 must be consumed. So if there are 2 moles of Cl_2 at equilibrium, there must be $6 - 2 = 4$ atm of PCl_5.

Now we can use the K_p expression.

$$K_p = \frac{\left(P_{PCl_3}\right)\left(P_{Cl_2}\right)}{\left(P_{PCl_5}\right)} = \frac{(2)(2)}{(4)} = 1$$

72. $Co^{2+} + 2\,e^- \rightarrow Co$ $\qquad E^\circ = -0.28$ V

$Cr^{3+} + e^- \rightarrow Cr^{2+}$ $\qquad E^\circ = -0.41$ V

According to the information shown above, which of the following statements is true regarding the reaction below?

$$2\,Co + Cr^{3+} \rightarrow 2\,Co^{2+} + Cr^{2+}$$

- (A) Cr^{3+} acts as the reducing agent.
- (B) The reaction would take place in a galvanic cell.
- (C) K_{eq} for the reaction is less than 1.
- (D) Cobalt metal is reduced in the reaction.
- (E) Chromium metal will plate out in the reaction.

72. C For a spontaneous reaction, ΔG° is negative.

From the equation $\Delta G^\circ = -nFE^\circ$, if ΔG° is negative, E° must be positive.

From the equation $\Delta G^\circ = -RT\ln K$, if ΔG° is negative, $\ln K$ must be positive, so K must be greater than 1.

73. What is the pH of a solution made by mixing 200 milliliters of a 0.20-molar solution of NH_3 with 200 milliliters of a 0.20 milliliters of an NH_4Cl solution? (The base dissociation constant, K_b, for NH_3 is 1.8×10^{-5}.)

- (A) Between 3 and 4
- (B) Between 4 and 5
- (C) Between 5 and 6
- (D) Between 8 and 9
- (E) Between 9 and 10

73. E This solution is a buffer with equal concentrations of acid (NH_4^+) and conjugate base (NH_3). For a buffer with equal concentrations of acid and conjugate base, pH = pK_a and pOH = pK_b.

We are given K_b, so we can estimate pOH. $K_b = 1.8 \times 10^{-5}$, so pK_b and pOH are between 4 and 5 (actually it's 4.7). But we need the pH.

pH + pOH = 14

pH + (between 4 and 5) = 14

pH is between 9 and 10

74. $$POCl_3(l) \rightarrow POCl_3(g)$$

For the reaction above, ΔH is 50 kilojoules per mole and ΔS is 100 joules per mole. What is the boiling point of $POCl_3$? (Assume that ΔH and ΔS remain constant with changing temperature.)

- (A) 0.5 K
- (B) 2 K
- (C) 50 K
- (D) 200 K
- (E) 500 K

74. E At the boiling point, the reaction is at equilibrium. Remember that, at equilibrium, $\Delta G = 0$. Use the relationship $\Delta G = \Delta H - T\Delta S$ and don't forget to convert kilojoules into joules.

$\Delta G = \Delta H - T\Delta S$

$0 = 50{,}000 - (T)(100)$

$T = \dfrac{50{,}000}{100}$ K $= 500$ K

75. When fluorine gas is bubbled through a concentrated aqueous solution of potassium chloride at room temperature, which of the following would be expected to occur?

- (A) A reaction will occur, and chlorine gas will be produced.
- (B) A reaction will occur, and potassium fluoride will precipitate.
- (C) A reaction will occur, and solid chlorine will precipitate.
- (D) A reaction will occur, and solid potassium will precipitate.
- (E) No reaction will occur, and the fluorine gas will bubble through.

75. A Fluorine gas is an extremely strong oxidizing agent. When F_2 gas is bubbled through a solution containing any other halogen anions, F^- ions will replace the other halogen ions in the solution. Chlorine ions can be oxidized to form Cl_2 gas, which will bubble out of the solution.

QUESTIONS	EXPLANATIONS

SECTION II—FREE RESPONSE

1. At 25°C the solubility product consant, K_{sp}, for nickel hydroxide, $Ni(OH)_2$, is 1.6×10^{-14}.

(a) Write a balanced equation for the solubility equilibrium for $Ni(OH)_2$ in water.

(a) $Ni(OH)_2 \leftrightarrow Ni^{2+} + 2\ OH^-$

(b) What is the molar solubility of $Ni(OH)_2$ in pure water at 25°C?

(b) Use the following K_{sp} expression:
$K_{sp} = [Ni^{2+}][OH^-]^2$

One Ni^{2+} is produced for every $Ni(OH)_2$ that dissolves, so the molar solubility of $Ni(OH)_2$ will be the same as the concentration of Ni^{2+}. Two OH^-s will be produced for every Ni^{2+}, so $[OH^-]$ will be twice as big as $[Ni^{2+}]$.

Let $x = [Ni^{2+}]$

$K_{sp} = [Ni^{2+}][OH^-]^2$

$1.6 \times 10^{-14} = (x)(2x)^2 = 4x^3$

$x = 1.6 \times 10^{-5}\ M$

So the solubility of $Ni(OH)_2$ at 25°C is 1.6×10^{-5} moles/liter.

(c) A 1.0-molar NaOH solution is slowly added to a saturated solution of $Ni(OH)_2$ at 25°C. If excess solid $Ni(OH)_2$ remains in the solution throughout the procedure, what is the concentration of Ni^{2+} ions in the solution at the moment when the pH is equal to 11?

(c) Use the following K_{sp} expression again:
$K_{sp} = [Ni^{2+}][OH^-]^2$

First let's find $[OH^-]$.

If the pH is 11, then from pH + pOH = 14, we know that the pOH is 3.

If the pOH is 3, then from pOH = –log $[OH^-]$, we know that $[OH^-] = 1.0 \times 10^{-3}$.

Now we can use the K_{sp} expression to find $[Ni^{2+}]$.

$K_{sp} = [Ni^{2+}][OH^-]^2$

$1.6 \times 10^{-14} = [Ni^{2+}](1.0 \times 10^{-3})^2$

$1.6 \times 10^{-14} = [Ni^{2+}](1.0 \times 10^{-6})$

$[Ni^{2+}] = 1.6 \times 10^{-8}\ M$

QUESTIONS	EXPLANATIONS

QUESTIONS

(d) Predict whether a precipitate will form when 200.0 milliliters of a 5.0×10^{-5}-molar KOH solution is mixed with 300.0 milliliters of a 2.0×10^{-4}-molar $Ni(NO_3)_2$ solution at 25°C. Show calculations to support your prediction.

EXPLANATIONS

(d) First we need to find the concentrations of the Ni^{2+} and OH^- ions.

Moles = (molarity)(volume)

Moles of $OH^- = (5.0 \times 10^{-5}\ M)(0.200\ L)$
$= 1.0 \times 10^{-5}\ mol$

Moles of $Ni^{2+} = (2.0 \times 10^{-4})(0.300\ L)$
$= 6.0 \times 10^{-5}\ mol$

Remember to add the two volumes: $(0.300\ L) + (0.200\ L) = 0.500\ L$

$$Molarity = \frac{moles}{liters}$$

$$[Ni^{2+}] = \frac{\left(6.0 \times 10^{-5}\ mol\right)}{(0.500\ L)} = 1.2 \times 10^{-4}\ M$$

$$[OH^-] = \frac{\left(1.0 \times 10^{-5}\ mol\right)}{(0.500\ L)} = 2.0 \times 10^{-5}\ M$$

Now test the solubility expression using the initial values, to find the reaction quotient.

$Q = [Ni^{2+}][OH^-]^2$

$Q = (1.2 \times 10^{-4})(2.0 \times 10^{-5})^2 = 4.8 \times 10^{-14}$

Q is greater than K_{sp}, so a precipitate will form.

(e) At 25°C, 100 ml of a saturated $Ni(OH)_2$ solution was prepared.

  (i) Calculate the mass of $Ni(OH)_2$ present in the solution.

  (ii) Calculate the pH of the solution.

  (iii) If the solution is allowed to evaporate to a final volume of 50 ml, what will be the pH? Justify your answer.

(e) (i) Use the result from (b). First convert moles/L to grams/L. The formula weight of $Ni(OH)_2$ is 92.69 g/mol.

Grams/L = (moles/L)(MW) = $(1.6 \times 10^{-5}\ mol/L)$ (92.69 g/mol) = 0.0015 g/L

Now you can find the grams in 0.100 L. (0.0015 g/L)(0.100 L) = 0.00015 grams.

  (ii) From (b), $[OH^-]$ will be twice as large as $[Ni^{2+}]$. So $[OH^-] = 3.2 \times 10^{-5}\ M$

pOH = $-log[OH^-] = -log(3.2 \times 10^{-5}) = 4.49$

pH = 14 − pOH = 14 − 4.49 = 9.51

  (iii) Since the solution is saturated, evaporation will not change the concentrations, so the pH will remain 9.51.

QUESTIONS	EXPLANATIONS

2. In two separate experiments, a sample of an unknown hydrocarbon was burned in air, and a sample of the same hydrocarbon was placed into an organic solvent.

(a) When the hydrocarbon sample was burned in a reaction that went to completion, 2.2 grams of water and 3.6 liters of carbon dioxide were produced under standard conditions. What is the empirical formula of the hydrocarbon?

(a) All of the hydrogen in the water and all of the carbon in the carbon dioxide must have come from the hydrocarbon.

First we'll find out how many moles of hydrogen there are.

$$\text{Moles} = \frac{\text{grams}}{\text{MW}}$$

$$\text{Moles of } H_2O = \frac{(2.2 \text{ g})}{(18.0 \text{ g/mol})} = 0.12 \text{ moles}$$

Every mole of water contains two moles of hydrogen, so there are 0.24 moles of hydrogen.

Now let's find out how many moles of carbon dioxide there are.

$$\text{Moles} = \frac{\text{liters}}{22.4 \text{ L/mol}}$$

$$\text{Moles of } CO_2 = \frac{(3.6 \text{ L})}{(22.4 \text{ L/mol})} = 0.16 \text{ moles}$$

Every mole of CO_2 contains one mole of carbon, so there are 0.16 moles of carbon.

The ratio of H to C is $\frac{0.24}{0.16} = \frac{1.5}{1} = \frac{3}{2}$

So the empirical formula of the hydrocarbon is C_2H_3.

QUESTIONS	EXPLANATIONS

(b) When 4.05 grams of the unknown hydrocarbon was placed in 100.0 grams of benzene, C_6H_6, the freezing point of the solution was measured to be 1.66°C. The normal freezing point of benzene is 5.50°C and the freezing point depression constant for benzene is 5.12°C/m. What is the molecular weight of the unknown hydrocarbon?

(b) First we'll find the molality of the solution. The freezing point depression, ΔT, is

$$5.50°C - 1.66°C = 3.84°C.$$

$$\Delta T = kmx$$

$x = 1$ because hydrocarbons don't dissociate in solution. Now solve for m.

$$m = \frac{\Delta T}{k} = \frac{(3.84°C)}{(5.12°C/m)} = 0.750\ m$$

From the molality of the solution, we can find the number of moles of the unknown hydrocarbon.

$$\text{Molality} = \frac{\text{moles of solute}}{\text{kg of solvent}}$$

Solve for moles.

Moles = (molality)(kg of solvent)

Moles of hydrocarbon = (0.750 m)(0.100 kg)
$$= 0.0750\ \text{moles}$$

Now we can find the molecular weight of the hydrocarbon.

$$\text{MW} = \frac{\text{grams}}{\text{moles}} = \frac{(4.05\ \text{g})}{(0.0750\ \text{mol})} = 54.0\ \text{g/mol}$$

(c) What is the molecular formula and name of the hydrocarbon?

(c) From (a) we know that the empirical formula is C_2H_3. For C_2H_3, the molecular weight would be 27 g/mol.

From (b) we know that the molecular weight of the compound is 54 g/mol. That's twice as large as the weight for the empirical formula, so we can just double the numbers in the empirical formula to get the molecular formula, C_4H_6, butyne.

(d) Write the balanced equation for the combustion reaction that took place in (a).

(d) $2\ C_4H_6 + 11\ O_2 \rightarrow 6\ H_2O + 8\ CO_2$

(e) Draw two isomers for the hydrocarbon.

(e)

1-butyne   2-butyne

QUESTIONS	EXPLANATIONS

3. A strip of Ni metal is placed in a 1-molar solution of $Ni(NO_3)_2$ and a strip of Ag metal is placed in a 1-molar solution of $AgNO_3$. An electrochemical cell is created when the two solutions are connected by a salt bridge and the two metal strips are connected by wires to a voltmeter.

(a) Write the balanced chemical equation for the overall reaction that occurs in the cell, and calculate the cell potential, $E°$.

(a) The half-reactions in the cell are as follows:

$$Ni \rightarrow Ni^{2+} + 2\ e^- \qquad E° = +0.25\ V$$

$$Ag^+ + e^- \rightarrow Ag \qquad E° = +0.80\ V$$

You can get the overall reaction by adding the half-reactions and balancing the electrons exchanged. You get the reaction potential by adding the potentials for the half-reactions and ignoring the coefficients in the balanced equation.

$$Ni + 2\ Ag^+ \rightarrow Ni^{2+} + 2\ Ag \quad E° = +1.05\ V$$

(b) Calculate how many grams of metal will be deposited on the cathode if the cell is allowed to run at a constant current of 1.5 amperes for 8.00 minutes.

(b) At the cathode, Ag^+ ions are reduced to $Ag(s)$, so the metal deposited at the cathode will be solid silver.

First we need to find out how many electrons were supplied by the current.

Coulombs = (amperes)(seconds)

Coulombs = (1.50 A)(8.00 min)(60 sec/min)
    = 720 C

$$\text{Moles of electrons} = \frac{\text{coulombs}}{96.500} = \frac{720}{96.500}$$
$$= 0.00746\ \text{moles}$$

From the silver half-reaction, we know that for every mole of electrons consumed, one mole of $Ag(s)$ is produced. So we have 0.00746 moles of $Ag(s)$.

Grams = (moles)(MW)

Grams of $Ag(s)$ = (0.00746 mol)(107.9 g/mol)
    = 0.805 grams

QUESTIONS	EXPLANATIONS
(c) Calculate the value of the standard free energy change, ΔG°, for the cell reaction.	(c) Use the following expression: $\Delta G^\circ = -nFE^\circ$ $n = 2$, because two moles of electrons are exchanged in the redox reaction. $E^\circ = 1.05$ V, from part (a) $F = 96{,}500$ C/mol $\Delta G^\circ = -(2)(96{,}500 \text{ C/mol})(1.05 \text{ V})$ $= -202{,}650 \text{ C-V/mol} = -202{,}650 \text{ J/mol}$ $= -203 \text{ kJ/mol}$
(d) Calculate the cell potential, E, at 25°C for the cell shown above if the initial concentration of $Ni(NO_3)_2$ is 0.100-molar and the initial concentration of $AgNO_3$ is 1.20-molar.	(d) Use the Nernst equation for 25°C: $E_{cell} = E^\circ - \dfrac{0.0592}{n} \log Q$ $E_{cell} = E^\circ - \dfrac{0.0592}{n} \log \dfrac{\left[Ni^{2+} \right]}{\left[Ag^+ \right]^2}$ $E_{cell} = 1.05 \text{ V} - \dfrac{0.0592}{2} \log \dfrac{(0.100)}{(1.20)} \text{ V}$ $E_{cell} = 1.05 \text{ V} - (-0.034 \text{ V}) = 1.08 \text{ V}$
(e) Is the reaction in the cell spontaneous under the conditions described in part (d)? Justify your answer.	(e) The cell potential is positive, so the reaction is spontaneous.

QUESTIONS	EXPLANATIONS

4.

(a) Solutions of sodium acetate and nitric acid are mixed.

   (i) Balanced equation:

   (i) $H^+ + C_2H_3O_2^- \Rightarrow HC_2H_3O_2$

   (ii) Identify the spectator ion in this reaction.

   (ii) Nitrate (NO_3^-) is not included in the balanced equation, so it is the spectator ion.

(b) A piece of solid calcium is heated in oxygen gas.

   (i) Balanced equation:

   (i) $2\ Ca + O_2 \Rightarrow 2\ CaO$

   (ii) Is the entropy change for this reaction positive or negative? Briefly explain.

   (ii) Entropy is decreasing during this reaction because 3 moles of reactants combine to make 2 moles of products and because a solid reacts with a gas to form a solid.

(c) Ammonia and boron trichloride gases are mixed.

   (i) Balanced equation:

   (i) $NH_3 + BCl_3 \Rightarrow H_3NBCl_3$

   (ii) Identify the Lewis acid in this reaction.

   (ii) Boron trichloride accepts the electron pair from ammonia, so BCl_3 acts as the Lewis acid.

QUESTIONS	EXPLANATIONS

5. Use your knowledge of chemical principles to answer the following questions.

    (a) The table above shows the first and second ionization energies for potassium and sodium

        (i) Explain the difference between the first ionization energies of potassium and sodium.

        (i) The first ionization energy is lower for potassium than for sodium. Ionization energy is the energy required for the removal of an electron. Potassium's single valence electron is located in the fourth shell, while sodium's single valence electron is in the third shell. So potassium's electron is farther from the nucleus and shielded by an extra shell, making it easier to remove.

        (ii) Explain why the second ionization energy for potassium is so much larger than the first ionization energy.

        (ii) Potassium's first electron is the only valence electron in the fourth shell, so it is easy to remove. The second electron must be removed from a completed, stable low energy third shell, which is much more difficult.

    (b) The table above shows the atomic radii of several atoms and ions.

        (i) The atomic radius of bromine is smaller than the atomic radius of potassium. Explain.

        (i) Potassium and bromine each have four shells, but bromine has many more protons in its nucleus than does potassium. Bromine's extra protons exert greater attractive force on its electrons, pulling them closer to the nucleus and making the atomic radius smaller.

        (ii) The potassium and chlorine ions have the same electronic structure. Explain the difference in their atomic radii.

        (ii) Both K^+ and Cl^- have the same electron configuration, that of neutral argon. But K^+ has two more protons in its nucleus. The extra protons exert greater attractive force on the potassium ion's electrons, pulling them closer to the nucleus and making the ionic radius smaller.

        (iii) When potassium bromide and sodium chloride are compared, which will have the higher melting point? Explain.

        (iii) KBr and NaCl solids are held together by ionic bonds. Ionic bond strength depends on two things: the size of the charges and the distance between them. All of the charges in KBr and NaCl are +1 or −1, so the difference in melting points must be based on distance. NaCl is composed of smaller ions, so the charges will be closer together, making the electrostatic attractions greater, the bonds stronger, and the melting point higher.

    (c) The table above shows the boiling points for several organic compounds. Explain the trend in increasing boiling point.

    (c) Methane (CH_4) is nonpolar and its molecules have only weak van der Waals attractions, so it has the lowest boiling point. Methoxymethane (CH_3OCH_3) is larger and slightly polar, so its stronger intermolecular attractions make for a higher boiling point. Ethanol (CH_3CH_2OH) exhibits hydrogen bonding, so it has the strongest intermolecular forces and the highest boiling point of the three.

QUESTIONS	EXPLANATIONS

6. An experiment is to be performed involving the titration of a solution of benzoic acid, $HC_7H_5O_2$, of unknown concentration with a standardized 1.00-molar NaOH solution. The pK_a of benzoic acid is 4.2.

(a) What measurements and calculations must be made if the concentration of the acid is to be determined?

(a) The volume of the acid must be measured before the titration begins. Then the volume of 1.00-molar NaOH solution required to reach the equivalence point must be measured.

Once the two volumes are known, the concentration of the acid can be found with the following calculation:

$$(M_{NaOH})(V_{NaOH\ added}) = (M_{acid})(V_{acid})$$

(b) An indicator must be chosen for the titration. If methyl red (pH interval for color change: 4.2–6.3) and thymol blue (pH interval for color change: 8.0–9.6) are the only indicators available, which one should be used? Explain your choice.

(b) Thymol blue should be chosen. When a weak acid reacts with a strong base, the equivalence point will be greater than 7 in the basic region. The color change for thymol blue occurs in the basic region, making it the more appropriate choice for this experiment.

(c) Describe how you could use the titration apparatus and benzoic acid and NaOH solutions to create a buffer solution with a pH of 4.2 after you measured the volume of NaOH solution required to reach the equivalence point for a given volume of benzoic acid solution.

(c) Start the titration over again with a fresh sample of benzoic acid solution of the same volume that was used to determine the equivalence point. Add exactly half the volume of NaOH solution this time. At this point, called the half-equivalence point, exactly half of the benzoic acid has been neutralized and the concentration of benzoic acid is equal to the concentration of its conjugate base, benzoate ion. At this point, $pH = pK_a$.

(d) Describe a way to test whether the neutralization of benzoic acid by a strong base is an exothermic process, using readily available school laboratory equipment.

(d) Put a thermometer in the calorimeter used for the titration. Measure the temperature of the acid solution before the titration. If the temperature increases during the titration, then energy is being released and the reaction is exothermic.

Index

ABOUT THE AUTHOR

Paul Foglino has taught for The Princeton Review for more than a decade. He has written and edited course materials for the SAT, GRE, GMAT, and MCAT courses. He is author of *Math Smart Jr. II* and coauthor of *Cracking the CLEP*. Foglino studied English and electrical engineering at Columbia University, but he remains convinced that he learned everything he ever needed to know in junior high school.

Completely darken bubbles with a No. 2 pencil. If you make a mistake, be sure to erase mark completely. Erase all stray marks.

1.

YOUR NAME: _____
(Print)
                  Last   First   M.I.

SIGNATURE: _____ DATE: _____ / _____ / _____

HOME ADDRESS: _____
(Print)
                        Number and Street

_____
           City   State   Zip Code

PHONE NO.: _____

IMPORTANT: Please fill in these boxes exactly as shown on the back cover of your test book.

2. TEST FORM

3. TEST CODE

4. REGISTRATION NUMBER

5. YOUR NAME

First 4 letters of last name				FIRST INIT	MID INIT
Ⓐ	Ⓐ	Ⓐ	Ⓐ	Ⓐ	Ⓐ
Ⓑ	Ⓑ	Ⓑ	Ⓑ	Ⓑ	Ⓑ
Ⓒ	Ⓒ	Ⓒ	Ⓒ	Ⓒ	Ⓒ
Ⓓ	Ⓓ	Ⓓ	Ⓓ	Ⓓ	Ⓓ
Ⓔ	Ⓔ	Ⓔ	Ⓔ	Ⓔ	Ⓔ
Ⓕ	Ⓕ	Ⓕ	Ⓕ	Ⓕ	Ⓕ
Ⓖ	Ⓖ	Ⓖ	Ⓖ	Ⓖ	Ⓖ
Ⓗ	Ⓗ	Ⓗ	Ⓗ	Ⓗ	Ⓗ
Ⓘ	Ⓘ	Ⓘ	Ⓘ	Ⓘ	Ⓘ
Ⓙ	Ⓙ	Ⓙ	Ⓙ	Ⓙ	Ⓙ
Ⓚ	Ⓚ	Ⓚ	Ⓚ	Ⓚ	Ⓚ
Ⓛ	Ⓛ	Ⓛ	Ⓛ	Ⓛ	Ⓛ
Ⓜ	Ⓜ	Ⓜ	Ⓜ	Ⓜ	Ⓜ
Ⓝ	Ⓝ	Ⓝ	Ⓝ	Ⓝ	Ⓝ
Ⓞ	Ⓞ	Ⓞ	Ⓞ	Ⓞ	Ⓞ
Ⓟ	Ⓟ	Ⓟ	Ⓟ	Ⓟ	Ⓟ
Ⓠ	Ⓠ	Ⓠ	Ⓠ	Ⓠ	Ⓠ
Ⓡ	Ⓡ	Ⓡ	Ⓡ	Ⓡ	Ⓡ
Ⓢ	Ⓢ	Ⓢ	Ⓢ	Ⓢ	Ⓢ
Ⓣ	Ⓣ	Ⓣ	Ⓣ	Ⓣ	Ⓣ
Ⓤ	Ⓤ	Ⓤ	Ⓤ	Ⓤ	Ⓤ
Ⓥ	Ⓥ	Ⓥ	Ⓥ	Ⓥ	Ⓥ
Ⓦ	Ⓦ	Ⓦ	Ⓦ	Ⓦ	Ⓦ
Ⓧ	Ⓧ	Ⓧ	Ⓧ	Ⓧ	Ⓧ
Ⓨ	Ⓨ	Ⓨ	Ⓨ	Ⓨ	Ⓨ
Ⓩ	Ⓩ	Ⓩ	Ⓩ	Ⓩ	Ⓩ

TEST CODE columns:

			REGISTRATION NUMBER						
⓪	Ⓐ	Ⓙ	⓪	⓪	⓪	⓪	⓪	⓪	⓪
①	Ⓑ	Ⓚ	①	①	①	①	①	①	①
②	Ⓒ	Ⓛ	②	②	②	②	②	②	②
③	Ⓓ	Ⓜ	③	③	③	③	③	③	③
④	Ⓔ	Ⓝ	④	④	④	④	④	④	④
⑤	Ⓕ	Ⓞ	⑤	⑤	⑤	⑤	⑤	⑤	⑤
⑥	Ⓖ	Ⓟ	⑥	⑥	⑥	⑥	⑥	⑥	⑥
⑦	Ⓗ	Ⓠ	⑦	⑦	⑦	⑦	⑦	⑦	⑦
⑧	Ⓘ	Ⓡ	⑧	⑧	⑧	⑧	⑧	⑧	⑧
⑨			⑨	⑨	⑨	⑨	⑨	⑨	⑨

6. DATE OF BIRTH

Month	Day		Year	
◯ JAN				
◯ FEB	⓪	⓪	⓪	⓪
◯ MAR	①	①	①	①
◯ APR	②	②	②	②
◯ MAY	③	③	③	③
◯ JUN		④	④	④
◯ JUL		⑤	⑤	⑤
◯ AUG		⑥	⑥	⑥
◯ SEP		⑦	⑦	⑦
◯ OCT		⑧	⑧	⑧
◯ NOV		⑨	⑨	⑨
◯ DEC				

7. GENDER
◯ MALE
◯ FEMALE

1. Ⓐ Ⓑ Ⓒ Ⓓ Ⓔ
2. Ⓐ Ⓑ Ⓒ Ⓓ Ⓔ
3. Ⓐ Ⓑ Ⓒ Ⓓ Ⓔ
4. Ⓐ Ⓑ Ⓒ Ⓓ Ⓔ
5. Ⓐ Ⓑ Ⓒ Ⓓ Ⓔ
6. Ⓐ Ⓑ Ⓒ Ⓓ Ⓔ
7. Ⓐ Ⓑ Ⓒ Ⓓ Ⓔ
8. Ⓐ Ⓑ Ⓒ Ⓓ Ⓔ
9. Ⓐ Ⓑ Ⓒ Ⓓ Ⓔ
10. Ⓐ Ⓑ Ⓒ Ⓓ Ⓔ
11. Ⓐ Ⓑ Ⓒ Ⓓ Ⓔ
12. Ⓐ Ⓑ Ⓒ Ⓓ Ⓔ
13. Ⓐ Ⓑ Ⓒ Ⓓ Ⓔ
14. Ⓐ Ⓑ Ⓒ Ⓓ Ⓔ
15. Ⓐ Ⓑ Ⓒ Ⓓ Ⓔ
16. Ⓐ Ⓑ Ⓒ Ⓓ Ⓔ
17. Ⓐ Ⓑ Ⓒ Ⓓ Ⓔ
18. Ⓐ Ⓑ Ⓒ Ⓓ Ⓔ
19. Ⓐ Ⓑ Ⓒ Ⓓ Ⓔ
20. Ⓐ Ⓑ Ⓒ Ⓓ Ⓔ
21. Ⓐ Ⓑ Ⓒ Ⓓ Ⓔ
22. Ⓐ Ⓑ Ⓒ Ⓓ Ⓔ
23. Ⓐ Ⓑ Ⓒ Ⓓ Ⓔ

24. Ⓐ Ⓑ Ⓒ Ⓓ Ⓔ
25. Ⓐ Ⓑ Ⓒ Ⓓ Ⓔ
26. Ⓐ Ⓑ Ⓒ Ⓓ Ⓔ
27. Ⓐ Ⓑ Ⓒ Ⓓ Ⓔ
28. Ⓐ Ⓑ Ⓒ Ⓓ Ⓔ
29. Ⓐ Ⓑ Ⓒ Ⓓ Ⓔ
30. Ⓐ Ⓑ Ⓒ Ⓓ Ⓔ
31. Ⓐ Ⓑ Ⓒ Ⓓ Ⓔ
32. Ⓐ Ⓑ Ⓒ Ⓓ Ⓔ
33. Ⓐ Ⓑ Ⓒ Ⓓ Ⓔ
34. Ⓐ Ⓑ Ⓒ Ⓓ Ⓔ
35. Ⓐ Ⓑ Ⓒ Ⓓ Ⓔ
36. Ⓐ Ⓑ Ⓒ Ⓓ Ⓔ
37. Ⓐ Ⓑ Ⓒ Ⓓ Ⓔ
38. Ⓐ Ⓑ Ⓒ Ⓓ Ⓔ
39. Ⓐ Ⓑ Ⓒ Ⓓ Ⓔ
40. Ⓐ Ⓑ Ⓒ Ⓓ Ⓔ
41. Ⓐ Ⓑ Ⓒ Ⓓ Ⓔ
42. Ⓐ Ⓑ Ⓒ Ⓓ Ⓔ
43. Ⓐ Ⓑ Ⓒ Ⓓ Ⓔ
44. Ⓐ Ⓑ Ⓒ Ⓓ Ⓔ
45. Ⓐ Ⓑ Ⓒ Ⓓ Ⓔ
46. Ⓐ Ⓑ Ⓒ Ⓓ Ⓔ

47. Ⓐ Ⓑ Ⓒ Ⓓ Ⓔ
48. Ⓐ Ⓑ Ⓒ Ⓓ Ⓔ
49. Ⓐ Ⓑ Ⓒ Ⓓ Ⓔ
50. Ⓐ Ⓑ Ⓒ Ⓓ Ⓔ
51. Ⓐ Ⓑ Ⓒ Ⓓ Ⓔ
52. Ⓐ Ⓑ Ⓒ Ⓓ Ⓔ
53. Ⓐ Ⓑ Ⓒ Ⓓ Ⓔ
54. Ⓐ Ⓑ Ⓒ Ⓓ Ⓔ
55. Ⓐ Ⓑ Ⓒ Ⓓ Ⓔ
56. Ⓐ Ⓑ Ⓒ Ⓓ Ⓔ
57. Ⓐ Ⓑ Ⓒ Ⓓ Ⓔ
58. Ⓐ Ⓑ Ⓒ Ⓓ Ⓔ
59. Ⓐ Ⓑ Ⓒ Ⓓ Ⓔ
60. Ⓐ Ⓑ Ⓒ Ⓓ Ⓔ
61. Ⓐ Ⓑ Ⓒ Ⓓ Ⓔ
62. Ⓐ Ⓑ Ⓒ Ⓓ Ⓔ
63. Ⓐ Ⓑ Ⓒ Ⓓ Ⓔ
64. Ⓐ Ⓑ Ⓒ Ⓓ Ⓔ
65. Ⓐ Ⓑ Ⓒ Ⓓ Ⓔ
66. Ⓐ Ⓑ Ⓒ Ⓓ Ⓔ
67. Ⓐ Ⓑ Ⓒ Ⓓ Ⓔ
68. Ⓐ Ⓑ Ⓒ Ⓓ Ⓔ
69. Ⓐ Ⓑ Ⓒ Ⓓ Ⓔ

70. Ⓐ Ⓑ Ⓒ Ⓓ Ⓔ
71. Ⓐ Ⓑ Ⓒ Ⓓ Ⓔ
72. Ⓐ Ⓑ Ⓒ Ⓓ Ⓔ
73. Ⓐ Ⓑ Ⓒ Ⓓ Ⓔ
74. Ⓐ Ⓑ Ⓒ Ⓓ Ⓔ
75. Ⓐ Ⓑ Ⓒ Ⓓ Ⓔ

The Princeton Review

Completely darken bubbles with a No. 2 pencil. If you make a mistake, be sure to erase mark completely. Erase all stray marks.

1.

YOUR NAME: _____
(Print)
Last   First   M.I.

SIGNATURE: _____   DATE: __ / __ / __

HOME ADDRESS: _____
(Print)
Number and Street

_____
City   State   Zip Code

PHONE NO.: _____

IMPORTANT: Please fill in these boxes exactly as shown on the back cover of your test book.

2. TEST FORM

3. TEST CODE

0	A	J	0	0
1	B	K	1	1
2	C	L	2	2
3	D	M	3	3
4	E	N	4	4
5	F	O	5	5
6	G	P	6	6
7	H	Q	7	7
8	I		8	8
9			9	9

4. REGISTRATION NUMBER

0	0	0	0	0	0
1	1	1	1	1	1
2	2	2	2	2	2
3	3	3	3	3	3
4	4	4	4	4	4
5	5	5	5	5	5
6	6	6	6	6	6
7	7	7	7	7	7
8	8	8	8	8	8
9	9	9	9	9	9

6. DATE OF BIRTH

Month	Day		Year	
○ JAN				
○ FEB	0	0	0	0
○ MAR	1	1	1	1
○ APR	2	2	2	2
○ MAY	3	3	3	3
○ JUN		4	4	4
○ JUL		5	5	5
○ AUG		6	6	6
○ SEP		7	7	7
○ OCT		8	8	8
○ NOV		9	9	9
○ DEC				

7. GENDER
○ MALE
○ FEMALE

The Princeton Review

5. YOUR NAME

First 4 letters of last name				FIRST INIT	MID INIT
A	A	A	A	A	A
B	B	B	B	B	B
C	C	C	C	C	C
D	D	D	D	D	D
E	E	E	E	E	E
F	F	F	F	F	F
G	G	G	G	G	G
H	H	H	H	H	H
I	I	I	I	I	I
J	J	J	J	J	J
K	K	K	K	K	K
L	L	L	L	L	L
M	M	M	M	M	M
N	N	N	N	N	N
O	O	O	O	O	O
P	P	P	P	P	P
Q	Q	Q	Q	Q	Q
R	R	R	R	R	R
S	S	S	S	S	S
T	T	T	T	T	T
U	U	U	U	U	U
V	V	V	V	V	V
W	W	W	W	W	W
X	X	X	X	X	X
Y	Y	Y	Y	Y	Y
Z	Z	Z	Z	Z	Z

1. A B C D E
2. A B C D E
3. A B C D E
4. A B C D E
5. A B C D E
6. A B C D E
7. A B C D E
8. A B C D E
9. A B C D E
10. A B C D E
11. A B C D E
12. A B C D E
13. A B C D E
14. A B C D E
15. A B C D E
16. A B C D E
17. A B C D E
18. A B C D E
19. A B C D E
20. A B C D E
21. A B C D E
22. A B C D E
23. A B C D E

24. A B C D E
25. A B C D E
26. A B C D E
27. A B C D E
28. A B C D E
29. A B C D E
30. A B C D E
31. A B C D E
32. A B C D E
33. A B C D E
34. A B C D E
35. A B C D E
36. A B C D E
37. A B C D E
38. A B C D E
39. A B C D E
40. A B C D E
41. A B C D E
42. A B C D E
43. A B C D E
44. A B C D E
45. A B C D E
46. A B C D E

47. A B C D E
48. A B C D E
49. A B C D E
50. A B C D E
51. A B C D E
52. A B C D E
53. A B C D E
54. A B C D E
55. A B C D E
56. A B C D E
57. A B C D E
58. A B C D E
59. A B C D E
60. A B C D E
61. A B C D E
62. A B C D E
63. A B C D E
64. A B C D E
65. A B C D E
66. A B C D E
67. A B C D E
68. A B C D E
69. A B C D E

70. A B C D E
71. A B C D E
72. A B C D E
73. A B C D E
74. A B C D E
75. A B C D E